SYSTEMATICS OF FLUID INCLUSIONS IN DIAGENETIC MINERALS

SEPM SHORT COURSE 31

by

Robert H. Goldstein
THE UNIVERSITY OF KANSAS
Department of Geology
120 Lindley Hall
Lawrence, KS 66045
USA

T. James Reynolds
FLUID INC.
P.O. BOX 6873
Denver, CO 80206
USA

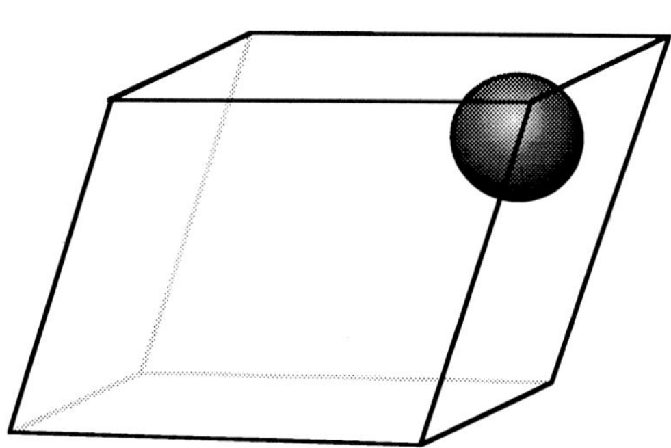

SEPM (Society for Sedimentary Geology)

ISBN #1-56576-008-5

Additional copies of this publication may be ordered from SEPM.
Send your order to

SEPM
Post Office Box 4756
Tulsa, Oklahoma 74159-0756
U.S.A.

Printed in the United States of America

CONTENTS

Chapter 3
PHASE CHANGES IN FLUID INCLUSIONS: THE BASICS

Chapter 4
ARE FLUID INCLUSIONS REPRESENTATIVE SAMPLES OF DIAGENETIC FLUIDS?

Chapter 5
THE PHILOSOPHY OF CONDUCTING A FLUID INCLUSION STUDY

Chapter 6
FLUID INCLUSION PETROGRAPHY

Chapter 7
FLUID INCLUSION MICROTHERMOMETRY

Chapter 8
DATA PRESENTATION

Chapter 9
PRACTICAL ASPECTS OF GEOTHERMOMETRY

Chapter 10
PRACTICAL ASPECTS OF GEOBAROMETRY

Chapter 11
CASE HISTORIES

Chapter 12
OTHER ANALYTICAL APPROACHES

PREFACE

The writing of this book evolved from our excitement about the potential uses of the fluid inclusion technique and our dismay over its common misuse in diagenetic minerals. The past decade has revealed significant advantages to using fluid inclusions as a means of understanding the physical and chemical history of fluids in sedimentary basins, but it also has revealed important limitations which have required that a new approach must be employed to effectively use fluid inclusions. Our main stimuli have been interactions with colleagues and students, in which we observed many having great difficulties extracting the basic fluid inclusion methodologies for diagenetic systems from the literature. Many of these people were misapplying fluid inclusions altogether, and had devoted many months of work to produce data of little utility. The end result was significant frustration among many workers and ineffective use of a potentially powerful technique.

In this book we concentrate on teaching the reader what must be known to use fluid inclusion techniques properly for applications in diagenetic minerals. In addition to those working in sedimentary rocks, this text will be useful to workers in other fields: the principles, philosophy, and procedures expressed herein are applicable to any fluid inclusion study, no matter what the emphasis. The text is organized as would be a practical course in the use of fluid inclusions in diagenetic minerals. It is not intended to be a comprehensive encyclopedia of all fluid inclusion work that has been done in sedimentary rocks. The book begins with what fluid inclusions are and what geologic history they are capable of recording. Following are the basic phase equilibria that must be known to understand the behavior of pore fluids and fluid inclusions in nature. Once this is explained, the question of validity of using fluid inclusions as records of ancient diagenetic systems is dealt with in such a way that the questions commonly asked about the limitations of the technique are addressed. These ideas set the stage for showing how to conduct a fluid inclusion study. We set out a new petrographically based approach for conducting fluid inclusion research that is logically followed by methods that allow for the interpretation of compositions of pore fluids that existed in sedimentary rocks, and methods of geothermometry and geobarometry. We then present selected case histories that are designed specifically to give the reader practice in evaluating fluid inclusion data from the diagenetic realm. Finally, we summarize briefly the arsenal of analytical techniques that may be applied to fluid inclusions to develop additional constraints on fluid inclusion composition.

R. H. Goldstein
T. J. Reynolds
March 1994

ACKNOWLEDGMENTS

The authors wish to thank the many individuals who contributed greatly to the completion of this book. Above all, we wish to thank Christina Baroth and Cindy Keeffe for countless hours of help and moral support. Linda Harris provided clerical help. Careful reviews by J. R. Allan, D. L. Hall, J. J. Irwin and E. Roedder greatly improved the manuscript. R. C. Burruss provided many helpful suggestions and aided in some of the calculations. S. M. Sterner provided samples of synthetic fluid inclusions. Some of the studies summarized in this book were supported by grants from Exxon Production Research, Texaco Research, and BP Research; the University of Kansas general research fund, NSF grants EAR-87-21229, and EAR-92-18463.

Chapter 1

INTRODUCTION TO FLUID INCLUSIONS AND THEIR APPLICATIONS

INTRODUCTION

For several decades geologists have been applying field, petrographic, and geochemical methods to study the diagenesis of limestones, dolomites, evaporites and sandstones. The most successful studies have integrated field and petrographic work with various geochemical methods. For most applications, the value of any one of the most commonly applied techniques has often been limited; however, when applied together they have proven very useful. Careful petrography has been the most important and reliable component of diagenetic studies. Trace and minor element analysis of diagenetic phases is limited by poor knowledge of distribution coefficients, unknown applicability of distribution coefficients, or unknown pore-fluid chemistry. Interpretation of values of stable isotopes ($\partial^{13}C$ and $\partial^{18}O$) may be plagued by unknown temperature, pore fluid composition, water-rock ratio, or unknown fractionation factors for some systems. All of the foregoing methods are indirect methods of interpreting diagenetic history in that they are the result of diagenetic processes rather than samples of the diagenetic systems themselves. Such indirect approaches often yield data that can easily be misinterpreted.

Fluid inclusions are fluid-filled vacuoles sealed within minerals. When trapped within diagenetic minerals, they provide the only direct means of examining the fluids present in ancient diagenetic environments. Fluid inclusions can be thought of as time capsules storing information about ancient temperatures, pressures, and fluid compositions. They may provide the following valuable information with simple petrographic observation, microthermometric analysis, or sophisticated geochemical analysis of inclusion contents.

Temperature of mineral precipitation — Fluid inclusions can be used to ascertain temperature of mineral formation at several different levels of precision, depending on the type of fluid inclusion study conducted and the fluid inclusion suite present. A temperature data set may commonly provide a minimum temperature of mineral formation. Other data may merely define conditions of precipitation during a particular monotonic increase or decrease in temperature of unknown magnitude. Or perhaps only a "ballpark idea" of low temperature versus high temperature may be determined. Sometimes the fluid inclusions may even yield the true temperature of mineral formation.

Pressure of mineral precipitation — The pressure of mineral precipitation can be determined using fluid inclusions, but our ability to make such an interpretation is entirely dependent on the particular suite of fluid inclusions present in the minerals studied. Fluid inclusions may be used to determine a minimum pressure of entrapment, or may be used to determine the true pressure of entrapment in certain sample suites.

Composition and origin of fluids of mineral precipitation — Given appropriate fluid inclusions, many aspects of aqueous fluid composition can be determined. The most common application is determining the salinity of the fluid of mineral precipitation. Other measurements can provide the identity and concentration of major ions in solution, presence of organics, major and minor ion ratios, concentration of particular dissolved components such as sulfate, dissolved gas identity and concentration, and even the isotopic composition of the fluid. Many other potential applications exist yet have not been exploited.

Later history of temperature, pressure, and fluid composition — Given the appropriate suite of fluid inclusions, there is potential for determining any of the above parameters from fluid inclusions trapped after mineral growth. Thus, there is potential for analyzing a more complete history for a sample suite than is available from precipitated minerals alone.

Of course, once these variables are constrained, there

are important geologic applications. These are but a few that have been aided by careful studies of fluid inclusions:

Improve understanding of diagenetic systems — The physical, chemical, and geological processes of ancient and modern diagenetic systems have proven difficult to understand. A major reason for this has been the indirect nature of diagenetic studies. Fluid inclusions are samples of the fluids responsible for diagenesis, and are a direct record of diagenetic systems. They have provided useful temperature, pressure, and fluid chemistry constraints on ancient diagenetic systems that cannot be obtained by other means.

Improve understanding of subsurface fluid evolution — Fluid inclusions provide a rather unique record of the pore fluids that were present in rocks during their burial and uplift history. When the data from fluid inclusions have been combined with other geological and paragenetic information, the evolution of subsurface brines has been elucidated.

Improve understanding of porosity evolution — The relative timing of reduction and occlusion of porosity has significant implications in exploration for petroleum and in development of petroleum reservoirs. Analyses of fluid inclusions have constrained the conditions at which the diagenetic phases precipitated, leading to a better understanding of the geologic controls and relative timing of the porosity-reducing events.

Improve interpretation of petroleum migration history — Fluid inclusions provide one of the best records of the history of migration of petroleum through a suite of rocks. The mere presence of petroleum fluid inclusions is a microscopic "oil show" that may carry great significance. When oil-filled inclusions have been placed in a paragenetic framework, the relative timing of petroleum migration has been determined. Temperature information from petroleum inclusions has been used to place the timing of petroleum migration into a geologic framework. In addition, the compositions of petroleum inclusions have been analyzed and fingerprinted, and used to trace the migration history of oils in a basin.

Improve reconstructions of thermal history — Fluid inclusions provide one of the most useful techniques for evaluating the temperatures experienced by sedimentary rocks. Because fluid inclusion data commonly are collected within a paragenetic framework, this technique has often provided detailed information on the thermal history experienced by sedimentary rocks.

Improve reconstructions of tectonic or stratigraphic history — The fluid inclusion technique has worked well for developing unique constraints on the temperatures and pressures present during deformation. Furthermore, the temperature and pressure data have been used in the reconstruction of burial and unroofing histories of basins.

Inherent in any of these applications is a requirement that the appropriate suite of fluid inclusions is present to answer the geologic question that has been posed. Most diagenetic minerals contain fluid inclusions, but every suite of rocks will not necessarily contain the fluid inclusions needed to answer a particular geologic question. Thus, although the fluid inclusion technique is an extremely valuable one, it should not be considered a panacea or a technique that must be routinely applied to every suite of rocks. Correct application of the technique requires initial formulation of the geologic problem to be solved, followed by careful study of samples in a geologic and paragenetic context to determine if fluid inclusions will prove useful in answering the particular geologic question posed. As samples of diagenetic systems, fluid inclusions trapped in minerals exist within a temporal or paragenetic context, so most fluid inclusion studies are based on careful petrographic observation. Perhaps the fluid inclusion approach would more appropriately be considered a petrographic method as opposed to a geochemical method, but in practice, it must be both to provide useful information.

This text will provide a framework for applying fluid inclusion techniques to diagenetic systems. Well-established concepts based on fundamental chemical and physical principles are stressed, but new approaches that cannot easily be gleaned from the literature are included as well. Also, many questions commonly asked about the applicability of fluid inclusions are addressed, and practical examples are provided to illustrate methods of fluid inclusion study.

HISTORICAL FRAMEWORK

In the 18th Century as the fundamental principles of geology were first being formulated, the presence of fluid inclusions in minerals was employed to support Neptunist theories. Dolomieu (1792) may have been the first to report inclusions in quartz filled with petroleum. The beginnings of the rigorous use of inclusions to decipher geologic processes came from

Sorby (1858) reasoning that bubbles within fluid inclusions were caused by differential thermal contraction, and that reheating inclusions would lead to disappearance of the bubbles and that the temperature at which this occurred could serve as an estimate of the temperature of mineral formation. Since that time fluid inclusion researchers have added many more applications, and have shown that Sorby was largely correct.

In the 20th Century Edwin Roedder has been a leader in fluid inclusion research and in bringing fluid inclusionists from diverse backgrounds together. His treatise on fluid inclusion research (Roedder, 1984) is an encyclopedia of fluid inclusion information, and is the embryo for all fluid inclusion research performed today. Prior to 1980, studies of fluid inclusions in sedimentary rocks were limited to those on evaporites and Mississippi Valley-type Pb-Zn ore deposits. This body of literature is a wealth of knowledge, and the lessons that can be learned provide a basis upon which this text is built. Nevertheless it has become increasingly apparent that "standard" procedures employed in these early works may lead to misinterpretations or improper applications of fluid inclusion data. In the early 1980s it became apparent that a more rigorous approach to the collection and interpretation of fluid inclusion data in sedimentary systems would be required to learn from inclusions in diagenetic phases. Works in the 1970s and early 1980s (Nelson, 1973; Klosterman, 1981; Moore and Druckman, 1981; Wagner and Matthews, 1982) led to controversies among diagenesis researchers which still rage today (Guscott and Burley, 1993; Osborne and Haszeldine, 1993). An SEPM research conference organized by Robert Burruss, Charles Barker, and Robert Halley convened in the mountains of Colorado in the fall of 1984 to discuss fluid inclusion research in sedimentary systems. Presentations by Terry O'Hearn, Dennis Prezbindowski, and Robert Goldstein left all participants with the distinct impression that if fluid inclusions were going to be employed to understand diagenetic processes, several problems would have to be resolved first.

During the past ten years, many workers have continued to apply an inappropriate methodology to fluid inclusion studies in sedimentary systems, and many of the published fluid inclusion studies in the 1990s are regarded with strong and well-deserved skepticism. Other workers have persevered in a determination to define the limitations of the fluid inclusion technique, and to derive a rigorous methodological approach that incorporates these limitations. So, over 200 years after what must have seemed like an obvious deduction for early geologists to think that the presence of water in crystals supported theories that all crustal materials had origins from ocean suspension or solution, we present to future inclusionists a text that should provide a foundation for making sound observations of fluid inclusions in sedimentary systems, and for knowing the limitations of the potential inferences from their observations.

Chapter 2

FLUID INCLUSIONS AND THEIR ORIGIN

FLUID INCLUSION APPEARANCE

When observed at room temperature using a transmitted light microscope, most fluid inclusions have a rather sharp outer boundary marking the edge of the inclusion cavity (Fig. 2.1). This is because of a significant difference in refractive index between inclusion fluids and their mineral hosts: most aqueous fluids have refractive indices between 1.33 and 1.45 whereas the minerals in which they are included have refractive indices from 1.43 to as high as 3.22. Hydrocarbon liquids, however, have refractive indices that may be similar to their mineral hosts (Burruss, 1981), and thus, are not all easily visible. The inclusion cavity generally contains a large amount of bright, clear liquid (Fig. 2.1A, D, E) and some may contain a small dark bubble of vapor or gas (although any liquid-to-vapor ratio is possible) that is dark because of internal reflection (Fig. 2.1D). However, as shown in Figure 2.1E, bubbles in flat inclusions may not appear that dark. Though most liquids appear colorless, some hydrocarbon liquids may have colors ranging from reddish-brown to yellow.

Inclusions smaller than 1 μm currently are not possible to study because of microscope optical limitations. The sizes of most inclusions readily studied in diagenetic phases are about 2 to 7 μm in longest dimension. For the most part, coarsely crystalline diagenetic minerals contain more workable-sized inclusions than fine-grained minerals, and smaller inclusions are typically much more abundant than larger inclusions in diagenetic phases. Because of the small size of the inclusions, petrographic study requires a good microscope properly adjusted, and well-polished samples (see Chapter 6 for detailed discussions).

INCLUSION ENTRAPMENT DURING CRYSTAL GROWTH

When a crystal precipitates from a fluid, the surface of crystal growth will, inevitably, be less than perfect. Typically, imperfections on the crystal surface will be engulfed by the surrounding crystal, producing a vacuole within the crystal, sealed-off by crystal growth and containing the fluid present at the moment of sealing. These trapped fluids are termed *primary* in that they were enclosed during growth of the crystal, and contain a sample of the fluid responsible for precipitation of the diagenetic phase. For many minerals studied, the exact cause of fluid inclusion entrapment during growth may not be known. However, experimental studies that produce inclusion entrapment in common diagenetic minerals at diagenetic temperatures and pressures (Sabouraud-Rosset, 1969; McLimans, 1987; Davis and others, 1990; Pironon and Barres, 1990; Kihle and Johansen, 1994) are advancing our understanding of the mechanisms of inclusion entrapment. Many of the potential mechanisms are summarized below.

Entrapment of fluid inclusions during crystal growth is normal and should be expected. Crystal growth can be thought of as occurring from lateral advancement of a series of step-like growth layers (Fig. 2.2). With kinks developed at their edges, it is possible for reentrants to be formed in the expanding growth layer, and eventually these reentrants are engulfed to form a hole. As succeeding growth layers seal off the holes, such imperfections in crystals are the sites of entrapment of fluid inclusions.

Roedder (1984) has described several mechanisms of entrapment known for primary fluid inclusions. Sometimes the centers of crystal faces become starved of nutrients relative to crystal edges, resulting in the formation of cavities which are sealed-off by later growth to form primary fluid inclusions (Fig. 2.3A). During major increases in the degree of supersaturation that might increase growth rate, voids can result from rapid growth, and when later overgrown, they enclose fluids which bathed the crystal (Fig. 2.3B). This mechanism has been demonstrated in experimentally grown calcite (Janssen-Van Rosmalen and Bennema, 1977; McLimans, 1987). Etching events are well known in diagenetic systems and have been described for many minerals. Such etching events create reentrants and channelways in the crystal surface that may be sealed by later growth to create fluid inclusions (Fig. 2.3C). Also, it is possible that fluid inclusions may be

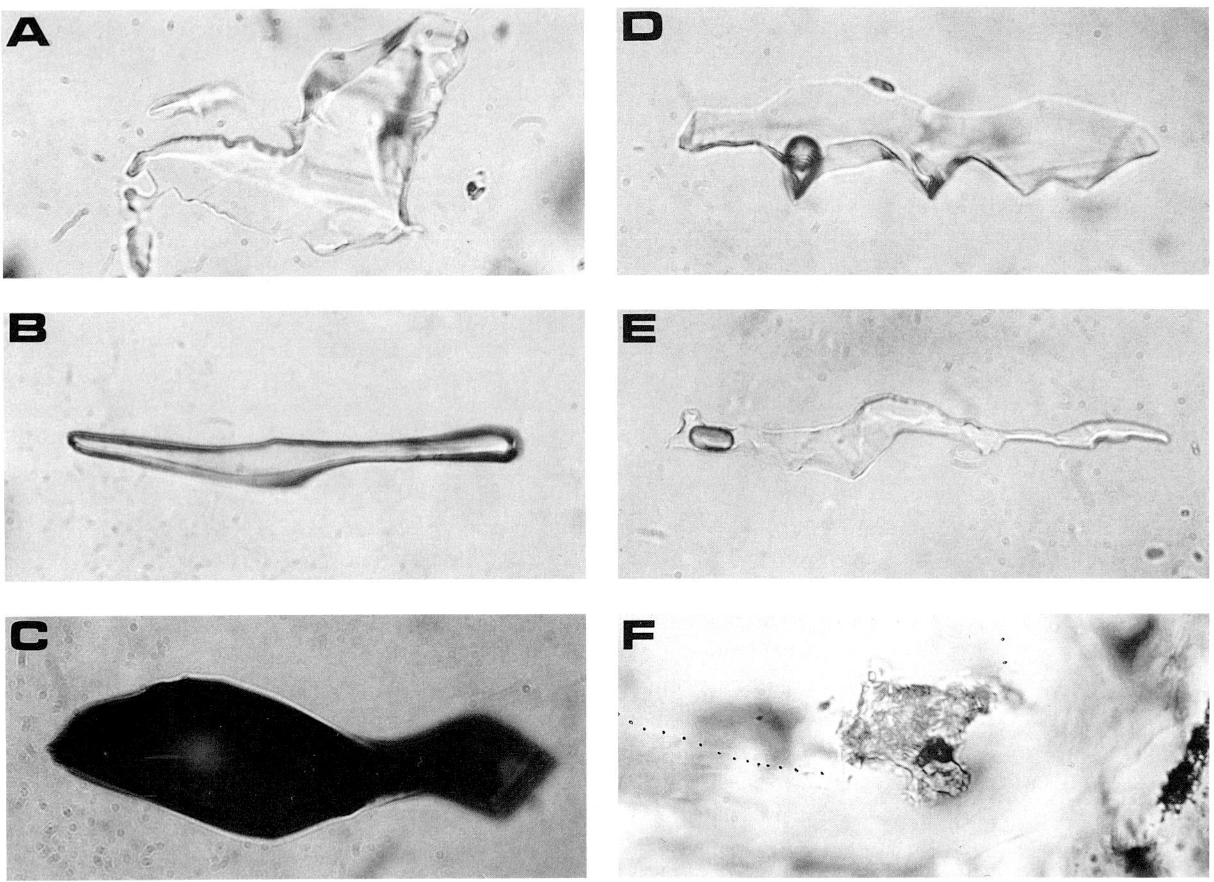

Fig. 2.1. Photomicrographs of fluid inclusions in authigenic quartz at room temperature. Left column (A, B, C) shows single-phase inclusions. Right column (D, E, F) shows two-phase inclusions. ***A)*** *Irregularly-shaped, liquid-filled aqueous inclusion.* ***B)*** *Clear, smooth-surfaced, single-phase methane inclusion.* ***C)*** *Dark, smooth-surfaced, single-phase methane inclusion.* ***D)*** *Irregularly-shaped, two-phase aqueous inclusion with spherical bubble.* ***E)*** *Irregularly shaped, two-phase fluid inclusion with elongate bubble.* ***F)*** *Highly irregularly-shaped, two-phase aqueous inclusion trapped in association with solid inclusions (clay?) at the boundary between a detrital quartz grain and quartz overgrowth. The boundary is not clear in the plane of this photomicrograph, so it is demarcated with dots. Most people having little experience with inclusions will typically ignore inclusions that look like those on the left (A, B, C), but, in fact, as will be stressed in this text, such inclusions should be scrutinized! The single-phase inclusions in A and B will look identical to most, but note the greater contrast between the edges of the inclusion in B with the surrounding quartz, as compared with A. This is an important clue that often indicates that a fluid of composition other than H_2O is present. The dark single-phase inclusion in C will appear to be empty to most novices; however, some inclusions that look like this may contain air, and for this inclusion shown, cooling to liquid nitrogen temperatures will prove that it contains methane. Also, many inexperienced inclusionists would disregard inclusions in D and E because they appear to be "necked." But as will be shown in this text, this is not sound reasoning: what is important for an inclusionist to note are the liquid-to-vapor ratios of inclusions within an individual fluid inclusion assemblage — shapes are not to be used for analyzing the results of necking down. Do not despair, as a complete comprehension and appreciation of this figure will follow upon a thorough reading of this text. For photogenic purposes, all of the inclusions selected for this figure are unusually large. Just pretend that the inclusions displayed are each <5 μm in longest dimension, as such a size is what realistically one can expect to find.*

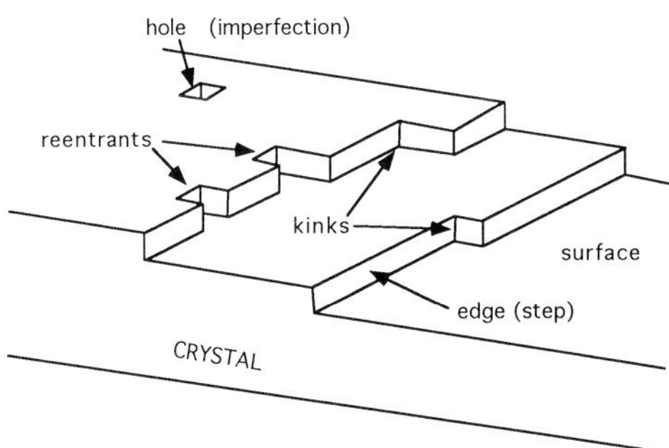

Fig. 2.2. *Schematic diagram showing process of inclusion formation from atomic-scale imperfections. Modified from McLimans (1987, figure 2).*

trapped preferentially at the contacts between growth twins, such as those common in gypsum.

It is probable that any poisoning or inhibition of growth on the growth surface could create a reentrant during later growth that is eventually sealed-off to form a fluid inclusion. For example, if a crack were to form within a crystal, subsequent growth may be disturbed at the discontinuity (Fig. 2.3D). Also, if a crystal of another substance were to fall or nucleate on the growing surface, a cavity could be formed in the wake of the growing crystal (Fig. 2.3E). Even a separate but mineralogically identical crystal nucleated on the growing surface could create a fluid inclusion cavity. Other substances adhering to the crystal surface could create growth discontinuities that would allow entrapment of fluid inclusions during subsequent growth. These could include bacterial bodies, other types of organic matter, globules of oil, or bubbles of gas. These are just a few of potentially dozens of mechanisms of fluid inclusion entrapment. The mechanisms discussed illustrate that fluid inclusion entrapment during crystal growth is a common process but not one that necessarily preserves a complete record of every phase of crystal growth.

Crystal growth may also occur through recrystallization of preexisting phases, and fluids may be formed during this process as well. Observations of inclusions in some recrystallized minerals have been recorded, but actual mechanisms are not well-understood. Recrystallized halite may contain large fluid inclusions which could have formed during recrystallization (Lazar and Holland, 1988; Horita and others, 1991; Bien and others, 1991). Recrystallization of aragonite to low-Mg calcite tends to exclude sizable fluid inclusions and this seems to be supported by IR spectroscopy analysis (Gaffey, 1988, 1990). However, on rare occasion we have found recrystallized aragonite that contains large fluid inclusions. Recrystallized dolomite and ankerite commonly contain fluid

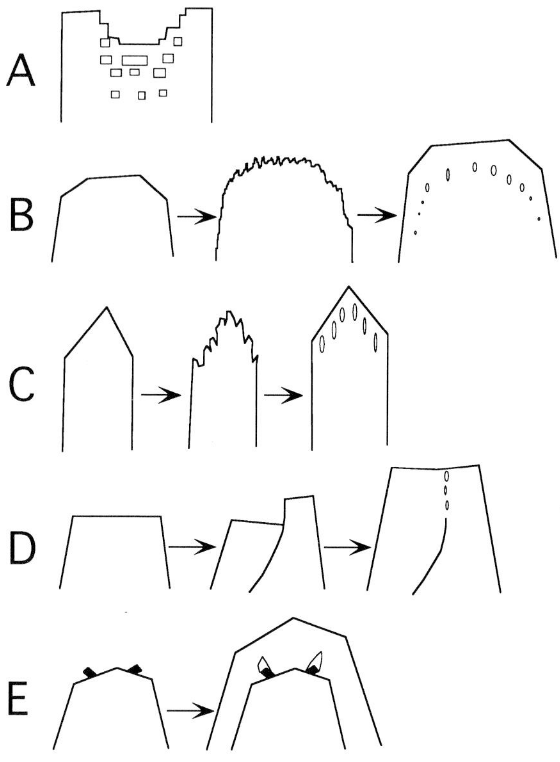

Fig. 2.3. *Schematic diagrams showing possible mechanisms of entrapment of primary fluid inclusions.* ***A)*** *Starvation of nutrients at centers of crystal faces cause imperfections that are propagated to form cavities that are later sealed by new growth.* ***B)*** *Irregularities formed by rapid growth are later sealed during subsequent growth stages.* ***C)*** *Irregularities formed by etching are later sealed by continued growth.* ***D)*** *A crack causes an imperfection which is propagated to form irregularities that eventually seal.* ***E)*** *Cavities form in the wake of materials foreign to the crystal that impede growth. Modified from Roedder (1984, figure 2.1).*

inclusions that appear to have been trapped during recrystallization (Abegg, 1990; Gregg and Shelton, 1990; Shelton and others, 1992; Wojcik and others, 1992, 1994). Low-Mg calcite that has recrystallized from high-Mg calcite preserves fluid inclusions (K. C. Lohmann, personal communication, 1988; James and Bone, 1992). Fluid inclusions trapped during processes of recrystallization will probably be employed as tools to enhance our understanding of these processes.

FLUID INCLUSION ENTRAPMENT AFTER CRYSTAL GROWTH

After mineral precipitation is complete, it is common for crystals to be deformed, both by brittle and

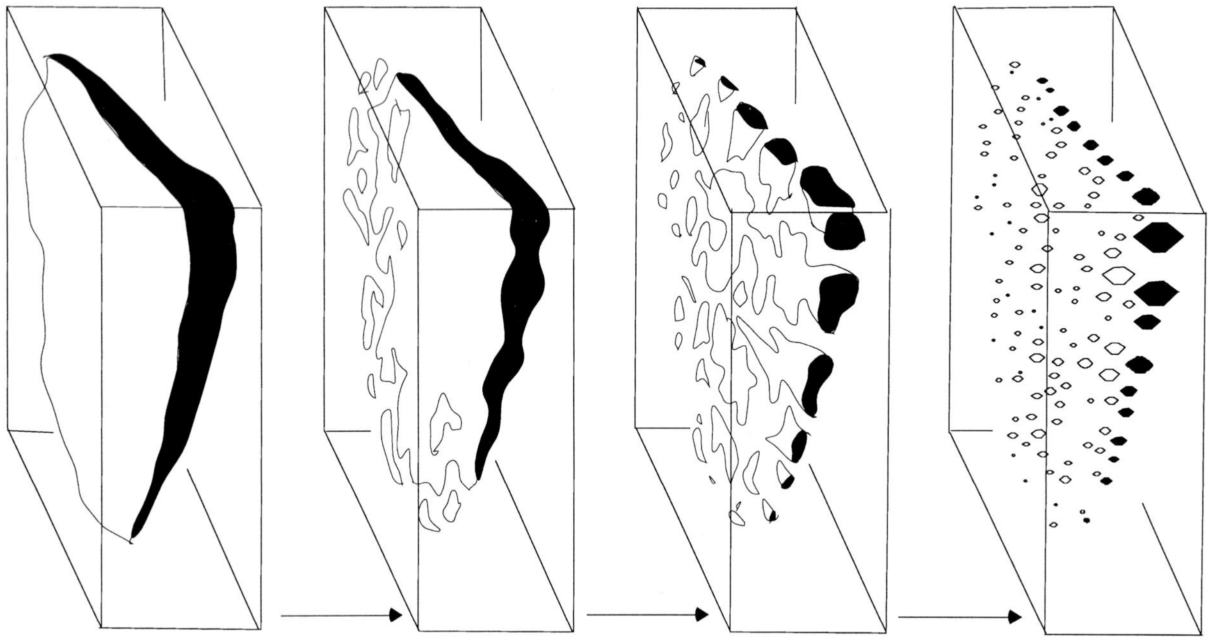

Fig. 2.4. *Schematic diagram showing stages in the healing of a microfracture in quartz. Note that the final shapes of the inclusions have crystal faces as surfaces. Such morphology is known as negative crystal shape. Generally, inclusions will evolve toward such a shape. This is the most stable state as it has the lowest surface free energy. However, whether or not negative crystal morphology is attained depends on many parameters — host mineral solubility in the fluid and time being of importance. Modified from Roedder (1962).*

ductile processes. Such deformation may result in the formation of μm-width microcracks, deformation twin surfaces, and shear planes. When such deformation features develop, it is likely that they will be filled with the fluid present during or after deformation. Fluid may be trapped between the deformed surfaces by either subsequent precipitation (on a submicroscopic scale), or by a dissolution-reprecipitation process. Once formed, planes of deformation do not simply "snap shut:" they must heal by either of these two aforementioned processes. The former requires transport of ions into the crack and supersaturation of the solution with respect to the precipitated mineral. The latter process will always occur (as long as the host mineral is soluble in the fluid) in a manner that reduces the high surface free energy of the surfaces of deformation (Roedder, 1984). As illustrated in Figure 2.4, the microcrack changes shape with time by dissolution and reprecipitation of the mineral to redistribute the surfaces of the crack into a lower surface free energy state. Such redistribution seals separate fluid inclusions along the position of the former crack. This process, known as *necking down*, does not require transport of new ions into the solution from outside the domain of the microcrack. These fluid inclusions in healed microcracks may provide a useful record of the fluids present after mineral growth during healing of cracks; they may contribute much information regarding a rock's history that is not preserved in primary inclusions formed during precipitation of diagenetic minerals. Fluid inclusions showing petrographic evidence indicative of entrapment after mineral precipitation are known as *secondary* fluid inclusions.

The process of necking down to form secondary fluid inclusions in some diagenetic minerals appears to be geologically relatively rapid, even at low temperature, and the presence of liquid water is not even necessary for necking to occur. Crack healing experiments in soluble minerals indicate that necking down can take place rapidly, on a scale of days and years (Lemmlein and Kliya, 1952). Our work has encountered rocks that have been uplifted into the vadose zone relatively late in the Neogene, and calcite cements from such settings may contain secondary inclusions enclosing vadose zone fluids. Furthermore, planes of secondary fluid inclusions commonly contain fluid inclusions with identical density and salinity throughout, indicating formation of all inclusions under the same set of conditions. It is a common misconception that secondary inclusions can only form from microfractures containing a liquid aqueous phase, but healed fractures containing oil-filled inclusions and gas-filled inclusions are not unusual (Burruss and Goldstein, 1980; Burruss, 1981; Horsfield and McLimans, 1984; Burruss and others, 1985; McLimans, 1987; Tilley and others, 1989; Lacazette, 1991). However, in such cases it is possible that an unobservable surface film of aqueous phase is present in these systems to facilitate necking down. In summary, healing of microcracks to trap secondary fluid inclusions appears to be a relatively rapid geologic phenomenon, and the fluid phase through which the necking down takes place need not be visible even at high magnification.

FLUID INCLUSION ENTRAPMENT BY MICROFRACTURING DURING CRYSTAL GROWTH

Microfractures and other deformation features can form during crystal growth. Fluid inclusions may be entrapped by processes of crack healing similar to secondary fluid inclusions within these features. If crystal growth continues after a deformational episode, then the fluid inclusions entrapped in microcracks retain a record of fluids present after the deformational event but before renewed crystal growth, or at the time the microfracture sealed upon renewed crystal growth. In each of these cases the fluids enclosed along the healed microfracture are fluids that bathed the crystal during a time interval of crystal growth. Such fluid inclusions are termed *pseudosecondary*, and like primary inclusions, record conditions that existed during a period of crystal growth.

CRITERIA TO DETERMINE ORIGIN OF FLUID INCLUSIONS

For most studies involving fluid inclusions, it is crucial for the inclusionist to determine the timing of entrapment of the inclusions relative to the detailed mineral paragenesis, and to accomplish this, an inclusionist should endeavor to locate unequivocal petrographic evidence. Many novices fail miserably in such an undertaking because they do not have the experience requisite for employing appropriate logic. A sound mental process is necessary because the frustrating truth for many samples is that the fluid inclusions required to answer a posed question are not present. Many researchers choose to carry-on in the face of such a fact in hope that "something will fall out." The danger of this approach is that the resulting data are not constrained, so a great potential exists for inappropriate inferences. For the vast majority of inclusions, the origin is either obvious or indeterminate. Thus, there are actually four choices for inclusion origin: primary, secondary, pseudosecondary, or unknown. Most importantly, an inclusionist should learn to report and to be able to defend the petrographic criteria used to determine fluid inclusion origin (discussed below) so that any critic could know the origin unequivocally.

Because petrographic criteria are used to determine *origin of the fluid inclusion vacuole,* and petrographic evidence of subsequent leakage and refilling of that vacuole may be lacking, this book will use the terms primary, secondary, and pseudosecondary only in the petrographic sense for describing the timing and origin of the original fluid inclusion vacuole, rather than using the terms in a genetic sense. The following discussion summarizes petrographic criteria that are useful for determining the origin of fluid inclusions.

Primary Fluid Inclusions

Primary fluid inclusions are best identified by their relationship to growth zonation of a crystal. Growth zonation is identified through variations in the distribution of fluid or solid inclusions that mimic crystal terminations, or compositional changes revealed by back-scattered electron images, cathodoluminescence, UV fluorescence, color variations in transmitted light, or other optical relationships. In the absence of demonstrating a relationship to growth zonation, evidence for primary origin is typically ambiguous at best in diagenetic minerals. Unfortunately, too many published papers fail to document their criteria for a primary origin — a fatal flaw — or have used inappropriate or ambiguous evidence for a primary origin, such as "large and isolated inclusions," "randomly distributed inclusions," or "negative crystal shape." In addition to simple relationships to growth zonation, there are other characteristics of primary fluid inclusions, specific to various minerals, that are useful for identifying them as primaries. The following summarizes some common petrographic characteristics of primary fluid inclusions specific to various minerals. This summary should not be viewed as "all inclusive," but should be employed only as an aid for identifying primary fluid inclusions. The figures are somewhat idealized and do not include the myriad of non-diagnostic occurrences; they should be supplemented with the researcher's logic of "what defines a relationship to growth" for each mineral as it is studied.

Calcite.—
Many of the diagnostic characteristics of primary fluid inclusions in calcite cement are illustrated in Figure 2.5. Note that the most common useful criterion is the occurrence of fluid inclusions confined by growth-zone boundaries (Fig. 2.5A-U). growth zones that are each defined by a single sheet of fluid inclusions (Fig. 2.5H-P), a crystal that is cloudy.

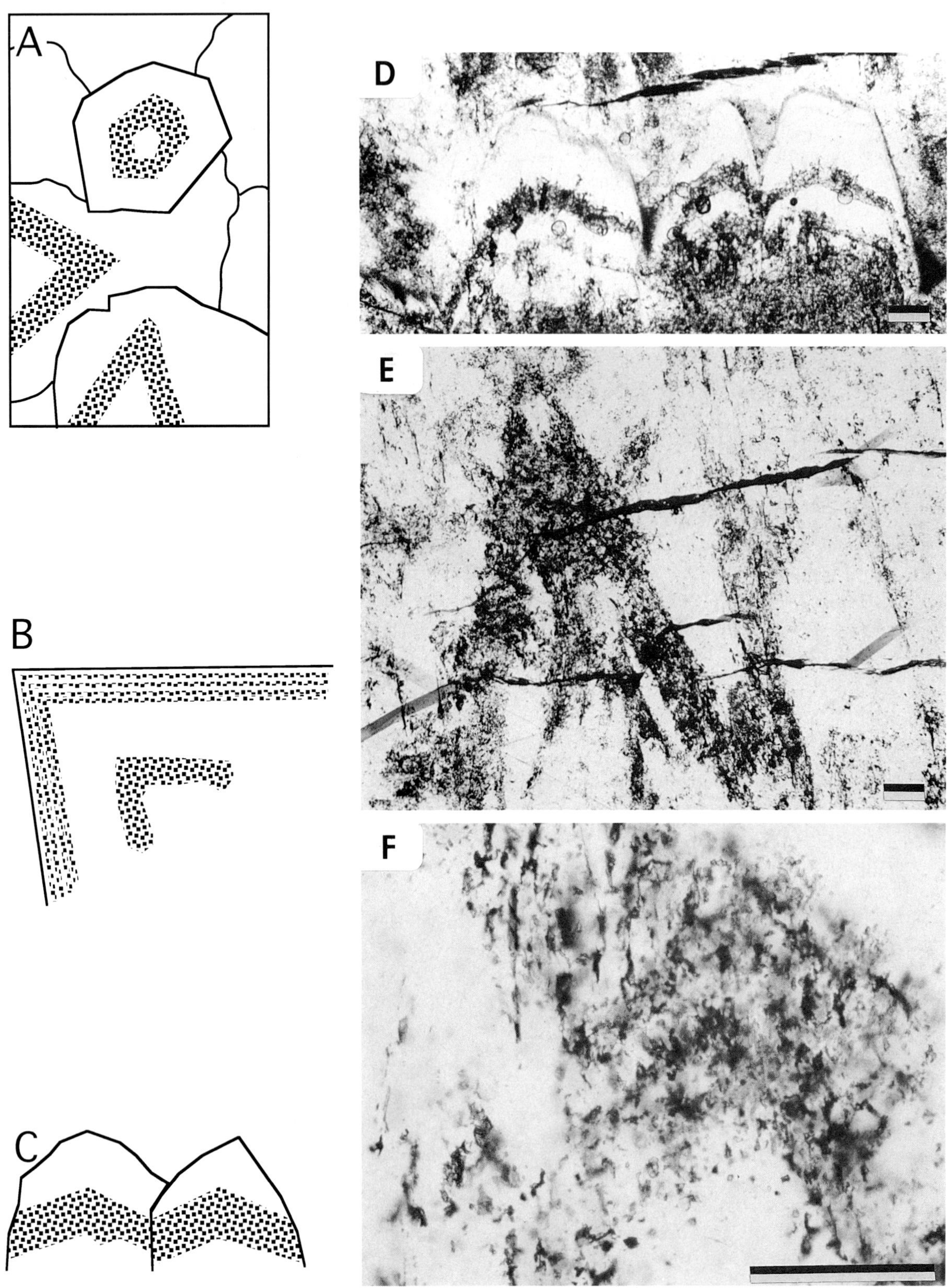

Fig. 2.5. Master diagram of occurrence of primary fluid inclusions in calcite (A-Z). Photomicrographs are in plane polarized light with scale of 100 µm. See text for explanation of each.

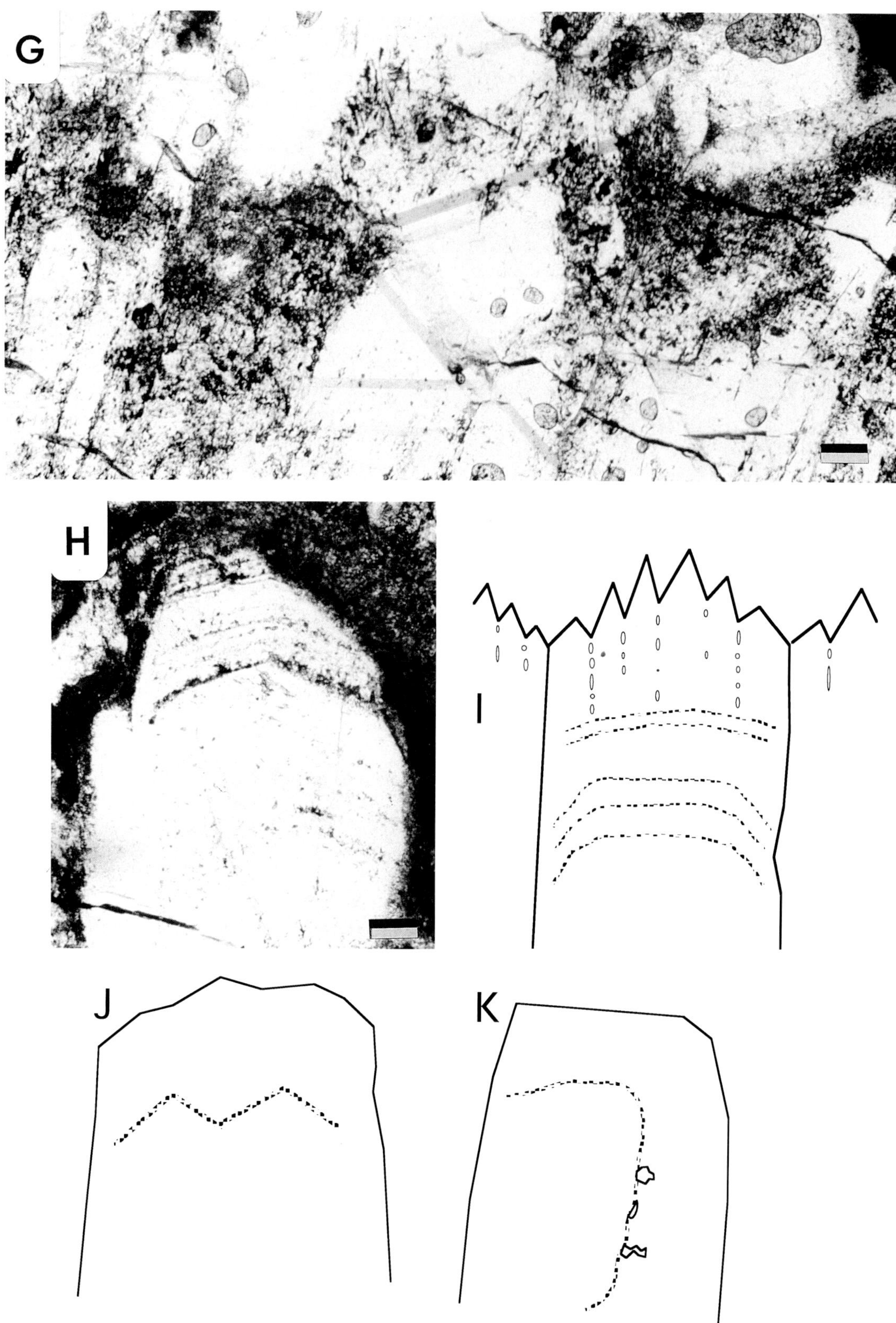

Fig. 2.5. Continued from previous page.

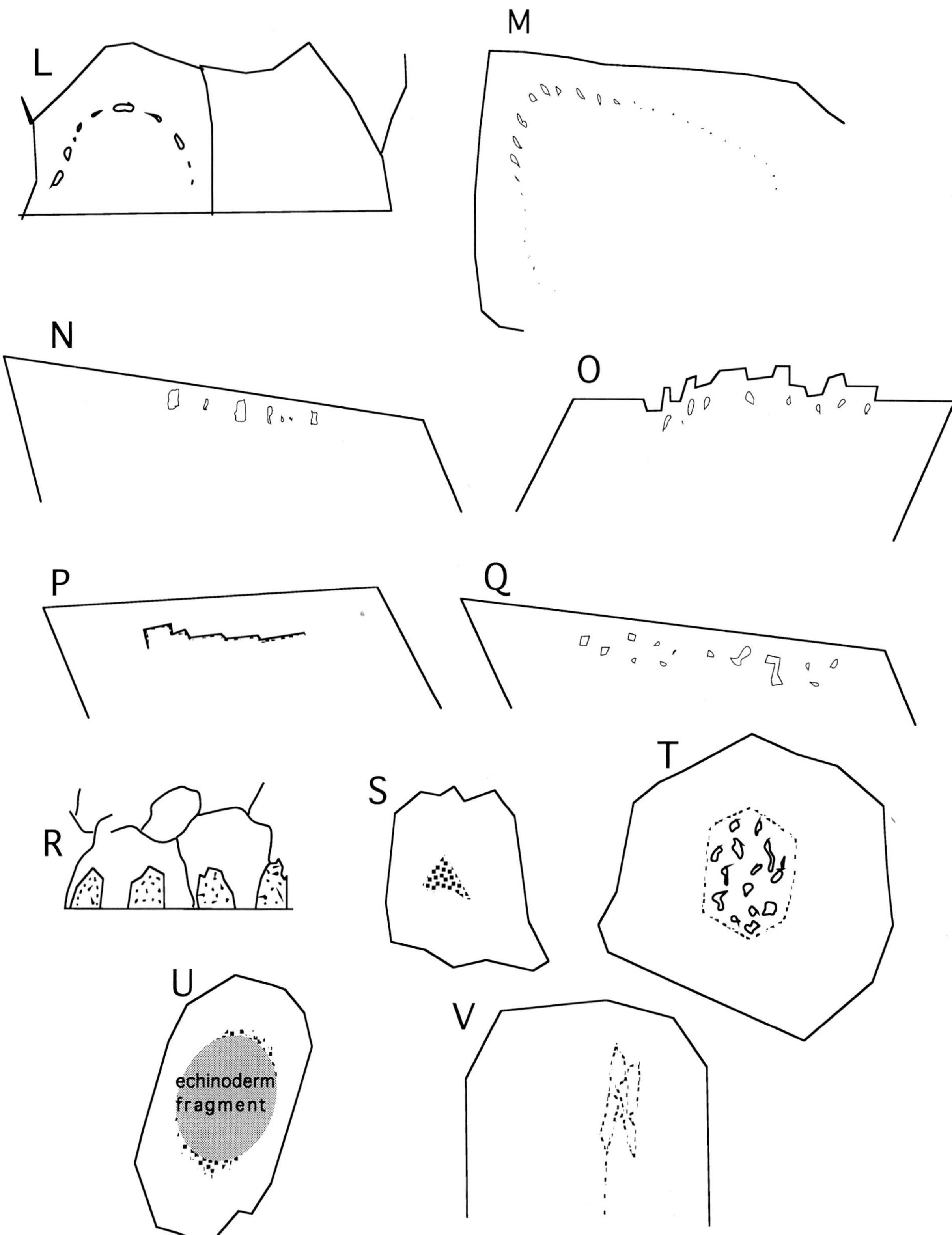

Fig. 2.5. *Continued from previous pages.*

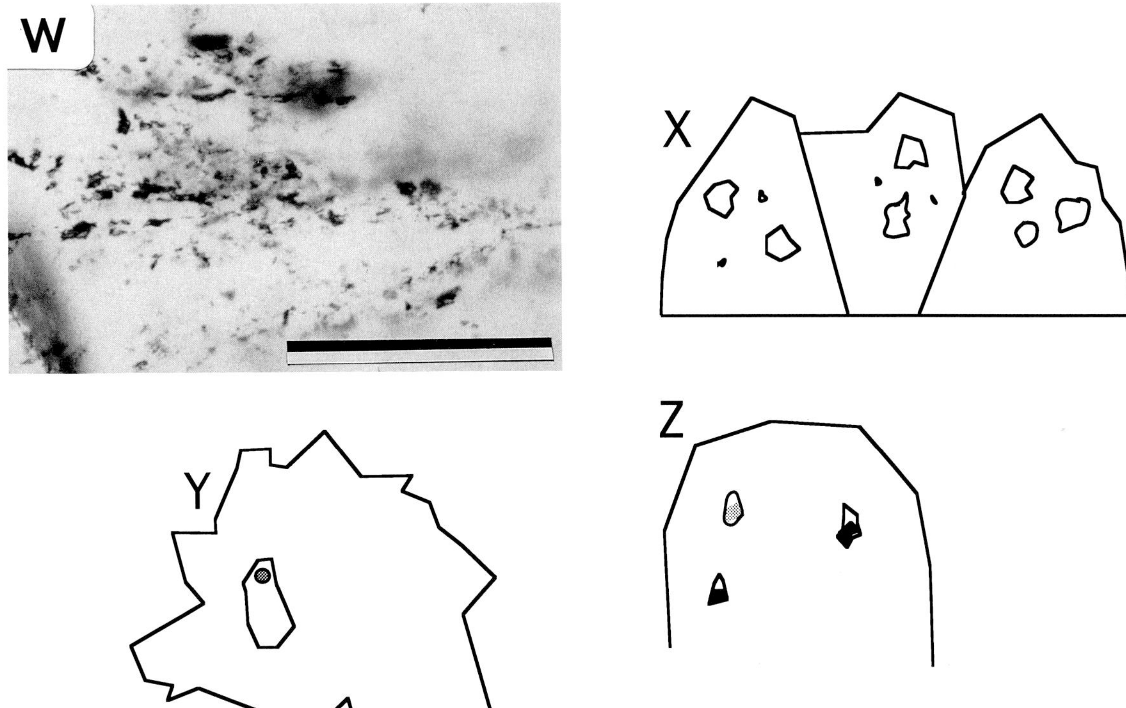

Fig. 2.5. Continued from previous pages.

These could include thick growth zones containing thousands of randomly distributed fluid inclusions (Fig. 2.5A-G), with fluid inclusions but with variation in inclusion abundances that can be ascribed to growth zonation (Fig. 2.5B, F), multiple inclusions that exclusively originate on, or are confined to, a single growth zone (Fig. 2.5A-U), or a crystal rich in randomly distributed fluid inclusions that becomes deficient in fluid inclusions at a growth zone boundary (Fig. 2.5D, R, T, U). Sometimes complex growth forms precipitate with multiple terminations, and linear trails of inclusions below termination re-entrants are generated (Fig. 2.5E, G, I). Composite crystals of many optically continuous subcrystals may trap planar and three-dimensional arrays of fluid inclusions at boundaries between subcrystals of a composite crystal (Fig. 2.5E, G, V, W). Rarely, some calcite crystals may contain single or several fluid inclusions that are extremely large compared to the crystal size (Fig. 2.5X, Y); beware, for some apparently isolated inclusions may be secondary in origin and should be evaluated with extreme skepticism. During crystal growth, contamination of the growth surface with another crystal of the same or a different mineral, or contamination with an immiscible fluid phase, can result in entrapment of a primary fluid inclusion in the growth wake of the contaminant (Fig. 2.5Z). These contaminant crystals commonly are called *accidental* solid inclusions and are here termed *accidentals*. Accidentals can be distinguished from *daughter minerals* that have precipitated from the inclusion fluid after entrapment: inclusions trapped at the same time and that later formed daughter minerals should have the same daughter mineral/fluid ratio (if the daughter has been able to nucleate), whereas inclusions with accidentals will lack such consistent phase ratios. Some calcite crystals may have cloudy cores rich in fluid inclusions randomly distributed in all three dimensions; at least two faces of the core should terminate against a growth zone boundary to demonstrate a primary origin (Fig. 2.5S). It may be impossible to distinguish between cloudy cores of primary origin versus those of pseudosecondary origin.

Primary fluid inclusions in calcite may have almost any shape ranging from negative crystal to globular (smooth-surfaced) to extremely irregular. In some samples inclusions of a particular origin may have a particular shape, and this shape may be helpful in distinguishing between primary and secondary fluid inclusions in that sample; however, there is no general correlation between inclusion shape and origin. Most flat inclusions are probably more likely secondary than primary, but some primary fluid inclusions are flat as well. Some fluid inclusions may retain a particularly unique morphological feature that may prove helpful in determining origin. For example, some primary fluid inclusions originate along the same growth surface and retain a flat base on that surface (Fig. 2.5K). Other inclusions may retain shapes and sizes that mimic fibrous or bladed growth textures. Some primary fluid inclusions become larger and more elongate along certain crystallographic directions (Fig. 2.5M).

Dolomite and ankerite.—

Dolomite and dolomite-like minerals preserve a variety of fabrics indicating primary origin of fluid inclusions. As in all minerals, the most important criterion for identifying primary fluid inclusions is inclusion distribution that is controlled by growth zonation (Fig. 2.6). It is important that secondary fluid inclusions trapped along the cleavage of carbonate minerals not be confused with growth zonation. Such a distinction is rather straightforward: growth zones can normally be traced to a point where they turn and parallel another crystal face, as is evident in Figure 2.6. The most common occurrence of primary fluid inclusions in dolomite is as cloudy, rhombohedral or baroque, inclusion-rich cores surrounded by clearer rims (Fig 2.6A, B). Other dolomites may have broad, inclusion-rich zones that are sharply bounded by clear zones (Fig 2.6C, D, E). Some thin growth zones may be defined by a single thickness of fluid inclusions along the growth zone (Fig. 2.6F, G). Dolomite crystals may be totally cloudy with fluid inclusions, but to ascribe a primary origin to those inclusions, the density of their distribution must vary and mimic growth zonation (Fig. 2.6H). However, in crystals with such a dense distribution of inclusions, selection of individual inclusions of definite primary origin may be impossible because of the potential presence of planes of secondary inclusions that are not visually discernible. In many dolomite and ankerite crystals, an internal area of elongate fluid inclusions is common and shows a divergent, "fir tree pattern" (Fig. 2.6I). Wojcik (1991) has used back-scattered electron imaging to show that such fluid inclusions are in fact related to replacement of internal zones with later dolomite or dolomite-like phases; therefore, the fir-tree pattern is a primary texture that may record one or multiple replacement events. Larger, box-shaped fluid inclusions

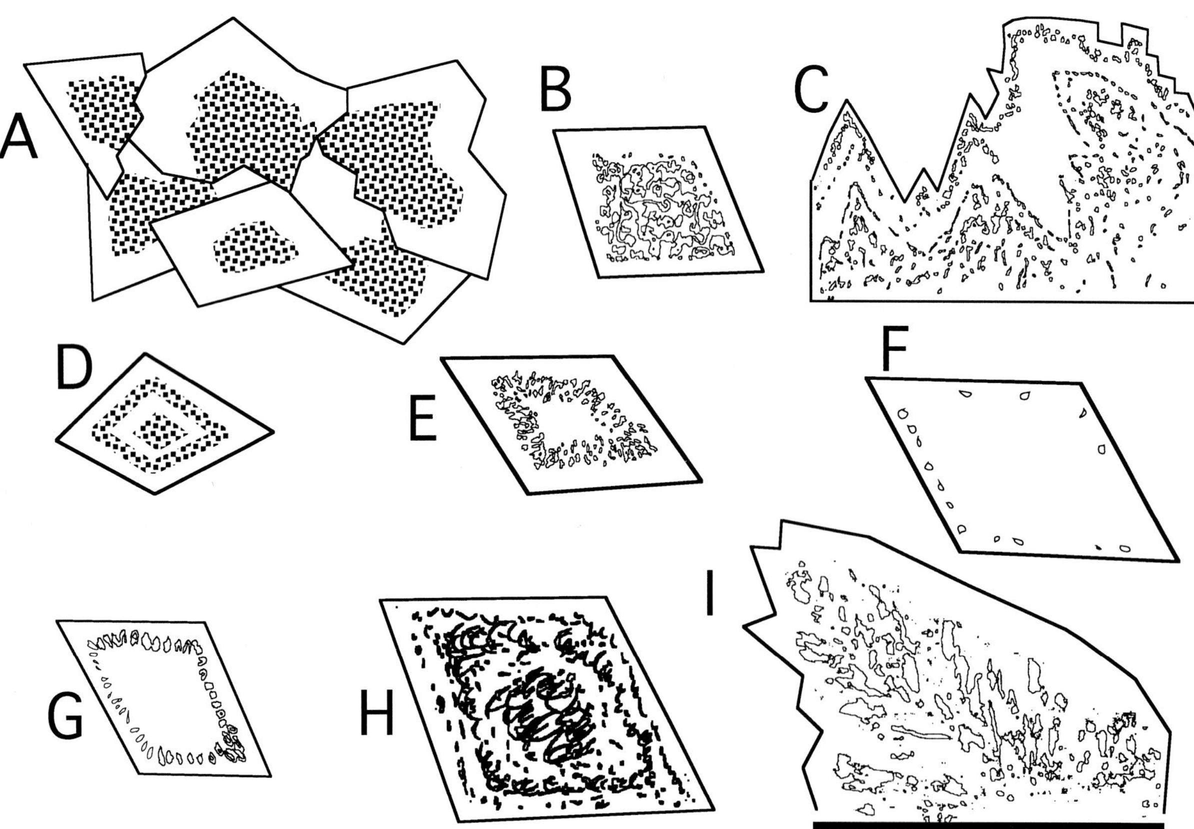

Fig. 2.6. Diagram illustrating occurrence of primary fluid inclusions in dolomite. See text for explanation.

are found in association with the elongate inclusions comprising the "fir tree pattern," and are also distributed in isolated areas of dolomite crystals. Wojcik (1991) has shown through back-scattered electron imaging that these box-shaped inclusions are entrapped during replacement events, but may not show a relationship to growth in normal transmitted light. Thus, a careful fluid inclusionist must treat such inclusions as being of unclear origin until a relationship to growth is identified through observation of compositional growth textures.

Fluid inclusions in dolomite may have any shape ranging from negative crystal to extremely irregular. Some may be elongate in the direction of growth, or elongate if trapped in a replaced internal zone. Most primary fluid inclusions in dolomite are <5 μm.

Quartz.—

Quartz overgrowths and coarse quartz cements (megaquartz) preserve a variety of fabrics indicative of primary origin for fluid inclusions. Studies of chalcedony and chert have not commonly identified fabrics indicative of primary fluid inclusions in the original unrecrystallized mineral. For quartz overgrowths and megaquartz cements, a relationship to growth of the crystal is generally easy to identify, and can be confirmed readily by use of SEM-cathodoluminescence (Guscott and Burley, 1993) or hot-cathodoluminescence techniques (Ramseyer and others, 1989; Burley and others, 1989; Walker and Burley, 1991).

Primary inclusions are common along the surface ("dust rim") separating the detrital grains from the authigenic overgrowth, and are sometimes found within the overgrowth as well (Fig. 2.7A, B). Inclusions along the detrital grain/overgrowth boundary typically parallel the boundary. They are commonly small (<2 μm), but can rarely be large (>10 μm). Shapes range from highly irregular to equant, and surfaces rough to smooth. Solid inclusions (e.g., clay minerals) that occur on the "dust rims" may act as sites of imperfections for entrapment of inclusions (Fig. 2.1F), but may also render the overgrowth boundaries weaker relative to the quartz on either side: physical weakness may allow the boundary to crack open, permitting new fluids to enter and be entrapped; chemical instability may allow later fluids to selectively dissolve the minerals at the boundary, enabling new fluids to permeate and be subsequently enclosed in new, younger fluid inclusions. Thus, it is a possibility that some inclusions at detrital grain/overgrowth boundaries do not contain the earliest diagenetic fluids from which authigenic quartz precipitated.

Primary fluid inclusions within authigenic overgrowths and within megaquartz cements are characterized by their alignment along concentric growth zones (Fig. 2.7C,D) that exist due to successive, discrete episodes of quartz precipitation (Guscott and Burley, 1993); or may be related to growth defects that propagate and cause a vacuole to be formed (Guscott and Burley, 1993). Again, sizes typically are small (<2 μm), but can exceed 10 μm in longest dimension. Shapes also range from irregular to equant, and surfaces are rough to smooth, to negative crystal-faceted. Fluid inclusions of larger sizes are more commonly found in coarse megaquartz cements and in quartz overgrowths in relatively coarse, matrix-poor sandstone. Primary inclusions may be parallel to the growth zonation, but are commonly parallel to growth direction (normal to the detrital grain substrate). Some quartz cements contain V-shaped reentrants that are rich in fluid inclusions that were trapped during crystal growth (Fig. 2.7E). Quartz crystals may contain extremely large inclusions (relative to the size of the enclosing crystal) that appear isolated (Fig. 2.7F). Some inclusionists misapply this criterion for primary origin by calling small, apparently isolated inclusions primary when they are actually secondary; such occurrences of large isolated primary inclusions are rare, but when they occur, they are obvious and easily documented.

Feldspar.—

Primary fluid inclusions in authigenic feldspar can be found in either replaced detrital grains or in overgrowths. In partially replaced detrital grains, the replaced part of the grain may contain primary fluid inclusions whereas the unaltered part of the grain remains relatively free of fluid inclusions or contains fluid inclusions of detrital origin (Fig. 2.8A, B). One must be able to show that the fluid inclusions are confined exclusively to the replaced part of the grain to identify a primary origin. Another potential problem is to prove that the timing of replacement is postdepositional. In overgrowths, primary origin may be ascribed to fluid inclusions entrapped within the overgrowth; these inclusions commonly begin at the outer margin of the detrital grain (Fig. 2.8B, C).

Primary fluid inclusions in authigenic feldspars may have elongate shapes. Some are relatively blocky in form, but may have a negative crystal shape on one side and an irregular to hackly appearance on the other side. Others have a negative crystal shape.

Halite.—

Primary fluid inclusions are common in halite. Many are identified by their occurrence as cloudy, three-dimensional arrays of fluid inclusions that end abruptly at growth zone boundaries and are overgrown by clear halite (Fig. 2.9A, B). Some primary fluid inclusions

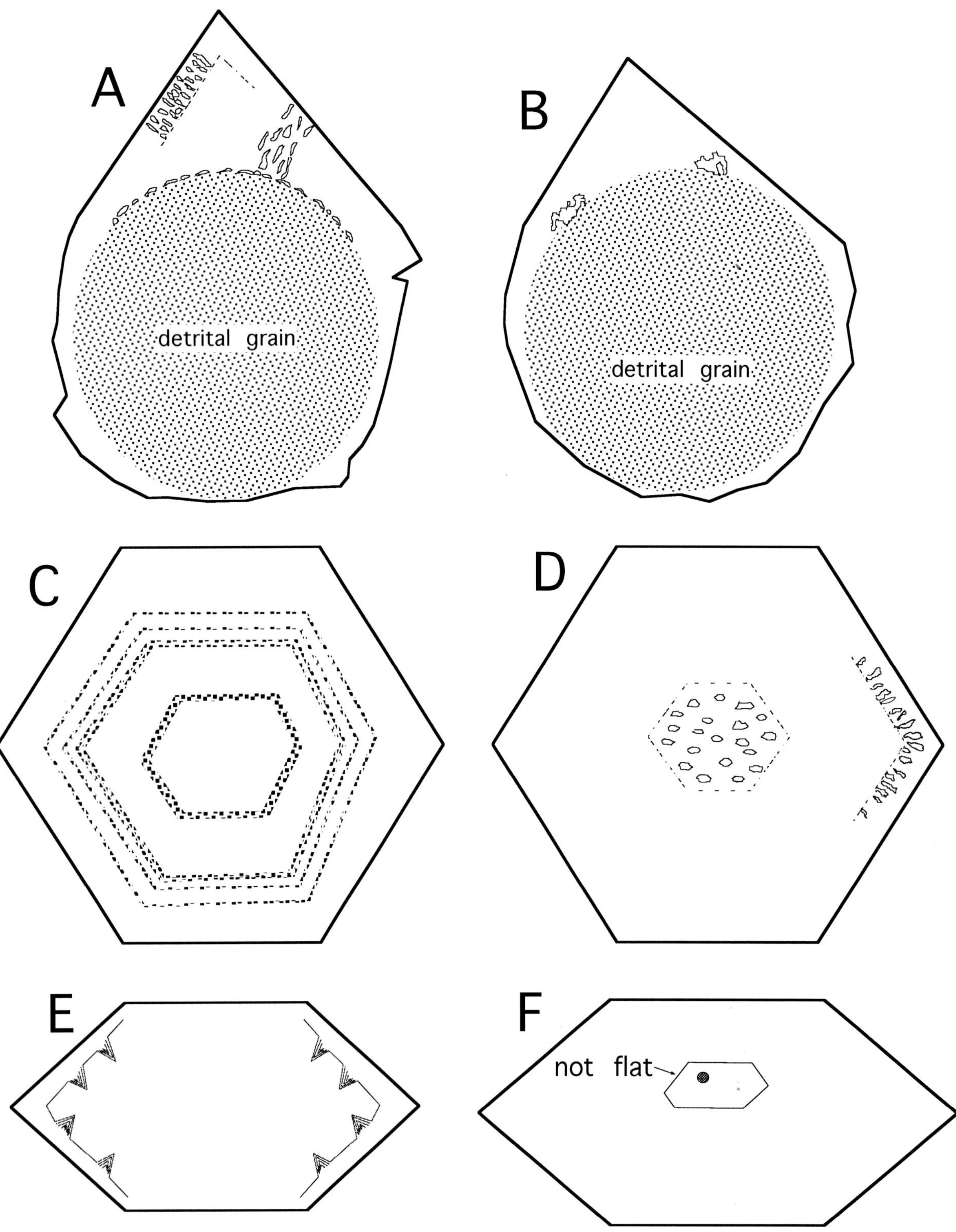

Fig. 2.7. *Sketches of primary fluid inclusions in authigenic quartz overgrowths (A, B) and in megaquartz cement (C-F). See text for details.*

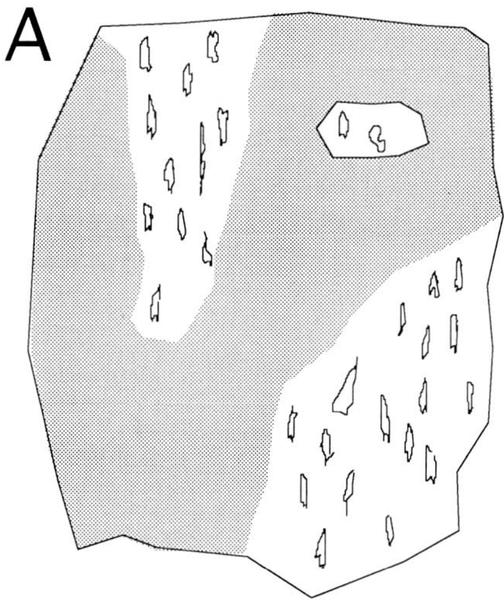

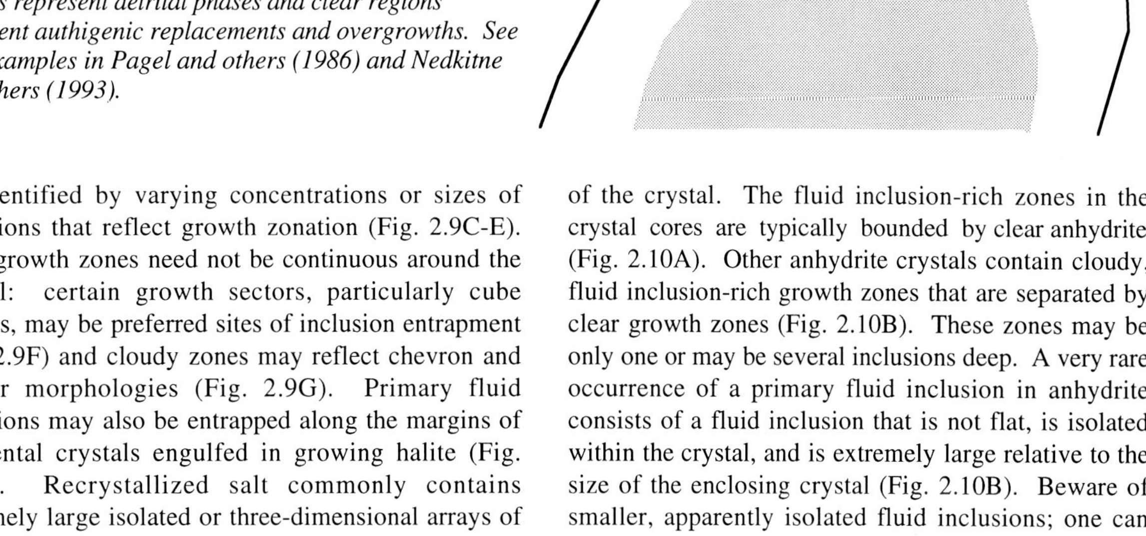

Fig. 2.8. *Diagram illustrating occurrence of primary fluid inclusions in authigenic feldspar. Shaded regions represent detrital phases and clear regions represent authigenic replacements and overgrowths. See also examples in Pagel and others (1986) and Nedkitne and others (1993).*

are identified by varying concentrations or sizes of inclusions that reflect growth zonation (Fig. 2.9C-E). Such growth zones need not be continuous around the crystal: certain growth sectors, particularly cube corners, may be preferred sites of inclusion entrapment (Fig. 2.9F) and cloudy zones may reflect chevron and hopper morphologies (Fig. 2.9G). Primary fluid inclusions may also be entrapped along the margins of accidental crystals engulfed in growing halite (Fig. 2.9H). Recrystallized salt commonly contains extremely large isolated or three-dimensional arrays of fluid inclusions that are also primary in origin (Fig. 2.9I), but are recording the conditions at which recrystallization occurred.

Most fluid inclusions in halite have a negative crystal morphology. The sizes of fluid inclusions in unrecrystallized salt vary from submicrometers to many tens of micrometers Many of the fluid inclusions in recrystallized salt are large, and fluid inclusions that are centimeters in size can be viewed in many museums.

Anhydrite.—

Primary fluid inclusions in anhydrite are best identified by a relationship to growth zone boundaries. The most common occurrence of primary fluid inclusions in anhydrite are three-dimensional arrays of tubular fluid inclusions that are confined to a core area of the crystal. The fluid inclusion-rich zones in the crystal cores are typically bounded by clear anhydrite (Fig. 2.10A). Other anhydrite crystals contain cloudy, fluid inclusion-rich growth zones that are separated by clear growth zones (Fig. 2.10B). These zones may be only one or may be several inclusions deep. A very rare occurrence of a primary fluid inclusion in anhydrite consists of a fluid inclusion that is not flat, is isolated within the crystal, and is extremely large relative to the size of the enclosing crystal (Fig. 2.10B). Beware of smaller, apparently isolated fluid inclusions; one can easily imagine ways by which some secondary fluid inclusions could appear isolated.

Most fluid inclusions in anhydrite have a tendency toward a rather tubular shape. However, rectangular, negative crystal shapes are also known.

Gypsum.—

Primary fluid inclusions in gypsum are identified by a relationship to growth surfaces. Surprisingly, gypsum has not been studied extensively for fluid inclusions, perhaps because of the obvious problem with its conversion to anhydrite. However, work by Sabouraud-Rosset (1972, 1976) summarized some aspects of occurrence and chemistry of fluid inclusions in gypsum. Elongate primary fluid inclusions in gypsum typically occur along the (001) plane (Fig.

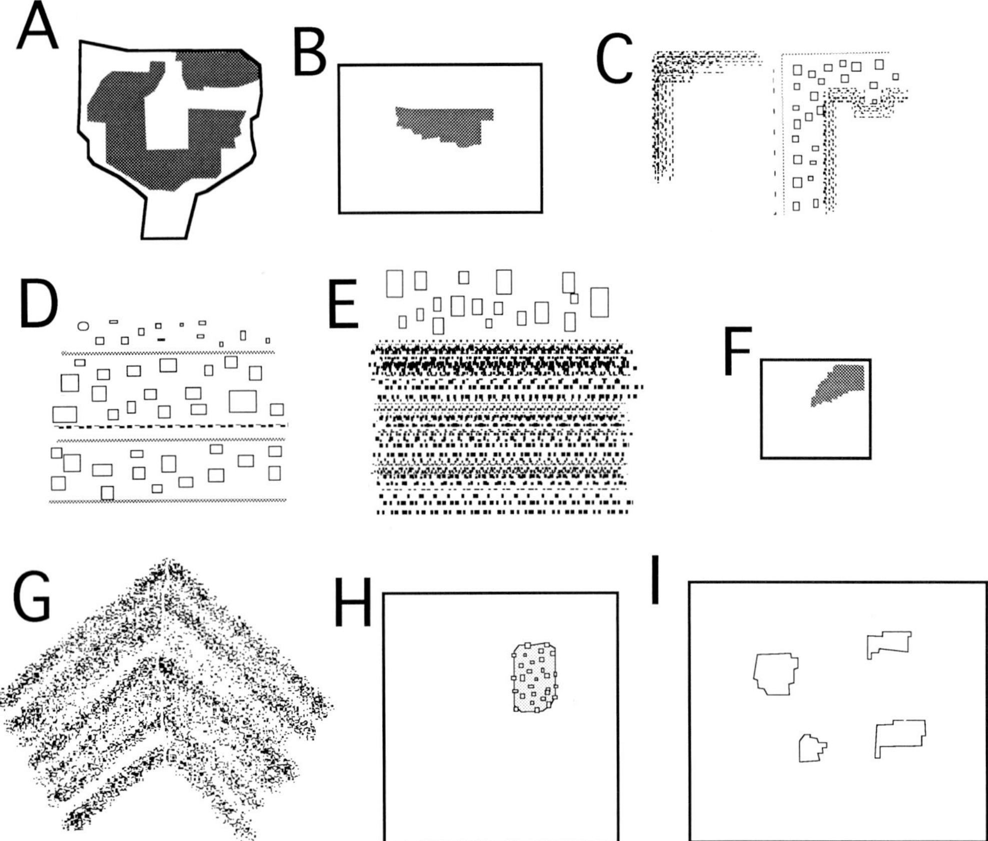

Fig. 2.9. Schematic illustration of primary fluid inclusions in halite. ***A, B, F)*** *Distribution of primary fluid inclusions in some individual crystals of halite.* ***C, D, E)*** *Growth zonation defined by distribution of fluid inclusions.* ***G)*** *A chevron structure defined by variation in density of primary fluid inclusions in growth zones. Notice the upward original orientation and septum that bisects the chevron.* ***H)*** *A single crystal of halite that has included a large crystal of anhydrite; fluid inclusions are trapped along the boundary between the anhydrite and halite.* ***I)*** *Large fluid inclusions in recrystallized halite.*

2.11A) and more equidimensional to elongate inclusions occur normal to the ($\bar{1}03$) plane (Fig. 2.11A-C). Many of these growth zones are defined by high concentrations of large inclusions and intervening areas of lower inclusion concentration.

Most fluid inclusions in gypsum preserve a spike-like shape or a negative crystal morphology. The tip of the spike typically points in the direction of growth. Inclusions are commonly large and may be tens of micrometers in size.

Fluorite.—

Fluorite is the one mineral found in diagenetic environments that may commonly contain primary inclusions that are identifiable by being truly isolated (Fig. 2.12A, B). Nevertheless, some apparently isolated inclusions in fluorite may be secondary inclusions and should be evaluated carefully. Primary inclusions are best identified by relationships to crystal growth zonation. This may include thick growth zones containing thousands of randomly distributed inclusions (Fig. 2.12C); thin growth zones that are defined by an array of inclusions only as wide as a single inclusion (Fig. 2.12B); inclusions that exclusively originate on, or are confined to, a single growth zone that is defined by features other than inclusions, such as color banding due to compositional differences (Fig. 2.12C); a crystal with abundant randomly distributed inclusions that becomes deficient of fluid inclusions at a growth boundary (Fig. 2.12D, E); and contamination of the growth surface with another crystal, or contamination with an immiscible fluid phase, that can result in entrapment of a primary fluid inclusion in the growth wake of the contaminant.

Primary fluid inclusions in fluorite can be any shape, ranging from negative crystal morphology to highly irregular. Most often, isolated inclusions are of negative crystal shape, but isolated inclusions in some

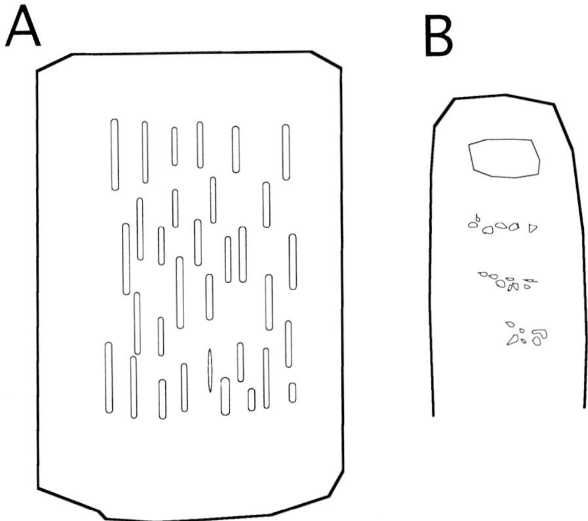

Fig. 2.10. *Sketches showing the distribution of primary fluid inclusions in anhydrite.* ***A)*** *Elongate fluid inclusions are confined to the growth-zone bounded core of the crystal.* ***B)*** *A single large inclusion is enclosed in the tip of the crystal and smaller inclusions are concentrated in growth zones (as illustrated by Dix and Jackson, 1982).*

samples can be highly irregular in shape as well. For example, primary inclusions trapped due to contamination of the growth surface can be highly irregular in shape. Oil-filled fluid inclusions may have a globular shape or even a bell-like or meniscus shape if been entrapped by contamination of the growth surface with an oil droplet. Some crystals of fluorite contain they have large tubular fluid inclusions that form normal to the growth direction (Fig. 2.12E). Inclusion-rich growth zones typically contain smooth-surfaced primary inclusions. Fluid inclusions in fluorite rank second only to halite in the abundance of large inclusions (>7 µm).

Secondary and Pseudosecondary Fluid Inclusions

After crystal growth, minute cracks, shear planes, or deformational twin planes may form in a crystal. The cracks and planes may heal to trap small samples of the fluid present during healing, providing a record of the fluids present after growth of a crystal. If such inclusions are trapped after crystal growth is complete, they are termed secondary; if they are trapped before crystal growth is complete, they are termed pseudosecondary. Secondary fluid inclusions may appear to cut across any or all growth zones of a crystal. Pseudosecondary inclusions terminate up against a growth-zone boundary. A single crystal may contain primary, secondary, and pseudosecondary fluid inclusions, as well as many inclusions of indeterminate origin.

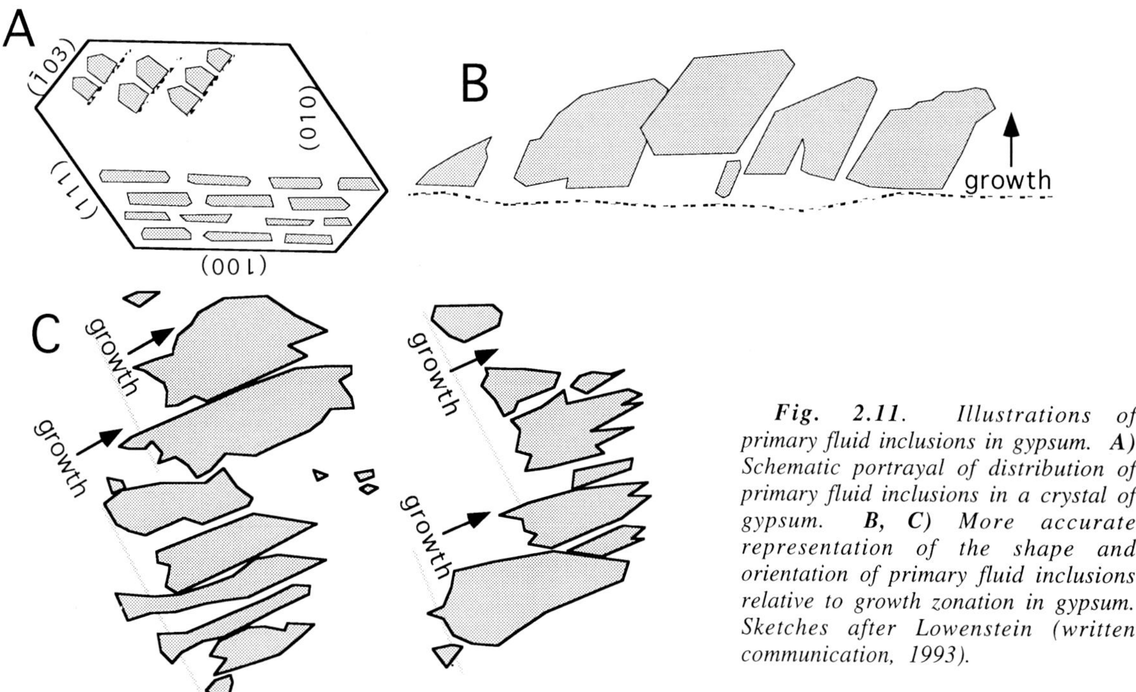

Fig. 2.11. *Illustrations of primary fluid inclusions in gypsum.* ***A)*** *Schematic portrayal of distribution of primary fluid inclusions in a crystal of gypsum.* ***B, C)*** *More accurate representation of the shape and orientation of primary fluid inclusions relative to growth zonation in gypsum. Sketches after Lowenstein (written communication, 1993).*

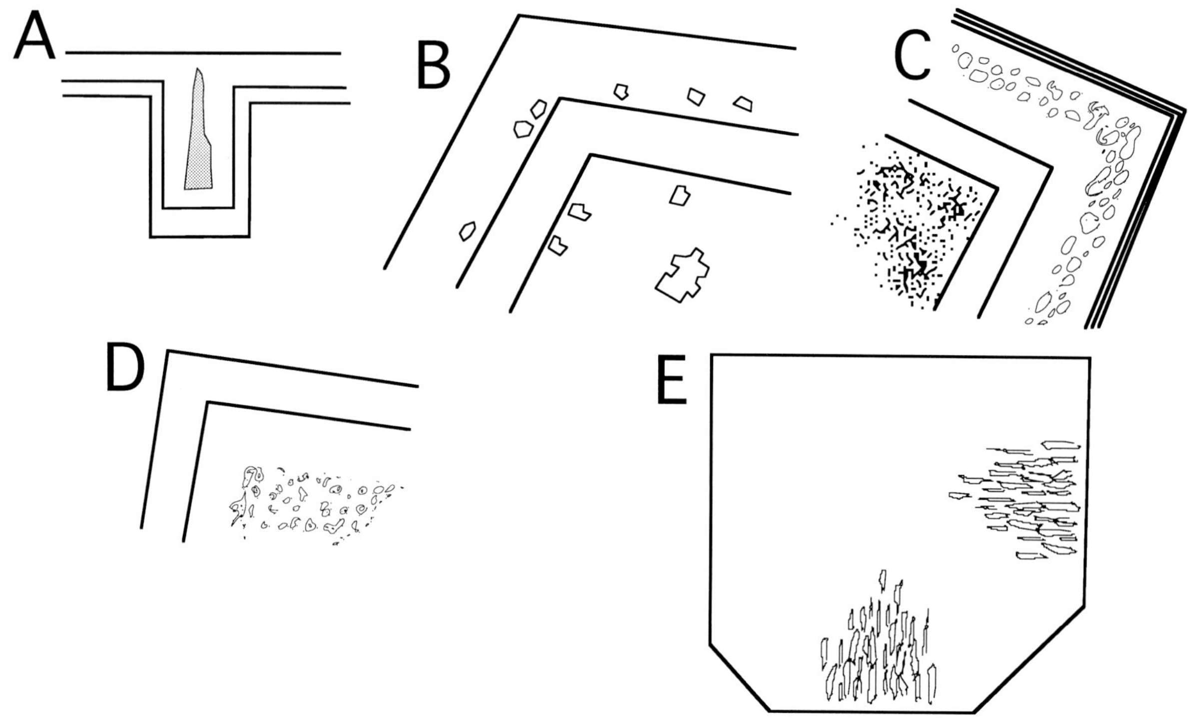

Fig. 2.12. Sketches showing primary fluid inclusions in fluorite. See text for discussion. A and E are modified from Roedder (1972, 1984).

Secondary fluid inclusions commonly occur in planar arrays or along curved surfaces that cut across growth zonation (Fig. 2.13). In many samples, the abundance of secondary fluid inclusions is so high, that individual planes cannot be discerned and the overall character appears to be a relatively random three-dimensional distribution of fluid inclusions. In such a case it may be impossible to distinguish between primary and secondary fluid inclusions, and it is best just to admit that the origin is unknown for such fluid inclusions. In other cases, the fluid inclusions trapped along planes may be so widely distributed that the planar arrays are not apparent, and the fluid inclusions may appear to be isolated. This type of distribution is commonly difficult to distinguish from the occurrence of some isolated primary fluid inclusions, so it is always appropriate to carefully view samples under low magnification (10X objective) to attempt to evaluate this possibility. However, rare isolated inclusions that are extremely large relative to the size of the crystal are more likely primary than secondary. Many secondary fluid inclusions are trapped along cleavage directions or along twin planes. One must be careful that the crystallographic control on these inclusions are not confused with primary fluid inclusions along growth zones. In some cases, the distinction is relatively easy because the cleavages and twin planes commonly do not parallel the growth zones or crystal terminations that are present (e.g., cleavages in fluorite versus cubic plane), and when planes of secondary fluid inclusions are crystallographically controlled, the planes cross one another, whereas primary inclusions in a growth zone would mimic crystal terminations.

Pseudosecondary fluid inclusions bear all of the characteristics of secondary fluid inclusions except that the planar arrays of fluid inclusions end abruptly at a growth zone boundary (Fig. 2.13). If there is one such occurrence, other examples of planes terminating at other localities of the same growth zone boundary should be located for unequivocal evidence of pseudosecondary origin. It is important to note that all fractures end somewhere, and that a single healed crack ending within the crystal rather than cutting across its entirety is not evidence of pseudosecondary origin. One must identify the consistent occurrence of termination of healed cracks at a growth zone boundary to call on a pseudosecondary origin. In general, evidence for pseudosecondary origin is much less frequently found than is evidence for primary origin in diagenetic phases.

SUMMARY

This chapter has shown that there are abundant and useful criteria for identifying the origin of the fluid

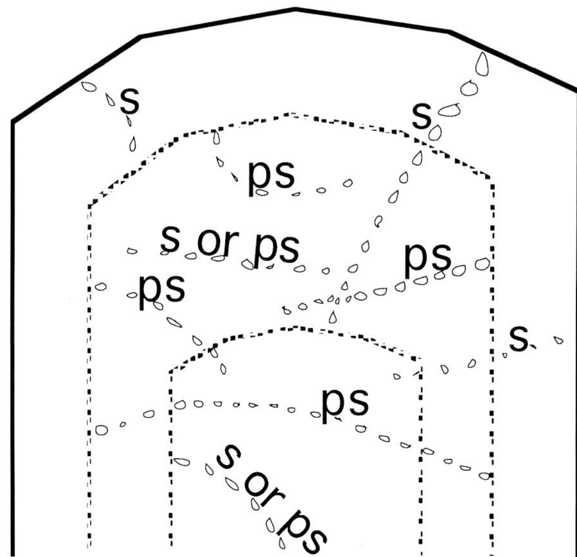

Fig. 2.13. *Sketch illustrating distinction between secondary (s) and pseudosecondary (ps) fluid inclusions. All inclusions occur in linear arrays because they were trapped along healed fractures. Secondary inclusions are trapped after crystal growth is complete, but pseudosecondary inclusions are trapped before final growth zones have formed. Notice that identification of pseudosecondary fluid inclusions requires that multiple healed fractures end at the same growth-zone boundary. A single healed fracture ending at a growth-zone boundary does not allow distinction between secondary and pseudosecondary fluid inclusions because a secondary healed fracture could end at a growth-zone boundary.*

studies, the petrographic relationships will be present inclusion vacuole in diagenetic minerals. For many and the study may proceed to answer the question posed. *The established criteria for fluid inclusion origin should always be carefully documented in the final report of the research; merely stating that the inclusions are primary or secondary is not sufficient.* However, never forget that the required criteria may often not be present in a sample suite, even after careful and exhaustive search of a large number of samples. In such a case, the researcher must either accept the conclusion that the origins of the inclusions are unknown, and that any data collected from the inclusions are limited to a composite record of various fluids that have been in the rock during its history (sometimes very useful information), or they must abandon the inclusion study altogether.

Chapter 3

PHASE CHANGES IN FLUID INCLUSIONS: THE BASICS

INTRODUCTION

This chapter considers the phase relations of simple systems applicable to aqueous and petroleum fluid inclusions in sedimentary environments. One cannot appreciate the potential power of fluid inclusion observations, or the inherent limitations of the technique of microthermometry, without a sound understanding of the phase equilibria and their representation as phase diagrams. The equilibrium phase relations provide the link between the laboratory measurements of temperatures at which phase transitions occur within inclusions when heated and cooled and the interpretations of the measurements. Without knowledge of the phase diagrams, one cannot be a careful, critical inclusionist because one will not know the appropriate phase transitions to look for, and one will not know the assumptions upon which interpretations of phase transitions are made. Even the most basic petrographic approach to the fluid inclusion technique begins with a strong foundation in phase equilibria. These fundamental principles are elucidated in this chapter by first evaluating the phase relations of the unary systems pure H_2O and pure CH_4. Then, the effects of adding NaCl to water are evaluated, followed by a treatment of the effects of adding CH_4 to water. Finally, a generalized treatment of the phase relations of petroleum fluids are considered. A more detailed presentation of phase behaviors that are observed within inclusions, as well as discussions of inherent observational and interpretative limitations are presented in Chapter 7.

SINGLE COMPONENT SYSTEM H_2O

A system of variable state is determined by the four quantities of pressure (P), volume (V), temperature (T), and composition (X). For an unary system, the transitions between the three basic physical states (solid, liquid, gas) may be affected by changing one or more of the parameters P, V, and T. Thus, a universal representation of the system can be shown in a P-V-T diagram. Figure 3.1A is such a diagram generalized for the system pure H_2O. Two useful projections, P-T, and T-V are shown to true scale in Figures 3.1B and 3.1C, respectively.

Basic Assumptions

In order to apply these graphical representations of the single component system pure H_2O shown in Figure 3.1 to fluid inclusions, several basic assumptions or requirements are necessary. First, it must be assumed that host crystals are impermeable to chemical changes and that an inclusion represents a chemically closed (isoplethic) system from the time of entrapment. Second, because of the rigidity of host crystals, and because fluids have much greater coefficients of thermal expansion than the enclosing host crystal, it must be assumed (as a first approximation) that the volume of an inclusion remains constant (isochoric) since entrapment. This assumption of constant volume is actually an approximation, as internal variations in pressure may cause minor elastic deformation, and removal of the crystal from higher P and T conditions may cause minor changes in the inclusion volume (Lacazette, personal communication, 1993). Also, it should be made clear that a careful inclusionist must never make assumptions without first attempting to evaluate whether or not they are justifiable for each inclusion measured, for nature is rarely as simple as our phase diagrams, and these two assumptions are sometimes not justifiable. Procedures for making such evaluations are presented in subsequent chapters. For the purpose of understanding the fundamental principles presented in this chapter, the two requirements that an inclusion represents an isoplethic (chemically closed) and isochoric (constant volume) system since entrapment will be assumed to be true. For the unary system pure H_2O, it would then follow that a fluid inclusion's specific volume (inverse of bulk density) would be constant throughout its history, and that the phase changes it experiences are those shown along sections of constant specific volume in Figure 3.1A. Furthermore, because the temperature of an inclusion is externally controlled, the pressure within

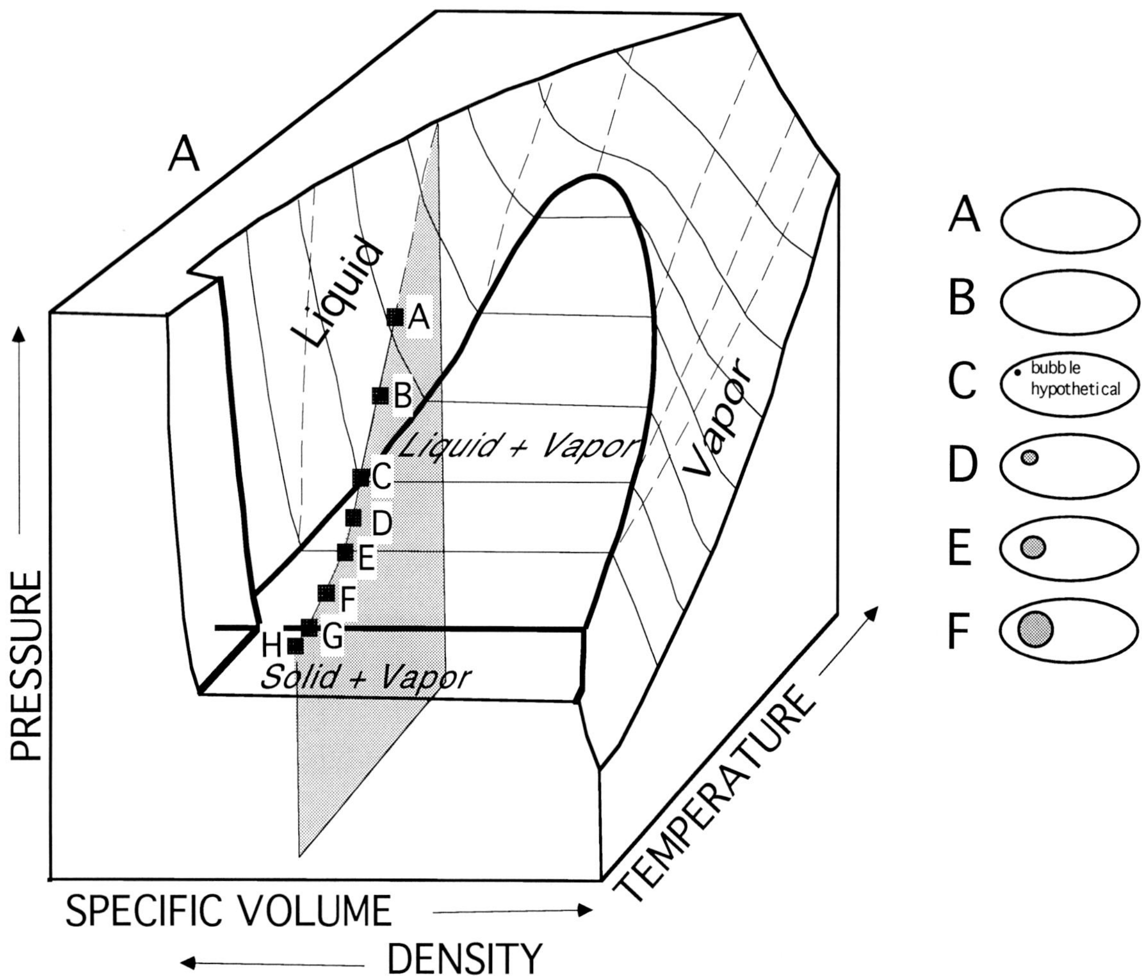

Fig. 3.1. *A) Generalized and schematic P-V-T diagram for the unary system H_2O (metastable surfaces have been omitted). The intersections of planes of constant volume with the surface representing stable single-phase equilibria are shown as dashed lines. The intersections of planes of constant temperature with the surface representing stable single-phase equilibria are shown as curved, light solid lines. The surface representing stable two-phase (liquid-vapor) equilibria is delimited by heavy solid lines. The intersection of these surfaces of stable equilibria with a plane of constant specific volume defines a curve known as an isochore. If an inclusion traps a fluid of specific volume (of that shown by the shaded plane) at the P-T conditions of point A, the inclusion will be constrained to the locus of points in P-V-T space described by the intersection of the surface of the shaded (constant volume) plane with the surfaces of stable equilibria (for cooling, A-B-C-D-E-F-G-H is shown; see text for additional discussion).*

The fluid inclusion sketches schematically show relative proportions of liquid (clear) and vapor (shaded) at the various labeled points shown on the phase diagrams. These diagrams are modified from similar figures in Roedder (1984) and Fisher (1976).

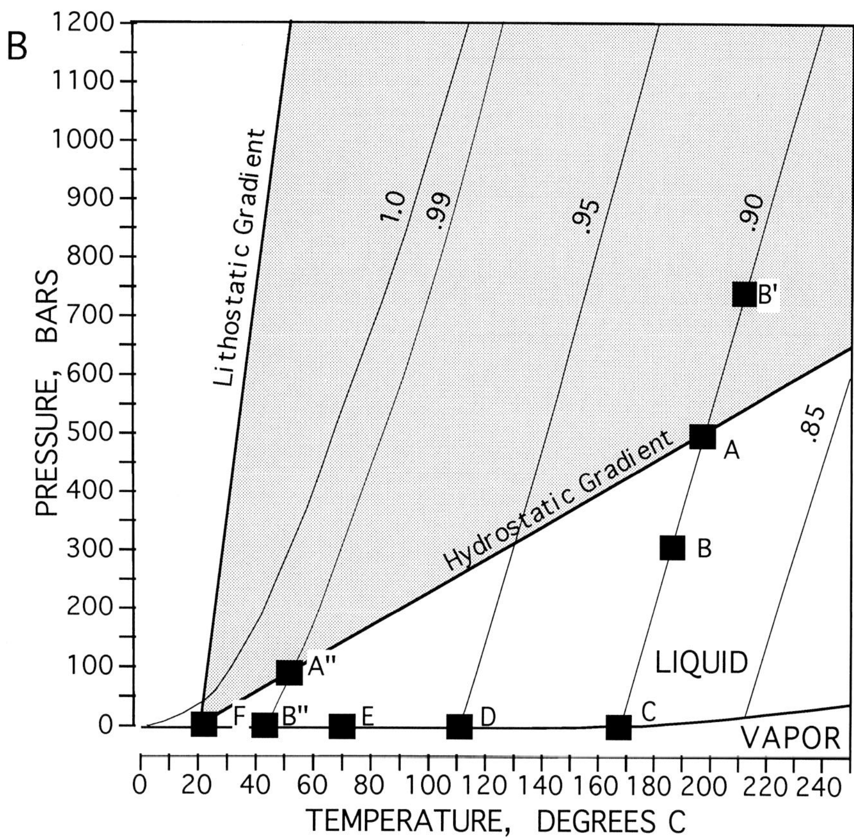

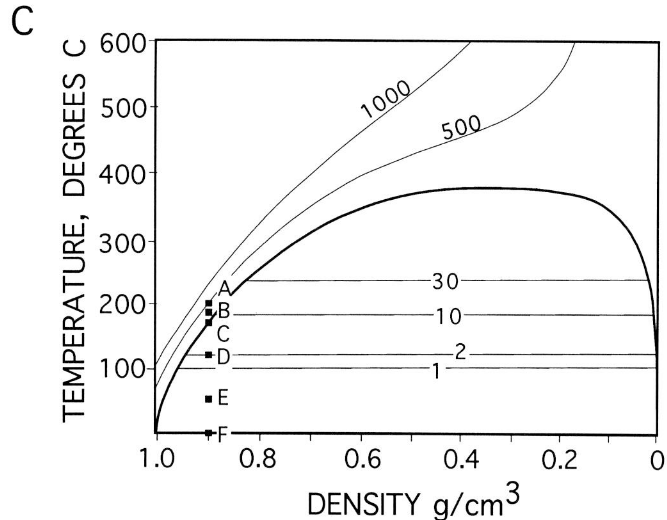

Fig. 3.1. *Continued from previous page.* ***B)*** *P-T projection of the H_2O system for diagenetic P-T conditions. Labeled contours are lines of constant density in g/cm^3. The shaded area shows P-T conditions that can be found in the diagenetic realm. See text for explanation.* ***C)*** *T-V (density) projection of the H_2O system, with contours of pressure in bars. The heavy curve separates the two-phase (liquid-vapor) region below from the one-phase region above. A fluid inclusion of a certain bulk density and temperature below the curve will have a unique pressure within the inclusion; the inclusion will have both a liquid and a vapor phase; the density of each of the two phases will be defined by the intersection of the pressure contour with the two sides of the two-phase region solvus; and the relative proportions of liquid and vapor within the inclusion will be defined by the lever rule.*

the inclusion is the dependent variable, meaning that the pressure within the inclusion will be specified by the density and temperature of the inclusion, as illustrated below.

P-T Conditions of Diagenesis

Two thermobaric gradients are presented in Figure 3.1B that define a range of P-T conditions that could occur in diagenetic systems. The geothermal gradient is normally within the range of 20°C to 50°C for each kilometer of burial depth. Pressure during burial normally is between a hydrostatic gradient of about 100 bars/km and a lithostatic gradient of about 226 bars/km. In Figure 3.1B, we have taken these pressure gradients and used an abnormally low geothermal gradient of 10°C/km and a common but high geothermal gradient of 35°C/km to illustrate likely ranges of natural pore fluid systems in the diagenetic realm. Thus, as can be seen in Figure 3.1B, simple (without gas) aqueous pore fluids will always be in the one-phase liquid stability field. Note that the condition of coexisting liquid and vapor H_2O (field boundary drawn between liquid water and water vapor — the boiling curve) could not occur in the diagenetic realm without abnormally high geothermal gradients (e.g., intrusion of a pluton in close proximity). As will be shown later in this chapter, the addition of salts makes the liquid stability field even larger; however, the addition of CH_4 to pure H_2O changes the phase relations considerably, so much so that a separate gas and liquid phase can easily coexist in pore fluids if CH_4 is present.

A Fluid Inclusion's Trek Through P-V-T Space

Consider a case where sediments had been buried along a common (but high) hydrostatic thermobaric gradient (35°C/km) shown in Figure 3.1B, and that at a temperature (197°C) and pressure (475 bars; point A) precipitation of a diagenetic mineral occurred. During precipitation, this mineral phase did not grow perfectly and enclosed some of the pore fluid due to the crystal's imperfections. The freshwater pore fluid at these P-T conditions is in the liquid stability field, and therefore the inclusion that traps this fresh water contains liquid H_2O with a density of .90 g/cm^3. Now, suppose that overlying rocks were progressively unroofed. The *crystal in the pore* would follow the thermobaric gradient path back toward surface conditions. However, if the inclusion satisfies the basic requirements stated above (remains a closed system and maintains a constant volume), then as the inclusion experiences a decrease in temperature, the P-T conditions *within the inclusion* are constrained to be at points on the line A-C (line of constant volume or isochore of a fluid with density of .90 g/cm^3) shown in Figure 3.1A and 3.1B. So at temperature B (187°C) on Figure 3.1, the inclusion experiences an internal pressure of 300 bars, and in P-T space the inclusion is in the liquid field at point B (Fig. 3.1A and 3.1B). As the temperature cools to point C (167°C and only 7 bars), the liquid-vapor curve is intersected and a minuscule vapor bubble is predicted to form in the inclusion. In actuality, this does not happen, as a bubble must be of a certain minimum diameter to reduce the effect of surface tension and to prevent the bubble from collapsing on itself, so the single phase liquid actually persists to lower temperatures in a "stretched" state (see Roedder, 1984, p. 293-296 for explanation). It is important to note that <1 km of overburden was removed to cause a fall in temperature large enough for the inclusion to have two phases (liquid and vapor) within it. With continued cooling of the inclusion (due to unroofing), the inclusion conditions remain on the liquid-vapor curve; the bulk density of the inclusion remains constant at .90 g/cm^3, but the proportions and densities of the liquid and vapor phases within the inclusion change progressively and are determined by the lever rule (Fig. 3.1C, points D, E, and F). As the inclusion cools from point C to point F, the volume of vapor grows in proportion to the volume of liquid and the internal pressure gradually falls so that at point D, the pressure is at 2 bars; between points D and E the fluid inclusion passes the 100°C and one-atmosphere pressure of boiling at the earth's surface, to approach about 0.5 bar at point E; and finally the inclusion reaches an internal pressure that is close to a vacuum at point F (Fig. 3.1B, C). So, the trek of the fluid inclusion through P-T space upon unroofing is precisely defined with knowledge of the phase relations. Furthermore, it is important to realize that during most of the unroofing history of a fluid inclusion, it contains two phases and its internal pressure is well below the surrounding pore fluid pressure.

Another scenario to consider for the inclusion trapped at point A in Figure 3.1 is a case that involves the deposition of more sediments. In this case, the crystal containing the inclusion is buried more deeply along the thermobaric gradient. Once again, if the inclusion satisfies the basic requirements of remaining a closed system and maintaining a constant volume, then the inclusion must stay on the .90 g/cm^3 isochore with increasing burial depth. This means that as the temperature increases with increasing burial depth, the pressure in the inclusion (point B' in Fig 3.1B) becomes higher than the pore fluid pressure that lies along the thermobaric gradient at the same temperature. The fact that inclusions heated beyond their initial conditions of entrapment develop internal pressures higher than the pore fluid pressure surrounding the crystal in which they are enclosed has significant consequences with respect to the approach that inclusionists must embrace to collect sound data and to make sound interpretations. These topics are discussed in detail in Chapters 4, 6, and 7.

For either of the examples above, it is important to note that the pressure within the inclusions at room conditions is much less than one atmosphere (760 mm Hg); namely, about 20 mm Hg (0.03 bar), which is virtually a vacuum. This fact has important implications for determining the presence of gases within inclusions, which is discussed in Chapter 6.

Inclusions trapped closer to earth surface conditions (e.g., A" in Fig. 3.1B) do not show a phase change as do the examples discussed above. Even though the inclusion trapped at A" will intersect the liquid-vapor curve at B" with cooling, commonly a vapor bubble will not nucleate, as contraction of the liquid caused by the change to ambient surface temperature is slight. Although lower temperatures of laboratory bubble disappearance (homogenization temperature) are known, for the most part, the lower temperature limit of aqueous homogenization temperatures is roughly 40 to 60°C (Roedder, 1979; Goldstein, 1986b, 1990; Anderson, 1989; Barker and Goldstein, 1990; Goldstein and others, 1990; Chapters 6 and 7 of this volume). Most aqueous fluid inclusions trapped below this temperature range remain as metastable or stable all-liquid inclusions at room temperature and may never nucleate a vapor bubble. Furthermore, it is not uncommon for small inclusions (<3 μm) in quartz overgrowths to remain as metastable all-liquid inclusions, even though the homogenization temperatures should be as high as 100°C.

A Fluid Inclusion's P-V-T Trek in the Laboratory

The natural cooling path through P-V-T space (A-B-C-D-E-F) of the fluid inclusion in Figure 3.1 may be reversed in the laboratory by heating while under observation with a microscope. The inclusion will follow the identical path except in reverse (F-E-D-C). From F to C the bubble will gradually shrink and at point C the bubble will disappear, and the inclusion once again becomes a single homogeneous liquid phase. The temperature at point C is thus known as the homogenization temperature (Th). This phase-change temperature is the most important in the technique of fluid inclusion microthermometry, as it provides an estimate of the temperature of entrapment of the fluid inclusion. Note that this temperature estimate is only a minimum, however. The inclusion that homogenizes at the temperature at point C (167°C) must have formed at conditions along the isochore intersecting point C (density of .90 g/cm^3), but is more likely to have formed at a higher temperature and pressure than point C, somewhere within the realm of diagenetic systems (Fig. 3.1B). Given an unknown sample with homogenization at point C, all that is definitely known from the homogenization temperature (as long as the basic assumptions are satisfied) is that the inclusion was entrapped at a temperature at least as high as the homogenization temperature; in other words, homogenization temperature is only a minimum estimate of the inclusion's entrapment temperature. Other information may also be obtained from the homogenization temperature. Since the phase relations and P-V-T properties of pure water are known from independent experimental data, then the density of the fluid within the inclusion is defined if the Th can be determined. (In fact, this follows for any isoplethic, isochoric inclusion, regardless of the composition, as shown later in this chapter). Also, if the pressure of entrapment can be determined (see Chapter 10) — in this case it was chosen at point A — then the true temperature of entrapment is obtained by proceeding up the isochore (C-B-A) to the known pressure, yielding the true temperature of entrapment. The difference between the temperature of entrapment (Tt) and the homogenization temperature (Th) is known as the *pressure correction* in the fluid inclusion literature.

As the pure H_2O inclusion is cooled to point G (Fig. 3.1A) in the laboratory, some of the liquid in the inclusion should freeze to ice as the inclusion crosses the line defining the triple point. Actually, this does not occur normally because the liquid persists metastably to lower temperatures before ice finally nucleates at point H on the solid (ice) + vapor surface. Upon warming, the ice will begin and end melting at the triple point (point G). This temperature of final melting of ice (Tm ice) is the second most important phase-change temperature for the technique of fluid inclusion microthermometry as it provides a powerful indication of the fluid's composition. In this case of a pure water inclusion, there are two important aspects of the solid to liquid H_2O phase change: the solid begins melting and the solid ends melting at the invariant triple point.

SINGLE COMPONENT SYSTEM CH_4

The unary system CH_4 is completely analogous to the system pure H_2O, as can be seen by comparing the CH_4 system P-T projection in Figure 3.2 with the H_2O system P-T projection in Figure 3.1B. The major difference is that the liquid-vapor phase boundary is at

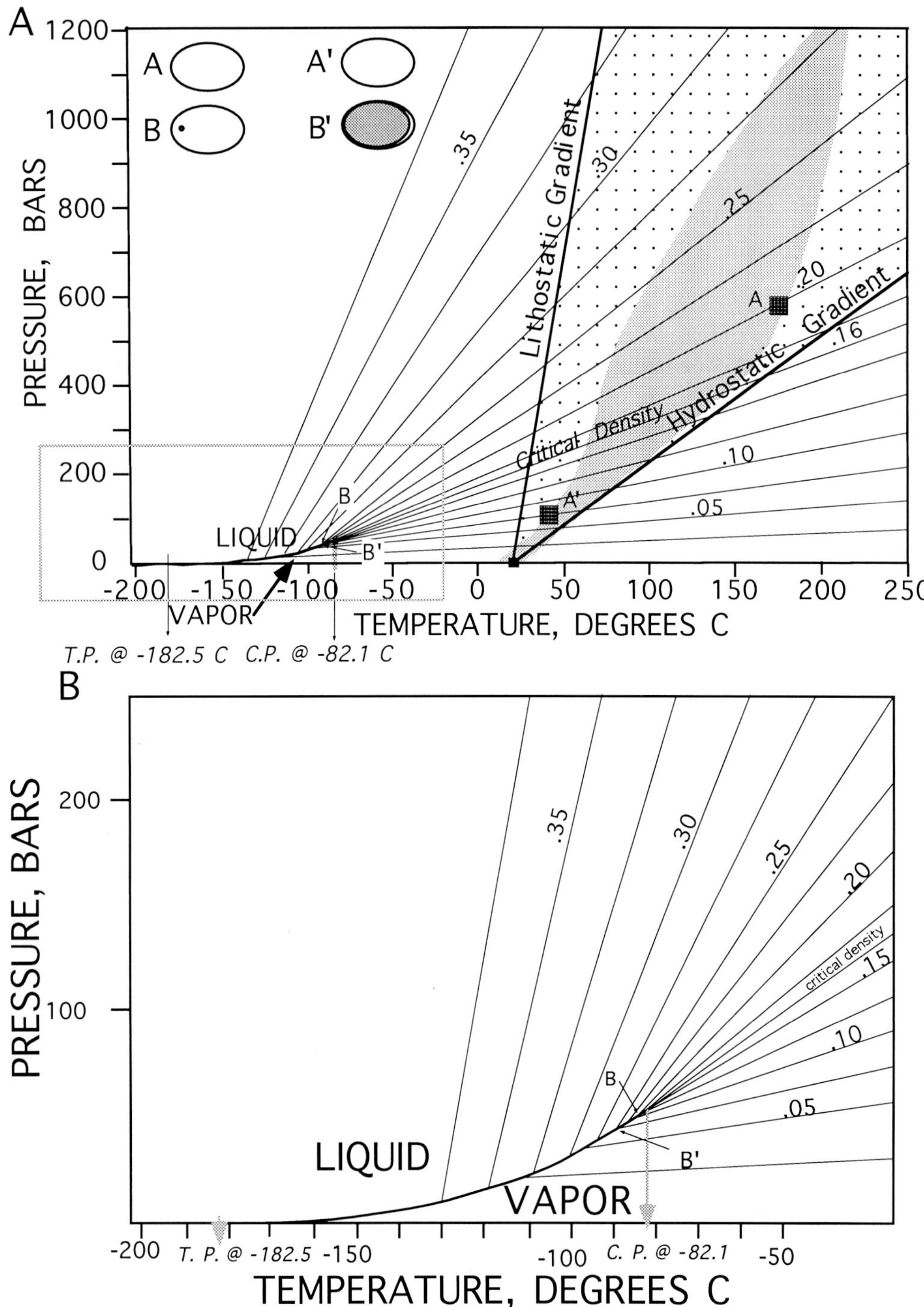

Fig. 3.2. *A) P-T plot of the CH_4 system. Labeled contours are lines of constant density in g/cm^3. The light shaded area shows the same P-T conditions found in the diagenetic realm as in Figure 3.1; for comparison, the darker shaded area shows the P-T regime present in nearshore Gulf basin sediments (from Hanor, 1980). Inset of fluid inclusion sketches schematically show relative proportions of liquid (clear) and vapor (shaded) at the various labeled points.* ***B)*** *Enlargement of area outlined by the rectangle in A. Critical point is at -82.1°C, 46.3 bars; triple point is at -182.5°C. Modified from Mullis (1979).*

much lower temperature conditions. Also shown on Figure 3.2 are the thermobaric gradients demarking P-T conditions that are predictable to exist in the diagenetic environment. There are several important implications of Figure 3.2. First, pure CH_4 inclusions will always contain only a single homogeneous phase under natural conditions. Natural diagenetic conditions exist at higher temperature and pressure conditions than those in which there is a distinction between gas and liquid CH_4; at such conditions CH_4 is referred to as a *fluid,* not gas or liquid. Second, most pure CH_4 inclusions formed in the diagenetic realm will have densities roughly .05 to .25 g/cm^3, and therefore, upon cooling in the laboratory to nucleate two phases, will normally homogenize between about -97 and -82°C (Fig. 3.2B) because that is where isochores of pure CH_4 inclusions formed in the diagenetic realm intersect methane's liquid-vapor field boundary. Lastly, CH_4 inclusions trapped on isochores above the critical density isochore (isochore extending from the critical point) will show phase changes identical to those discussed above for pure H_2O, except at different P-T conditions. So, for an inclusion trapped at point A, cooling during natural unroofing to the Earth's surface would cause the inclusion's internal pressure to drop along its isochore (the one at .20 g/cm^3). Notice that unlike the aqueous system discussed previously, this CH_4 inclusion at surface temperatures will contain an internal pressure greater than 200 bars whereas the pressure at the surface is only about one bar. As the inclusion is cooled to point B (about -84°C), a bubble of CH_4 gas is predicted to nucleate within an inclusion dominated by liquid CH_4. Additional cooling would cause the gas bubble to increase in size, and if the inclusion were subsequently warmed in the laboratory, the bubble would shrink and disappear just as the inclusion homogenized to liquid CH_4 at point B. Alternatively, a pure CH_4 inclusion trapped below the critical isochore (lower density fluids), such as at point A' (isochore with density of .075 g/cm^3) will instead nucleate a rim of liquid CH_4 within a gas dominated inclusion upon cooling the inclusion to point B' (about -88°C), at the position in P-T space in which the .075 g/cm^3 isochore intersects the liquid-vapor phase boundary. With subsequent rewarming, the inclusion would homogenize by expansion and filling by the vapor phase at point B'.

Like pure H_2O inclusions, pure CH_4 inclusions also react to cooling below the triple point (-182.5°C). Supercooling below the triple point is required to nucleate solid CH_4, and this typically happens below -190°C. Melting of pure CH_4 will begin and end at the triple point. If any impurities exist as dissolved components in the CH_4 phase, then melting occurs over a range of temperature, just like in the H_2O-NaCl system as discussed below.

TWO-COMPONENT SYSTEM H_2O-NaCl

The principles explained above for the unary systems are applicable to more complex mixtures. For the binary system H_2O-NaCl, the P-T projections of the liquid-vapor phase boundary for fluids of different salinities are similar to that of pure H_2O, except that the liquid phase equilibrium shifts with greater amounts of dissolved NaCl (Fig. 3.3). For the inclusionist, there are two important consequences of these shifts in P-T space. First is the decreasing slope and shift of lines of constant volume (isochores) in the liquid field (for temperature conditions prevalent in sedimentary environments). Thus, just as for the pure H_2O system discussed above, if an H_2O-NaCl inclusion remains isoplethic (constant composition) throughout its history from the time of entrapment, then these lines of constant volume become lines of constant density. And since the phase relations and P-V-T-X properties of the H_2O-NaCl system are known from independent experimental data, then the density of the fluid within the inclusion can be determined if the NaCl content and the Th of the inclusion can be obtained from microthermometry. The other important consequence of the addition of NaCl to H_2O is that the highest temperatures at which ice can exist are depressed to lower temperatures as salinities are increased (Fig. 3.3). Therefore, the final melting of the ice (Tm ice) correlates to the amount of NaCl in the fluid, and it is this fact which enables an inclusionist to approximate the salinity of the fluid inclusion.

For many aqueous systems containing dissolved salts, phase equilibria are known, so by combining Tm ice measurements and other techniques (see Chapter 7) to determine the composition of the fluid, the inclusion worker can determine which phase diagram might be applied to determine the density of the inclusion, and thus know the isochore (line of constant volume) in P-T space along which the inclusion was trapped. Nevertheless, as with the pure H_2O system discussed above, the Th of a suitable (see Chapter 6) inclusion will always provide a minimum temperature of entrapment, regardless of whether or not any compositional data has been obtained.

Another effect of adding NaCl to the pure water system is in the region of P-V-T-X space less than 0°C. A convenient and useful projection of the region is a T-X diagram (Fig. 3.4A) where P is the equilibrium vapor pressure at every point on the diagram; that is, vapor is present in every field. To explain the phase changes that may be observed upon freezing a fluid inclusion containing only NaCl and H_2O, four different inclusions

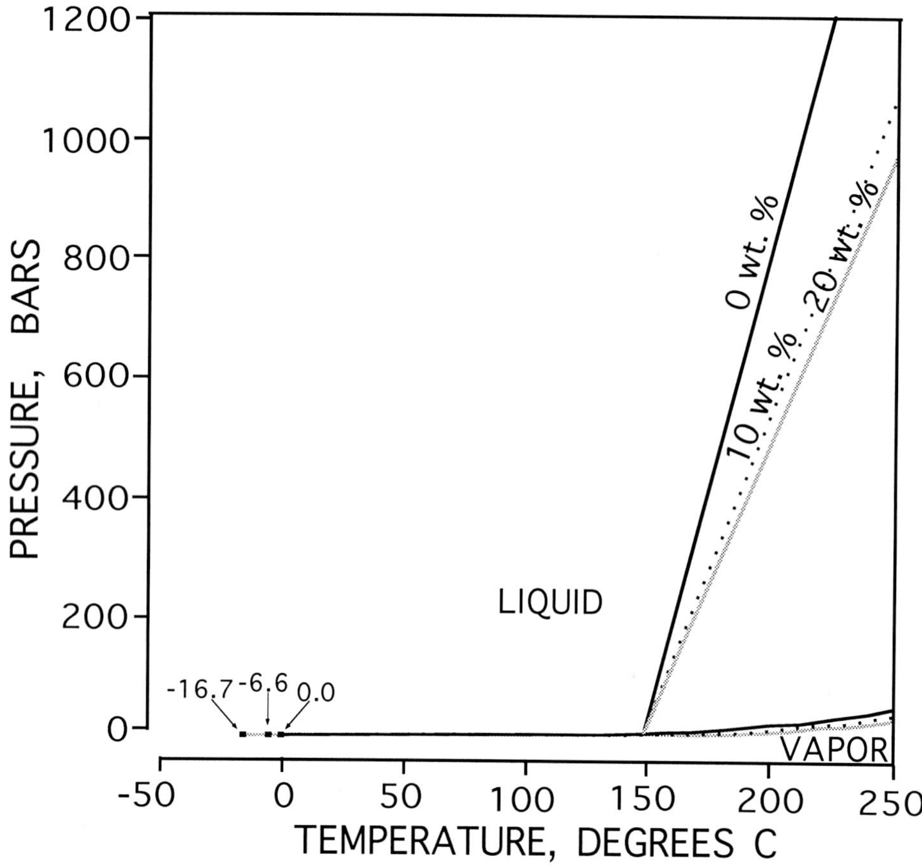

Fig. 3.3. *P-T plots for H_2O-NaCl systems showing similarity in positions of the liquid-vapor curves, differences in slopes of isochores, and differences in triple points between pure water and water with 10 and 20 wt.% NaCl dissolved. New P-V-T data suggest that the change in isochoric slope with salinity may be less than what is shown in this diagram (D. Hall, personal communication, 1994).*

of differing compositions are considered below.

Inclusion compositions #1 (10 wt.% NaCl), #2 (23.2 wt.% NaCl, eutectic composition) and #3 (25 wt.% NaCl) (Fig. 3.4A, B) at room conditions (20°C) will be undersaturated with respect to NaCl (room temperature saturation is about 26 wt.% NaCl) and will contain aqueous liquid and a vapor bubble (Fig. 3.4B). Inclusion composition #4 (27.5 wt.%) at room conditions is in the NaCl solid + liquid + vapor field, and therefore contains all three phases at room conditions (Fig. 3.4B). Except for inclusions in halite, such NaCl-saturated inclusions from sedimentary environments are relatively rare, but have been reported (Haynes, 1988).

When inclusion #1 is cooled in the laboratory, the equilibrium phase diagram of Figure 3.4A dictates that the inclusion should nucleate ice at some negative temperature at point E_1. This does not actually happen within the fluid inclusion because of the kinetic problem of nucleation of the first crystal of ice. So the inclusion is cooled to lower temperatures until the nucleation occurs, which could be at point D_1, but most commonly is at B_1, within the field vapor + ice + hydrohalite (hydrohalite is the solid, $NaCl \cdot 2H_2O$). Once nucleation occurs at B_1, the inclusion is now filled with vapor, ice and hydrohalite, and since ice has a lower density than liquid H_2O, it occupies enough space so that the "vapor" cavity is decreased in size as shown in Figure 3.4B (the vapor bubble is a cavity almost at vacuum pressure at these conditions). *For small inclusions with small bubbles, the entire vapor cavity could be filled by the two solid phases. If this were to occur, Figure 3.4A could not be used to evaluate the phase changes because a vapor bubble must be present in the inclusion to employ this phase diagram.*

Inclusions #2 and #3 will act basically the same as inclusion #1 on cooling. However, because the composition of ice is 0 wt.% NaCl there will be proportionally more hydrohalite (and less ice) in the inclusions with higher salinities, and therefore the decrease in volume of the vapor cavity upon nucleation of solids will become smaller with increasing salinity.

Upon warming of inclusions #1, #2, and #3 from temperatures at points B to C, the first liquid H_2O will form at the eutectic temperature (-21.2°C; Hall and others, 1988), and this liquid will have the composition at the eutectic point (e) of 23.2 wt.% NaCl. Continued addition of heat will cause further melting of the ice and the breakdown of hydrohalite, but no increase in temperature occurs because all four phases can occur together only at one (invariant) point — the eutectic. An increase in temperature will only occur when one or more phases has "melted" completely. In fluid inclusion #2, the last crystal of ice and the last crystal of hydrohalite will both disappear at the same time at the eutectic temperature. In fluid inclusion #1, all of the hydrohalite breaks down leaving ice, brine and vapor; only then can the inclusion rise in temperature with addition of heat. At the temperature of point D_1, there is ice, brine, and vapor in the inclusion. The relative proportions of ice and liquid are given by the lever rule: the ice proportion is given by the relative lengths of D_1-I/L-I, the liquid proportion is given by the relative lengths D_1-L/L-I. At point E_1 the last ice crystal in the inclusion will melt. It is this temperature of final melting of ice (Tm ice) that defines the composition of inclusion #1. Notice that on the ice + liquid + vapor/liquid + vapor field boundary, each temperature has a corresponding and unique composition. This temperature value, also commonly termed the freezing point depression, provides powerful data on the salinity of fluid inclusions. In fluid inclusion #2, the fact that the last ice and hydrohalite crystals disappear at the same temperature (-21.2°C) yields the eutectic composition for the fluid inclusion (23.2 wt.%). In fluid inclusion #3, temperature cannot increase above the eutectic temperature until all of the ice has melted completely. At the temperature of point D_3, there is hydrohalite, brine, and vapor in the inclusion. The relative proportions of hydrohalite and liquid are given by the lever rule: the hydrohalite proportion is given by the relative lengths of D_3-H/L-H, the liquid proportion is given by the relative lengths D_3-L/L-H. At point E_3, the last hydrohalite in the inclusion will break down. It is this temperature of final breakdown of hydrohalite (Tm hydrohalite) that defines the composition of inclusion #3 as containing 25 wt.% NaCl. It is very important for the reader to note that a novice inclusionist might accidentally mistake, or not be able to distinguish between, final melting of ice and final breakdown of hydrohalite (point E_1 versus point E_3); but note how important this distinction is, for the two resulting salinities can differ markedly for identical melting points. For instance, -0.1°C final "melting" could correspond to either a salinity close to 0 wt.% or a salinity close to 26 wt.%, depending on which phase (ice or hydrohalite) is the final phase to disappear in the inclusion. Thus, it should be obvious how important it is for the observer to always attempt to conclusively determine the identity of the last phase to disappear. Identification of ice and hydrohalite in H_2O-NaCl inclusions is discussed in Chapter 7.

Inclusion #4 contains a solid crystal of halite at room temperature and must be heated above point A_4 to determine its composition. At point B_4 the bubble has decreased in size and some of the halite crystal has dissolved (it becomes smaller and more rounded in shape). At point C_4 the NaCl crystal disappears at the temperature of C_4. This temperature of disappearance defines the bulk salinity of the fluid inclusion.

The above discussion considers the phase relations of the system that results from adding a single electrolyte component to water. But all natural waters have multiple salt components. Addition of Ca and Mg salts (see Chapter 7) affects the freezing point depression, the eutectic point, the position of the liquid-vapor phase boundary, and the position of the isochores slightly, but does not change the overall approach to the phase diagrams discussed above. However, addition of a gaseous component such as CH_4 has a marked effect on the phase equilibria and is now considered separately.

TWO-COMPONENT SYSTEM H_2O-CH_4

Methane is common in many sedimentary basins, so it is predictable that it will be a component of diagenetic fluid inclusions. Thus, it is essential to understand the phase behavior of the H_2O-CH_4 system. In the diagenetic realm, the most important implications of the phase relations of the H_2O-CH_4 system are those that occur above 30°C, and to understand these, a three-dimensional diagram with pressure, temperature, and composition axes is required, as the volume of any inclusion, once trapped, will be assumed to remain constant. This phase diagram is constructed in Figure 3.5A, B, and C, as presented by Diamond (personal communication, 1992). In Figure 3.5A, the liquid-vapor phase boundary for pure H_2O is shown on the T-P face (at X = 100% H_2O) , along with the critical point (C) for pure H_2O. On the X-T face (a plane at constant pressure) is a solvus that exists in the H_2O-CH_4 system, separating conditions where only one phase exists from conditions where two phases must coexist. The top of the solvus is the critical point at that pressure. The curve through the P-T-X volume connecting all critical points is called the critical curve (Fig. 3.5A). In three-dimensional (P-T-X) space, the solvus becomes a surface of immiscibility separating the field where only a single fluid phase exists, from the field where two fluid phases coexist (Fig. 3.5A).

Another way to visualize this surface is to show the curves that result by the intersection of the surface with constant temperature planes transecting the P-T-X volume as shown in Figure 3.5B. The important aspect to note from Figure 3.5B is that the surface of immiscibility separates two fields (volumes). Within the volume confined by the surface (and on the surface as well) is a field of P-T-X conditions where two immiscible phases will coexist: an H_2O-rich phase with CH_4 dissolved in it, and a CH_4-rich phase with H_2O dissolved. At P-T-X conditions outside the surface, only a single phase will exist. If the bulk composition of this phase is near the H_2O-rich end-member, then this single phase will be dominantly H_2O with CH_4 dissolved in it. Consider the case where such a single fluid phase is trapped within an inclusion (T-P-X conditions at the star shown in Figure 3.5C). Once trapped, the inclusion must remain on the I_2 isoplethic plane if the inclusion remains a chemically closed system throughout its history. Furthermore, since the volume is assumed constant, the inclusion will be constrained to an isochoric path on the isoplethic plane, which is shown on Figure 3.5C and labeled as an isochore. Thus, just as in the pure water system discussed above, as the fluid inclusion experiences cooling, the pressure within the inclusion will adjust so that the inclusion follows the isochore within the plane of the isopleth (I_2) to lower P-T conditions.

At point H_2 along the isochoric path of the inclusion the surface of immiscibility is intersected: an infinitesimally small amount of CH_4-rich fluid can coexist with the H_2O-rich fluid. Line V_2-H_2 in Figure 3.5C is a tie line at a given P-T condition, and this tie line intersects the surface of immiscibility at two points

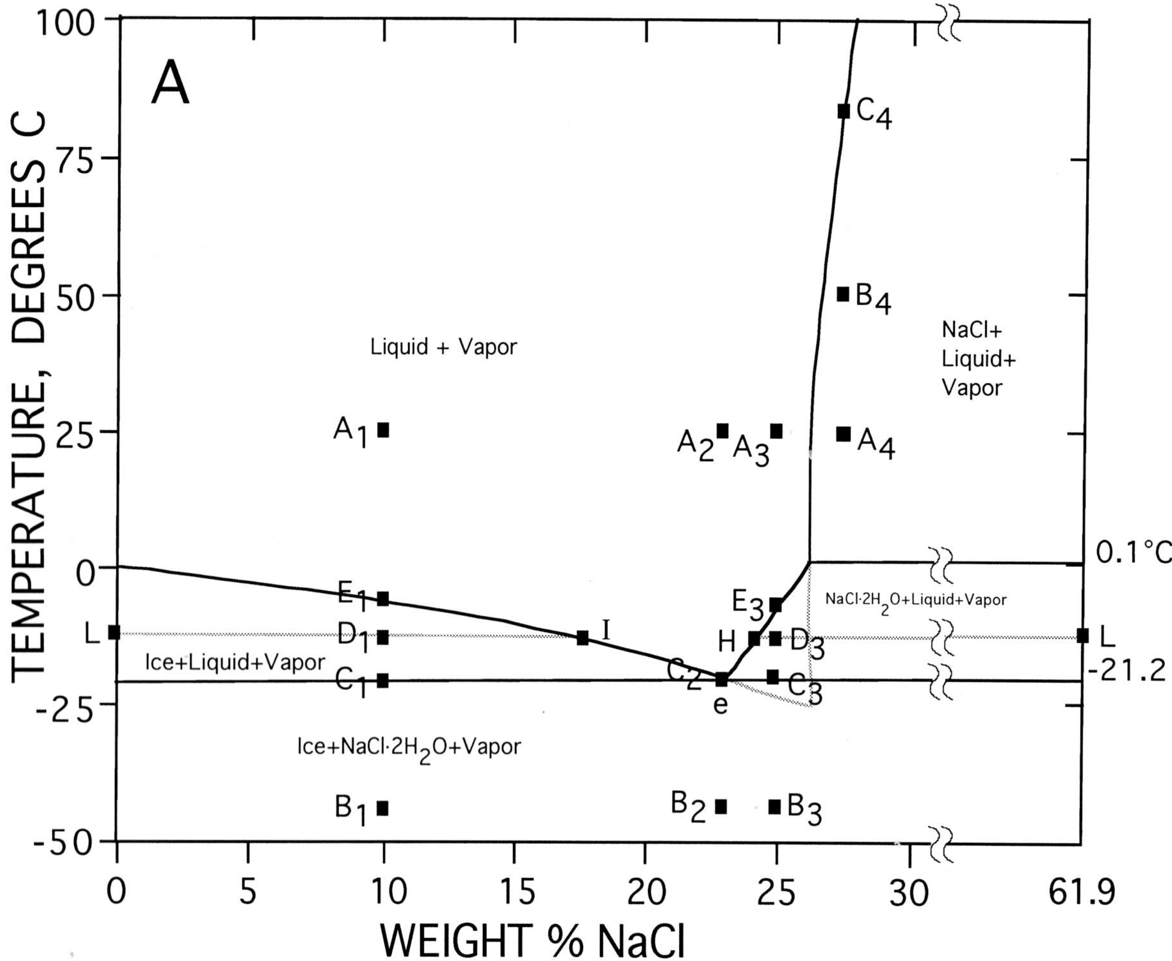

Fig. 3.4. A) T-X plot for the lower temperature, lower salinity portions of the system H_2O-NaCl. Each point on this diagram is at equilibrium vapor pressure. Produced from Crawford (1981), Roedder (1984), and Hall and others (1988). See text for complete discussion.

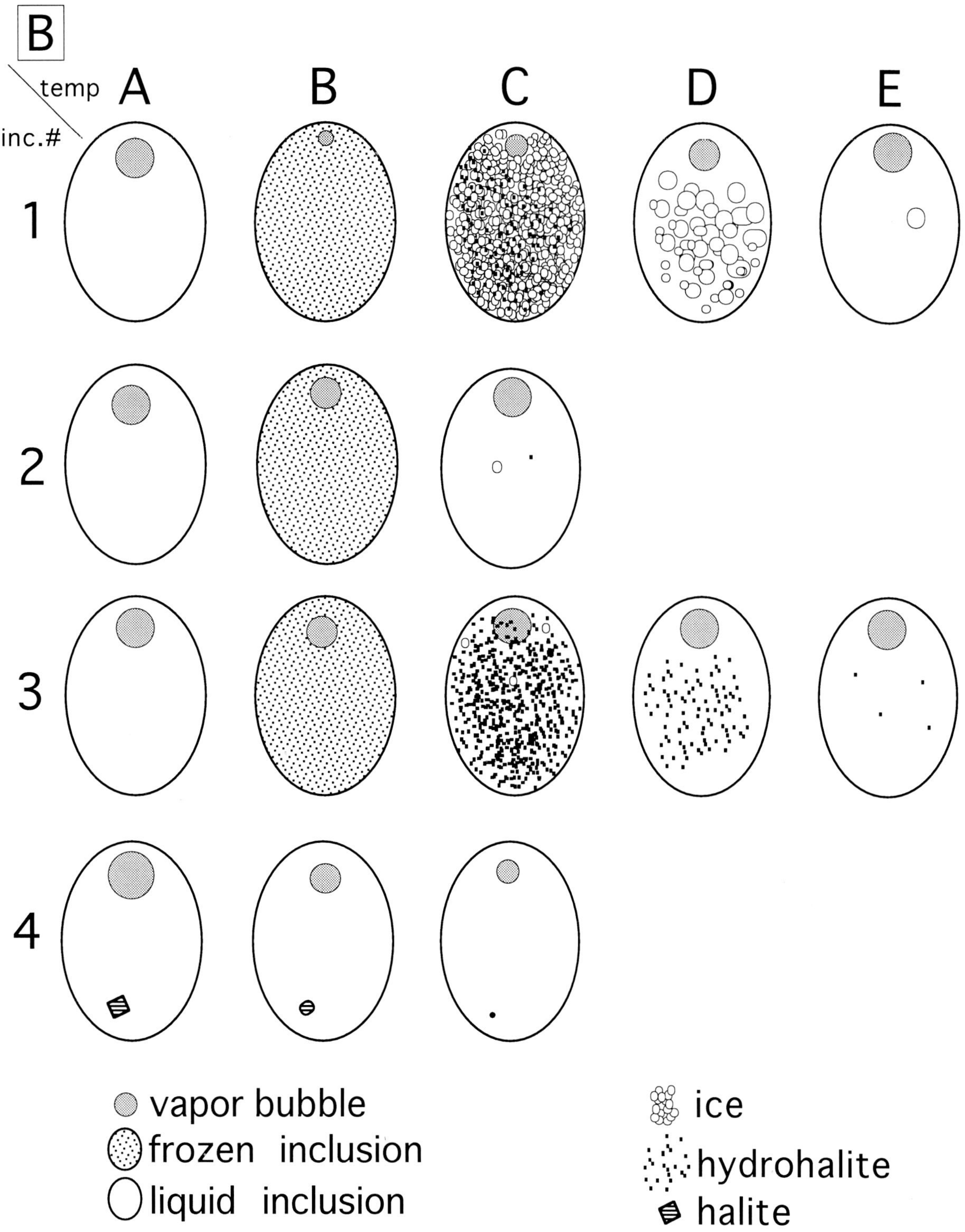

Fig. 3.4**. Continued from previous page.* ***B) *Schematic representation of the phases and their rough proportions in four different inclusions (1, 2, 3, and 4) of four different compositions (10 wt.% NaCl, 23.2 wt.% NaCl, 25 wt.% NaCl, and about 27 wt.% NaCl respectively) at the different temperatures shown in A. See text for complete discussion.*

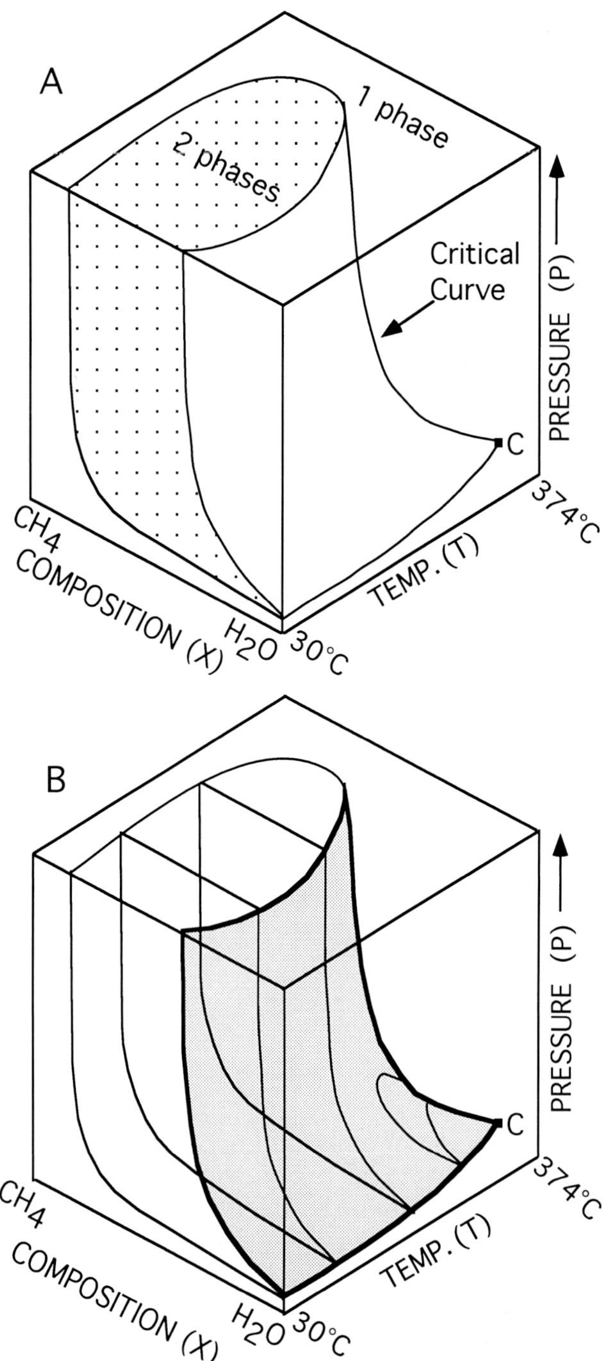

Fig. 3.5. ***A)*** *Schematic P-T-X diagram for the system H_2O-CH_4 showing the one-phase field separated from the two-phase field by a surface of immiscibility. On the P-T face at 100% H_2O the phase boundary for pure water separating the liquid and vapor fields is shown, as well as the position of the critical point (C) for pure H_2O.* ***B)*** *Same diagram as in A, except that the surface of immiscibility is better delimited by curves shown at the intersections of four constant temperature planes. Under this surface is a field of P-T-X conditions where two immiscible phases will coexist: an H_2O-rich phase with CH_4 dissolved in it, and a CH_4-rich phase with dissolved H_2O. At any P-T-X condition outside the surface, only a single phase will exist. The surface of immiscibility near the H_2O-rich end-member is shaded. The position of the intersection of the surface of immiscibility with the P-X face is not presented to scale to permit important relations to be readily shown in part C of this figure (in particular, the shaded part of the surface is grossly displaced toward the CH_4 end-member composition).*

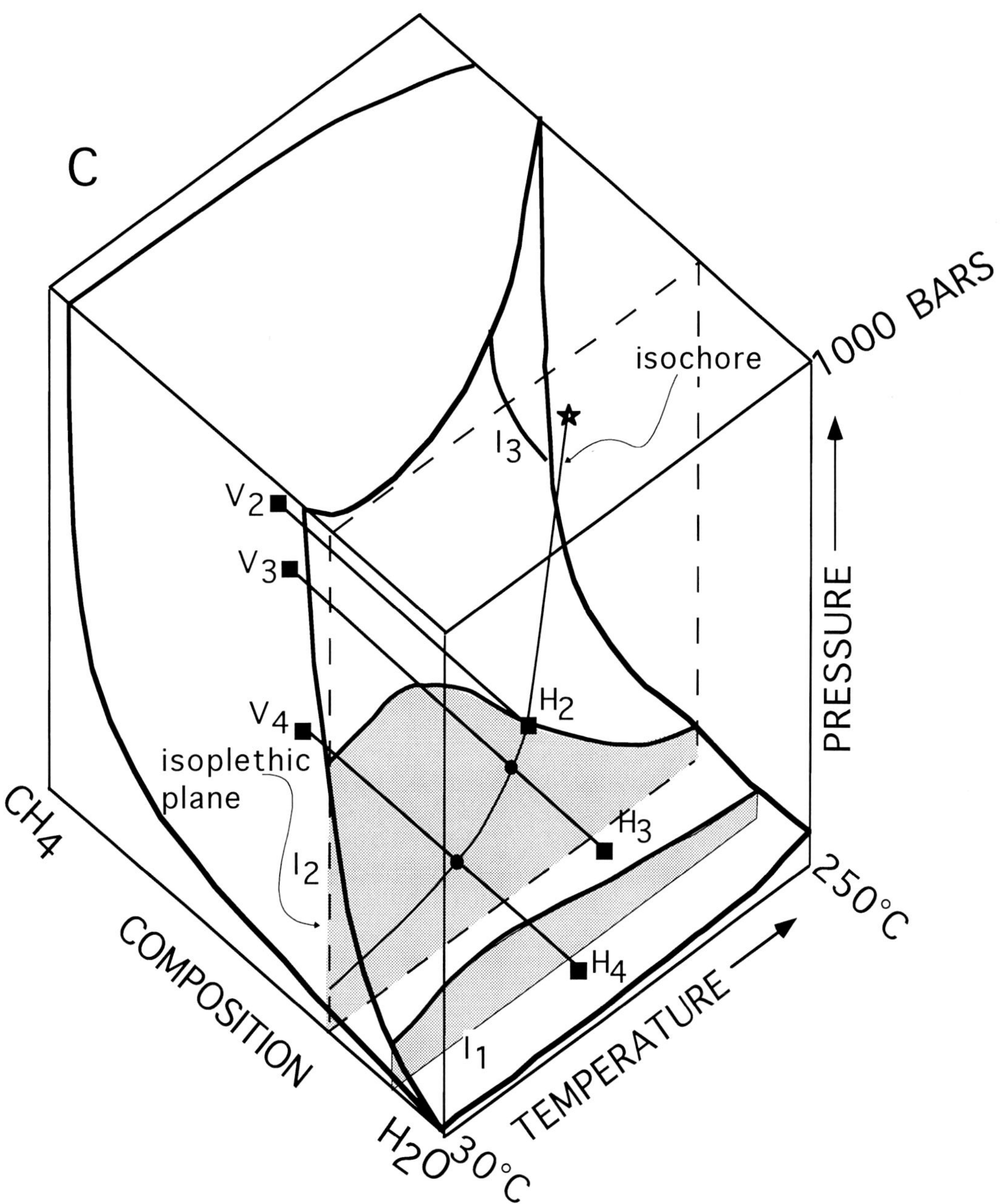

Fig. 3.5. Continued from previous page. ***C)*** *Same diagram as in A and B except that the surface of immiscibility near the H_2O-rich end-member is better delimited by its intersection with three isoplethic (constant composition) planes (I_1, I_2, I_3). The I_2 isoplethic plane is shown cutting the entire P-T-X volume with dashed lines, and the part of the plane that is beneath the surface of immiscibility is shaded (as is that for isoplethic plane I_1). A CH_4-bearing H_2O inclusion trapped on this plane at a P-T condition at the position of the star will be constrained to the plane as long as the assumption of remaining a chemically closed system is true. See text for discussion of tie lines V-H.*

V_2 and H_2. Note that V_2 gives the composition of the infinitesimally small bubble that exsolved from the water-rich fluid, just when the water-rich fluid became saturated with CH_4 at point H_2.

With continued decrease in temperature, the inclusion remains on the isoplethic plane (I_2) and follows an isochoric path through the two-phase field, but the composition of the two phases within the inclusion are given by the points of intersection of the tie lines (at a given P-T condition) with the surface of immiscibility (V_3 and V_4 are points of intersection on the CH_4-rich side of the surface, and H_3 and H_4 are points on the H_2O-rich side). Such a path continues to ambient temperatures.

Just as with the pure H_2O inclusion discussed above, the aqueous inclusion with CH_4 can be reheated in the laboratory, and the path reversed: the inclusion travels up the isochoric path and homogenizes at point H_2. Again, the homogenization temperature of an aqueous inclusion containing CH_4 is only a minimum temperature of entrapment, because the inclusion could have been trapped at any higher P-T condition along the isochore.

Figure 3.5C can also be studied to assess what phases could be present in the pore spaces in sedimentary environments. Even though the figure is schematic, one can see that in the diagenetic realm, at temperatures below 250°C and pressures below about 1000 bars, and with any of the many possible fluid compositions possible in the diagenetic realm, a point defining the P-T-X conditions of a pore fluid could be anywhere within the volume of Figure 3.5C. Therefore, H_2O containing CH_4 could exist as a single phase, CH_4 containing H_2O could exist as a single phase, or both could exist as two coexisting immiscible phases.

Consider a P-T condition where two immiscible phases coexist in a pore at H_3 and V_3 (Fig. 3.5C), and each phase is trapped separately in two individual fluid inclusions. The inclusion trapping the H_3 fluid contains H_2O saturated with CH_4, and once trapped is constrained to lie on the isopleth (not shown in Figure 3.5C) that passes through H_3. Likewise, the inclusion trapping the V_3 fluid contains CH_4 saturated with H_2O, and once trapped, is constrained to the isopleth (not shown) that passes through V_3. With cooling, the H_3 fluid inclusion instantly forms an infinitesimally small amount of CH_4-rich fluid of composition V_3, and the V_3 fluid inclusion instantly forms an even smaller amount of water-rich fluid with fluid composition H_3. As explained in examples above, both inclusions will continue to follow isochoric paths until surface conditions are reached. However, the differences between these cases and that described above is that when these two inclusions are heated in the laboratory, they should both homogenize at the temperature of entrapment; that is, no pressure correction is necessary. In actuality, one likely will not be able to view the V_3 inclusion homogenize (through disappearance of the water phase) because the amount of H_2O that can be dissolved in the CH_4 phase in the diagenetic realm is so small (<5-10 mol.%; Welsch 1973; Pichavant and others, 1982) that one would not only be unable to see this H_2O phase before homogenization, but one likely would be unable to see the H_2O phase wetting the walls of the inclusion at ambient conditions in the laboratory!

This example illustrates an important principle of fluid inclusions; namely, when a fluid inclusion entraps a single phase of an immiscible fluid, the homogenization temperature (if observable) yields the true temperature of trapping (Tt). If the composition of the inclusion fluids can be determined (see later chapters), and if the P-V-T-X phase relations are known for the system being studied, then the pressure of trapping (Pt) is also known once Tt is determined. However, remember that Th will not equal Tt if more than one phase of the immiscible fluids were trapped within the inclusion when it formed: a fundamental requirement for fluid inclusions to yield meaningful Th data is that only a single homogeneous fluid was trapped within an inclusion at the instant of its sealing.

In order to assess other important implications of the addition of CH_4 to H_2O with respect to its affect on pore fluids in natural systems and to the behavior of phases within fluid inclusions, it is necessary to consider the phase relations more quantitatively. To accomplish this objective, the curves defined by the intersections of the surface of immiscibility with various isoplethic planes of H_2O-rich compositions are projected onto a single P-T plane as shown in Figure 3.6A. These curves are known as bubble point curves, or CH_4 solubility curves. There are an infinite number of bubble point curves beginning just past the isoplethic plane of pure water, and continuing to the isopleth intersecting the apex of the immiscibility surface of Figure 3.5C. (This surface closes at higher pressure conditions not shown in Figure 3.5C.) Figure 3.6B is similar to 3.6A, but NaCl has been added to the H_2O. Note that although the overall solubility of CH_4 in H_2O appears to be relatively low, there is a marked increase in CH_4 solubility with increasing depth (increasing P and T) in both figures. Also, presence of dissolved salts in an aqueous fluid decreases the solubility of CH_4 (Haas, 1978): a salinity increase to 15 wt.% NaCl reduces CH_4 solubility to about half of what would be dissolved in water with no electrolytes. Thus, the addition of salt has the important effect of raising the surface of immiscibility (field of coexistence

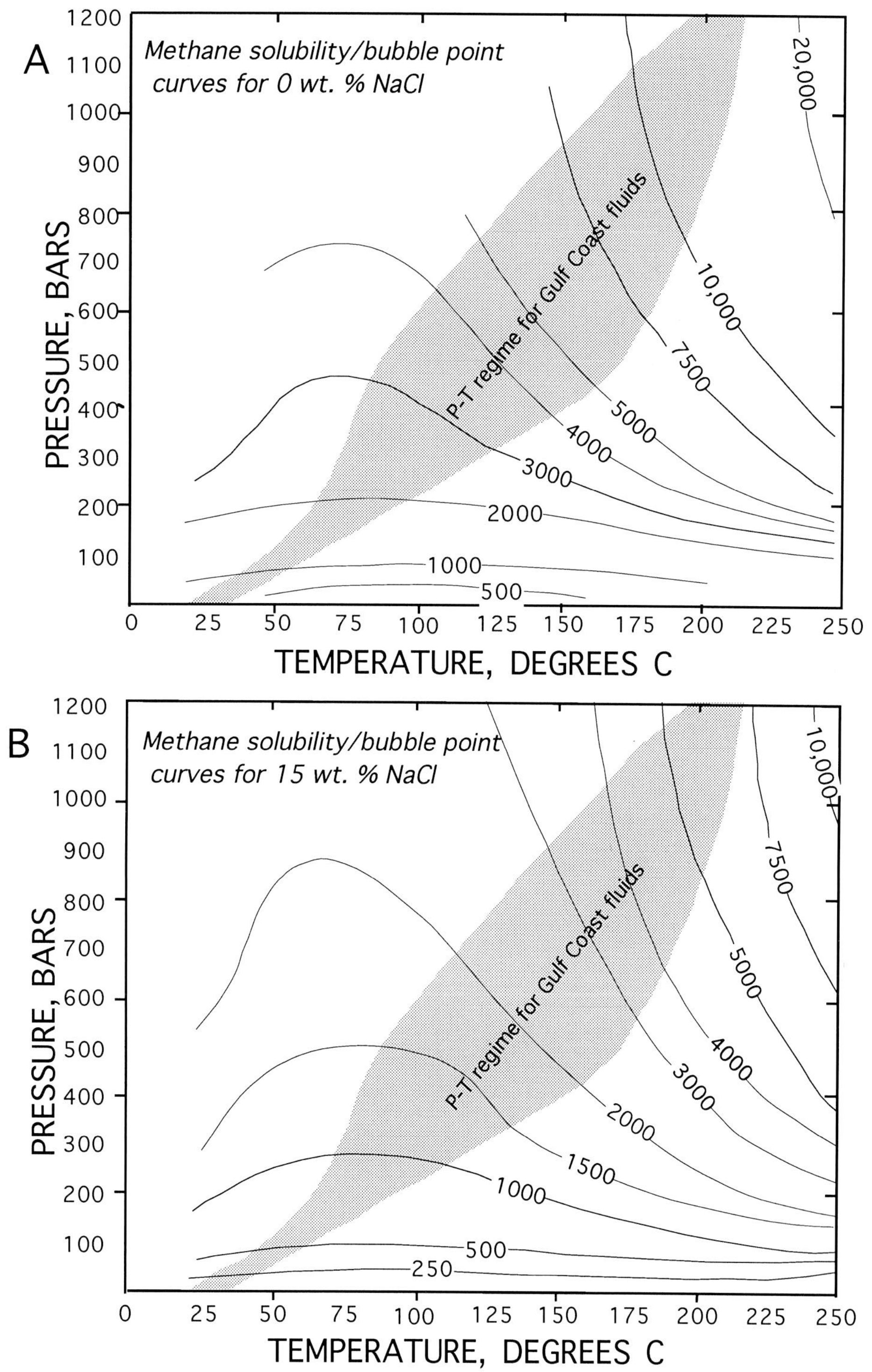

Fig. 3.6. Intersections of isoplethic planes for H_2O-rich fluids with various amounts (ppm) of CH_4 with the H_2O-CH_4 immiscibility surface projected onto a single P-T diagram. ***A)*** *H_2O-CH_4 fluids containing no NaCl.* ***B)*** *H_2O-CH_4 fluids containing 15 wt.% NaCl. Modified from Hanor (1980).*

of two immiscible fluids) to higher P-T conditions (compare Figures 3.6A and 3.6B). Moreover, the addition of other electrolytes to the H_2O-CH_4 solution (KCl, $CaCl_2$, $MgCl_2$, $SO_4^{=}$, HCO_3^-, etc.) will also affect the solubility of CH_4 (Stoessel and Byrne, 1982; Duan and others, 1992).

Shown in Figures 3.6A and 3.6B by shading are the P-T conditions present in the Gulf Coast Basin, USA. With these constraints, one can see that a deep-seated aqueous fluid that is undersaturated with respect to CH_4 (in the one-phase field) could reach saturation (hit the two-phase surface of immiscibility, shown in Figures 3.6A and 3.6B as bubble point curves) during upward flow of fluids to lower P-T conditions. Conversely, a pore fluid that is being buried along a thermobaric gradient, or that is flowing downward into a basin, will experience an increased capacity to dissolve CH_4. Because sedimentary basins are dynamic and complex fluid systems, and because CH_4 is produced at slow and variable rates, an assumption that all sedimentary fluids are saturated with respect to CH_4 is totally unjustifiable (Jones, 1976; Wallace and others, 1978; Hanor, 1980; R. Capuano, personal communication, 1992; Y. Kharaka, personal communication, 1992). A more appropriate outlook would be that only some subsurface waters are saturated with respect to CH_4 (that is, two immiscible phases coexist at the same P-T conditions in the same pore — an H_2O-rich phase saturated with CH_4, and a CH_4-rich phase saturated with H_2O), and that other subsurface waters are undersaturated with respect to CH_4.

Figure 3.7 shows a bubble point curve, or solubility curve, for a salt-free aqueous fluid with 3200 ppm dissolved CH_4. Also shown are the lines of constant volume (or constant bulk density; i.e., isochores) in P-T space. Note that these lines are straight in the one-phase field, and curved in the two-phase field. With this plot the fluid inclusion phase behavior occurring after entrapment can be assessed more quantitatively. If fluid inclusions remain closed systems throughout their history, then the lines of constant volume become lines of constant density, as labeled on Figure 3.7. If an inclusion were to trap this H_2O-CH_4 fluid in a diagenetic mineral at a burial depth of 4 km, assuming a

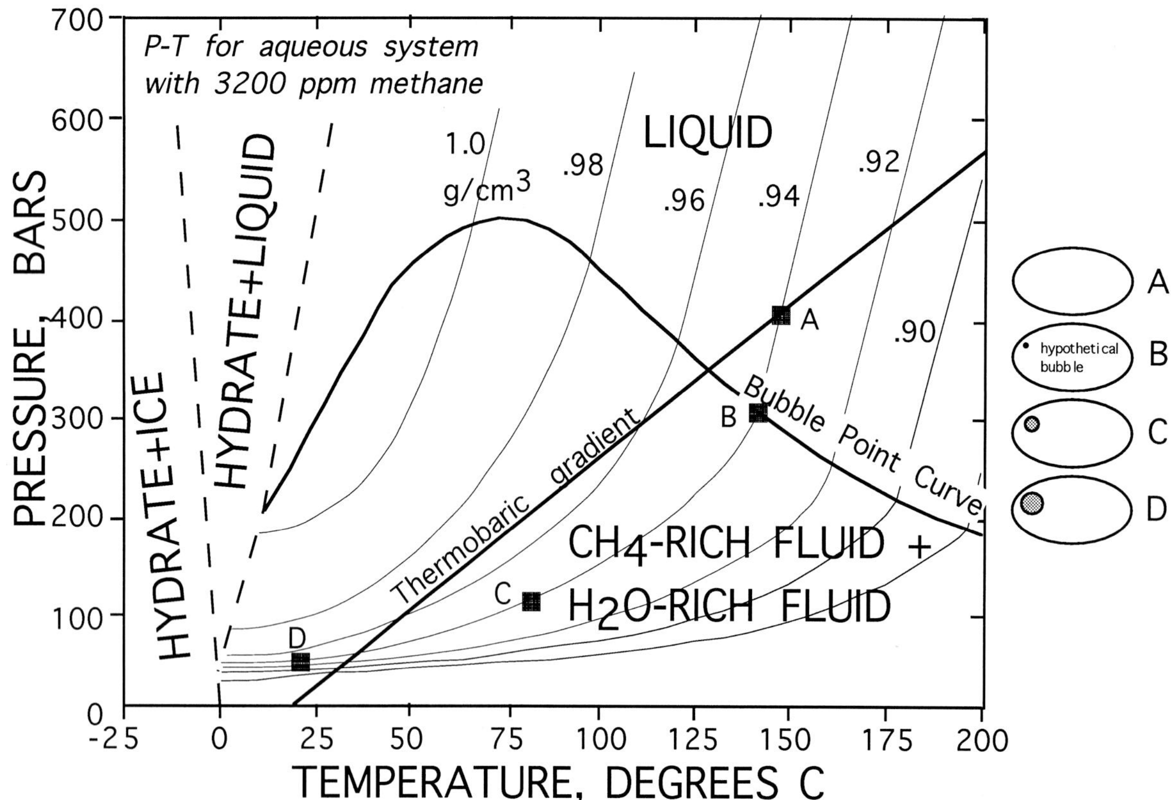

Fig. 3.7. P-T phase diagram for H_2O-CH_4 system with 3200 ppm CH_4. Sketches of inclusions to the right of diagram schematically show proportions of the phases in an inclusion at several P-T conditions (A, B, C, D). See text for explanation. Thermobaric gradient is 32°C/100 bars. Modified from Hanor (1980).

geothermal gradient of 32°C/km and a hydrostatic pressure gradient of 100 bars/km, the inclusion would be at point A (148°C and 400 bars) on the 0.94g/cm^3 isochore, and would of course be required to remain on this isochore throughout its history. If the inclusion were buried deeper, it would experience an increase in temperature, and the internal pressure within the inclusion would be constrained to follow the isochoric path, and would be higher than the pressure of the pore fluid, as defined by the thermobaric gradient (Fig. 3.7).

Alternatively, as with the pure H_2O inclusion previously discussed, upon unroofing and uplift, the inclusion would follow the isochoric path (A-B-C-D) to surface conditions. There are several important points to make about this cooling path. First, for the inclusion to intersect the bubble point curve (point B) and form a bubble of CH_4-rich fluid, only about 200 m of overburden need be removed to provide the 7°C drop in temperature required. The implication here is that an inclusion containing CH_4 will have two phases throughout most of its uplift history. Second, the inclusion is still under high pressure at surface temperatures (point D, Fig. 3.7), which is in marked contrast to inclusions without dissolved CH_4, which are nearly a vacuum at earth-surface temperature. These points will be of critical importance for interpreting fluid inclusions encountered in natural systems, as discussed in subsequent chapters.

As explained before for aqueous inclusions without CH_4, the cooling trend can be reversed in the laboratory by placing the sample with the inclusion in a heating stage. The inclusion will follow the path D-C-B on Figure 3.7. Disappearance of the bubble will occur at the temperature at B which is the temperature of homogenization (Th) and once again is only a minimum temperature of formation of the inclusion, as the inclusion could have been trapped at any P-T condition along the isochore extending from point B. However, in contrast to the systems without CH_4 considered earlier, the Th of an inclusion containing CH_4 more closely approaches the temperature of trapping (Tt) than the inclusions that do not contain CH_4 because the position of the bubble point curves of fluids containing CH_4 are at much higher pressures than the positions of those without CH_4. Therefore, pressure corrections for aqueous fluid inclusions containing CH_4 are typically much smaller than for aqueous inclusions that do not contain CH_4.

MULTICOMPONENT SYSTEM H_2O-PETROLEUM

There are numerous examples of the coexistence of petroleum with an immiscible aqueous phase in sedimentary basins throughout the world. As a first approximation, the phase behavior of the H_2O-petroleum system is analogous to the H_2O-CH_4 system. When the two phases coexist, there will be a certain amount of one dissolved in the other. If the aqueous phase is enclosed within an inclusion, this phase will be saturated with hydrocarbon components, so upon cooling, a fraction of these components will instantly exsolve. Similarly, if the petroleum phase is entrapped, it will be saturated with H_2O, and upon cooling should exsolve H_2O instantly. Therefore, in theory, as explained above in the discussion of the H_2O-CH_4 system, the Th of either type of inclusion should be the true temperature of entrapment (as long as only a single phase is initially entrapped). However, there is a significant practical problem that is rooted in the low mutual solubilities between water and hydrocarbons: the Th's may not be visually detectable. In fact, it is predictable that it would be *very unlikely* for one to be able to observe the tiny amount of H_2O that exsolves from a petroleum phase in a fluid inclusion. Whether or not the homogenization of the hydrocarbon phase in an aqueous inclusion would be visually detectable is still an open question, and is the subject of modeling studies at the time of this writing (R. Burruss, personal communication, 1994).

MULTICOMPONENT SYSTEM GAS-PETROLEUM

Some researchers have chosen to treat petroleum phase relations without the H_2O component, ignoring the fact that the inclusions were probably trapped by the growth of a mineral precipitating from an aqueous phase in equilibrium with the petroleum. Burruss (1992) used the Peng and Robinson (1976) equation of state to construct phase diagrams for petroleum of various types and compositions (Fig. 3.8). There are several important implications of this exercise. First note that there is a tremendous range in the position in P-T space of the boundaries between the one-phase petroleum fields and the two-phase petroleum fields for petroleum of different types: obviously, the composition of the petroleum is extremely important for defining phase relationships. Furthermore, note that for entrapment conditions of 600 bars and 165°C (point A), petroleum inclusions of different compositions will yield a tremendous range of homogenization temperatures from -51°C to 82°C (points B). Secondly, since the slopes of the isochores for most petroleum fluids are much shallower than those of aqueous fluids, the apparent Th of petroleum inclusions commonly do not provide as close of an estimate of minimum entrapment conditions as do aqueous inclusions (unless they were entrapped in equilibrium with a gas phase). On the other hand, because the petroleum fluid isochores are commonly equal to or shallower than thermobaric gradients present in sedimentary basins (Fig. 3.8), if a petroleum inclusion is buried after entrapment, the pressure within the inclusion is not likely to rise significantly above the external pore pressure, which is exactly the opposite of the effect in aqueous inclusions. This may have important implications in the interpretation of the petrography of fluid inclusions found in diagenetic phases (see Chapter 6). Finally, note that the bubble point curves for the various petroleum fluids occur at elevated pressures in comparison to aqueous fluid without gases. The implication is analogous to that for the H_2O-CH_4 system discussed above: immiscibility between petroleum and gas is predictable to exist over a wide range of conditions in natural diagenetic environments. In addition, if homogeneous entrapment of either phase from an immiscible gas-petroleum fluid system occurred, then the Th's of the petroleum and gas inclusions are the entrapment temperatures. True homogenization of the gas inclusions may be difficult to see, but the Th of a petroleum inclusion trapped when saturated with gas will be easily visible.

SUMMARY

In this chapter several relatively simple chemical systems were chosen to illustrate the basic principles of phase equilibria that are necessary to understand in order to apply fluid inclusion observations to problems concerning sedimentary basin processes. Understanding the phase relations of these simple systems has proven to be very fruitful. Nevertheless, the novice

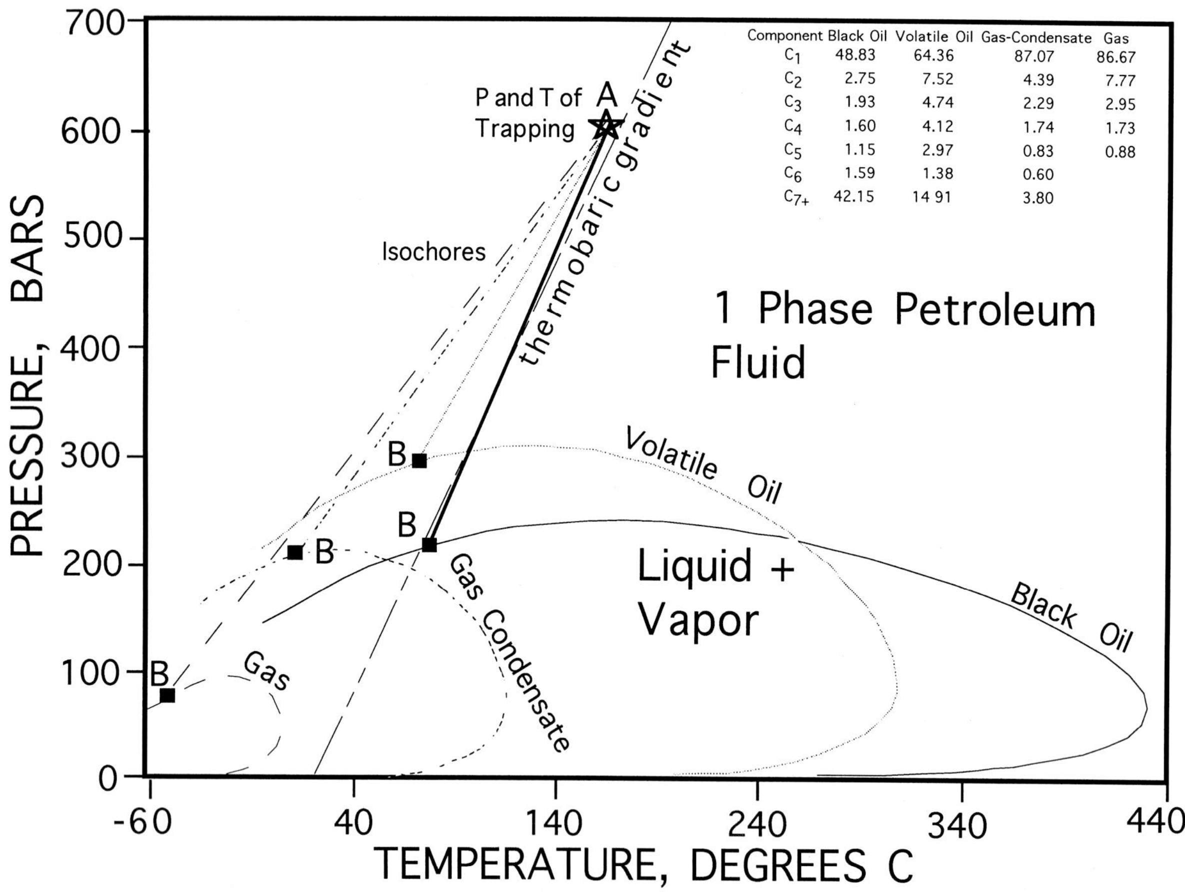

Component	Black Oil	Volatile Oil	Gas-Condensate	Gas
C_1	48.83	64.36	87.07	86.67
C_2	2.75	7.52	4.39	7.77
C_3	1.93	4.74	2.29	2.95
C_4	1.60	4.12	1.74	1.73
C_5	1.15	2.97	0.83	0.88
C_6	1.59	1.38	0.60	
C_{7+}	42.15	14 91	3.80	

Fig. 3.8. *P-T phase diagram showing immiscibility curves and isochores of four "generic" petroleum fluids without H_2O, derived from commercially available software (EQUI-PHASE: D.B. Robinson Research, Ltd., Edmonton, Alberta, Canada) modified from Burruss (1992). Mole percent of components modeled are presented in the upper right. Within the envelopes two petroleum fluids would coexist; outside the curves only a single petroleum fluid would exist. The thermobaric gradient shown is 25°C/100 bars. As pointed out by Burruss (1992), with decreasing mole percent CH_4 and with increasing mole percent of the high molecular weight components (C7 and higher), the curves shift to higher positions, and the slopes of the isochores increase (the fluids become less compressible). See text for additional discussion.*

inclusionist must realize that because natural fluids are much more complex than those covered here, the application of these principles is limited. Furthermore, because of the optical limitations imposed by the sizes of inclusions and microscope resolution, some of the important phase transitions discussed are impossible to observe. These theoretical and practical limitations will be considered in great detail in subsequent chapters.

Chapter 4

ARE FLUID INCLUSIONS REPRESENTATIVE SAMPLES OF DIAGENETIC FLUIDS?

INTRODUCTION

The phases trapped within fluid inclusions and any phase changes that may subsequently be observed in the lab are dependent not only on the micrometer and submicrometer scale physical and chemical processes active during entrapment, but also on processes that proceed after entrapment. If fluid inclusions are to be useful, it is important to know the degree to which their chemical composition and density are representative of the bulk of the diagenetic fluid from which they were entrapped. The degree to which an inclusion's fluid is representative of the ancient diagenetic fluid involves an assessment of several important questions. First, for any fluid trapped within an inclusion, a question arises as to its similarity with the major, minor, trace element, and isotopic composition of the ancient pore fluid that existed prior to entrapment. Second, if more than one fluid phase were present in a diagenetic system during inclusion formation, a question arises as to whether the phases trapped within inclusions are representative, proportionately and compositionally, to those phases present in the pore fluid. Finally, one must question the potential of natural processes to cause changes in fluid inclusion location, shape, volume, and compositions after initial entrapment and during subsequent uplift and/or burial. This chapter attempts to answer these essential questions in two major parts. First, we will determine the degree to which fluid inclusions (formed from both homogeneous and heterogeneous fluid systems) might differ from the bulk of the diagenetic fluid during the entrapment process. Second, any changes resulting from post entrapment processes are evaluated. Overall, it appears that although some differences from the bulk diagenetic fluid may occur, in many cases, the differences are relatively minor, or can be recognized, thus leaving fluid inclusions a useful tool for evaluating diagenetic environments.

DIAGENETIC FLUID VERSUS INCLUSION FLUID

Homogeneous Fluid

During growth of diagenetic minerals from aqueous solutions and entrapment of fluid inclusions containing those solutions, it is important to consider whether that fluid is truly representative of the bulk of the solution responsible for mineral precipitation. As a first approximation, the fluid within the inclusion should be in equilibrium with the pore fluid at the instant before sealing, because mineral growth and changes in fluid composition are relatively slow compared with the rates of diffusion through aqueous liquids. However, on a more detailed level, the inclusion fluid can never be exactly the same as the bulk fluid because of the "boundary layer effect:" there will always be a concentration gradient between the face of a crystal and the fluid. Worst case scenarios would involve a rapidly precipitating mineral, the composition of which involves a major ion in a viscous solution. A small fluid inclusion formed under such circumstances encompasses a significant amount of the "boundary layer" and thus could differ from the bulk composition of the fluid. Such a scenario has been evaluated experimentally by growing salt crystals from concentrated brines. These experiments show that there is no measurable difference between the measured bulk composition of the inclusion and the fabricated bulk composition of the fluid from which it was grown (Davis and others, 1990; Lowenstein and Spencer, 1990). In other experimental studies in which fluid inclusions have been synthesized in other minerals such as gypsum (Sabouraud-Rosset, 1969) and high-temperature quartz (Shelton and Orville, 1980; Sterner and Bodnar, 1984; Zhang and Frantz, 1987), fluid inclusion compositions were essentially representative of the composition of the bulk solution from which the minerals formed. Thus, the magnitude of the boundary layer effect on the concentration of major ions in fluid inclusions appears to be minor.

Empirical observations also support this view. The salinities determined from various cements that have precipitated in relatively well-constrained diagenetic environments seem to provide the correct result. Salinities of fluid inclusions trapped in "mixing-zone" cements from the Plio-Pleistocene Hope Gate Formation of Jamaica lie between fresh and marine values (Lehrmann and Goldstein, in prep.); and marine cements that have not recrystallized contain fluid inclusions with marine salinities (Johnson and Goldstein, 1993).

Assessing whether the boundary layer gradient affects the concentrations of minor and trace elements in aqueous fluid inclusion solutions is not as straightforward because experimental studies are lacking. Fortunately, other experimental data show that major ion compositions appear to be preserved in worst-case scenarios of rapid salt growth, and this suggests that minor ion compositions might also be preserved to some degree. Systems in which a mineral is growing from a minor component of the solution probably precipitate more slowly than evaporites; thus, such systems are more likely to approach equilibrium during much of their history. It was pointed out by Roedder (1984) that for systems at equilibrium, structural gradients in the solution above the mineral could extend for several hundred angstroms (Henniker, 1949). If true, there should be almost no effect on the composition of fluid inclusions in the workable size range. The same logic applies to components that are incorporated as trace elements in the growing crystal, but the effects could vary somewhat depending on partition coefficients and relative concentrations of ions in solution. Overall, from such simple reasoning, it appears that even minor and trace elements trapped in fluid inclusions may be relatively representative of the bulk diagenetic fluid. However, the magnitude of any minor differences are at this point unknown until careful experimental studies are undertaken.

Heterogeneous Entrapment

If fluid inclusions are trapped when immiscible fluid phases are present in a pore, the inclusions most probably will not be representative of the bulk composition of the diagenetic system. Several examples of heterogeneous systems are common in diagenetic environments: aqueous liquid phases coexisting with a liquid hydrocarbon or gaseous hydrocarbon phase; an aqueous liquid phase coexisting with air in the vadose zone; and in some diagenetic environments below the water table, various gas phases may outgas and coexist with an aqueous phase. Entrapment of fluid inclusions within such heterogeneous environments easily can yield fluid inclusion populations with variable proportions of the different phases in individual inclusions. Some phases may be preferentially entrapped. For instance, if a diagenetic system were dominated by aqueous fluid, but contained a few bubbles of methane gas, the bubbles might preferentially cling to the surfaces of growing crystals and could poison the surface where they are in contact with the crystal. These clinging bubbles could act as preferential sites for fluid inclusion entrapment to yield inclusions dominated by methane gas, even though methane gas might have been rare in the system. Similar phenomena might be expected during entrapment of petroleum, and a few experimental studies have demonstrated how it may occur (McLimans, 1987; Pironon and Barres, 1990; Kihle and Johansen, 1994). The reverse could just as easily occur as one phase may not adhere to the surface of a crystal due to surface tensional-wetting characteristics (Roedder, 1984). The recognition of fluid inclusions that have trapped heterogeneous mixtures is very important, for it alerts the inclusionist to obtain the appropriate measurements and phase equilibria to provide valuable constraints on temperature and pressure of fluid inclusion entrapment (see Chapter 3).

Some diagenetic systems precipitate solid phases different from the mineral in which inclusions are entrapped. If such a solid phase were to adhere to the surface of a growing crystal, a fluid inclusion might be formed in the wake of the solid phase. The resulting inclusion would contain the solid phase (an accidental mineral, see Chapter 2) as well as a sample of fluid from the diagenetic system. A population of fluid inclusions entrapped in this way would contain highly variable volumetric ratios of solid phase to fluid phase. This should be easy to distinguish from a population of fluid inclusions that contain crystals which have precipitated from the inclusion fluid after inclusion entrapment (daughter crystals). Evidence for true daughter crystals includes inclusions with consistent volumetric ratios of daughter crystal to inclusion fluid, and daughter crystals should redissolve upon heating at similar temperatures within each inclusion.

CHANGES AFTER ENTRAPMENT

There are potential processes, other than predictable phase changes, that should be expected in a fluid inclusion as it cools (such as nucleation of a vapor bubble or daughter crystal), that may affect fluid inclusions after entrapment. Some of these bear on the question of how representative fluid inclusions are of the diagenetic fluid. They are considered below.

Reaction With Host Crystal

Entrapment of a fluid inclusion occurs at a specific temperature and pressure. For the most part, the fluid is probably at equilibrium with the crystal in which it is entrapped. However, equilibrium conditions change as the thermal environment changes during later burial and unroofing: daughter crystals may precipitate, more of the crystal host may be added to the walls of an inclusion, or some of the crystal host may be dissolved from the walls of an inclusion. Concomitant with this change is a change in the volume of the inclusion cavity and a change in the composition of the inclusion fluid. For most inclusions in the diagenetic realm, most of the changes should be reversible. In other words, if equilibrium has changed because of cooling during uplift to laboratory temperature, this shift can be reversed by heating the fluid inclusion in the lab. However, there are probably changes by which reaction with the crystal host may alter the composition of the fluid inclusion in ways that are not easy to reverse.

There are two known mechanisms by which such irreversible reaction might occur: solid diffusion and dissolution-reprecipitation reactions. The solid diffusion reactions are probably not very significant given the low temperatures of diagenesis, as evidenced by trace element growth zoning preserved on a submicrometer scale in many diagenetic phases (Barker and Kopp, 1991), but the degree to which solid diffusion occurs for every element of interest is not well known. The other major mechanism of reaction with the crystal host involves dissolution and reprecipitation of mineral, generally to change the shape of a fluid inclusion to a state of lower surface free energy (necking down; see Chapter 2 and below). For this common phenomenon to occur, the fluid must be involved in multiple dissolution-reprecipitation reactions of the mineral enclosing, and in contact with, the fluid in the inclusion. Thus, the inclusion fluid will become "rock-dominated" for some elements of interest, whereas the original diagenetic system might have been water-dominated. Such rock domination can cause changes in trace element composition of inclusion fluids. The amount of change is determined by the distribution coefficients for the elements of interest, the ion ratios of the elements of interest in the inclusion fluid and in the mineral host, and the final water/rock ratio achieved from the dissolution-reprecipitation reactions. For many low-temperature systems, aqueous fluid inclusions retain an irregular shape and apparently have not necked down to a significant degree, but other fluid inclusions in diagenetic minerals have negative crystal shapes, some of which may have resulted from necking down. If such dissolution-reprecipitation reactions are a viable mechanism by which the composition of fluid inclusions may change, then it follows that non-breached inclusions that have not changed in shape (necked down) have the best potential of preserving the original trace element composition of the diagenetic fluid.

There is also potential for change in the oxygen isotopic composition of the inclusion water from reaction between the inclusion fluid and the mineral host during changing temperature. For many diagenetic minerals such as silicates, carbonates and sulfates, the oxygen present in the mineral host may dominate the isotopic composition of the fluid in the inclusion if there has been significant exchange. If isotopic exchange occurs between mineral host and inclusion fluid, the fluid merely may show equilibrium at laboratory temperature in a system dominated by the mineral composition, and this may be significantly different from the original isotopic composition of the fluid upon entrapment. Interestingly however, Vityk and others (1993) presented evidence that for one case, $\partial^{18}O$ composition of fluid inclusions in quartz had not reequilibrated with the quartz after entrapment, so reequilibration of oxygen isotopic ratios in oxygen-bearing minerals is not necessarily a problem for every study. One might expect preservation of the fluid's oxygen isotopic composition when inclusions are entrapped in oxygen-free mineral species such as sphalerite, halite, and fluorite.

It is important to evaluate the maximum effect of reaction between inclusion fluid and host mineral on the bulk salinity of the fluid inclusion, even though there is as yet no experimental data that will constrain such a calculation. For example, starting with a hypothetical crystal of calcite that precipitated from marine waters, the fluid inclusions in the sample would initially have approximately 3.5 wt.% total salinity. Removing trace elements in concentration of 1000 ppm from a volume of calcite equal to the volume of all fluid inclusions, adding these trace components to the inclusions, and calculating the change in salinity for the inclusions results in an increase in salinity of the inclusions of only 0.27 wt.% (2.7 parts per thousand) — a relatively small change. This simple calculation illustrates the relative stability of the total salinity of a fluid inclusion. In reality, the actual change in composition of the inclusion fluid would be controlled by equilibrium and electroneutrality considerations, and could be modeled for various systems using trace element distribution coefficients assuming a change from a water-dominated to a rock-dominated system. Moreover, data on seawater fluid inclusions in calcite from rocks as old as Cambrian show that the mode in the data occurs at the salinity of seawater and has not been altered (Johnson and Goldstein, 1993). Thus, the evidence we have so far argues against significant changes in salinity from scavenging of trace elements from the host crystal.

Diffusion Through the Host

Once fluid inclusions are entrapped, it is important to evaluate whether they act as sealed bottles or if the minerals enclosing them are permeable to some components to allow leakage of material in or out. In addition to physical opening of the fluid inclusion from a crack (discussed in detail below), the potential for diffusion of components through the crystal itself, and along dislocations in the crystal, must be evaluated. Experimental work (Blacic, 1975; Kekulawala and others, 1981; McLaren and others, 1983; Gratier and Jenaton, 1984; Pecher and Boullier, 1984; and Bakker and Jansen, 1990, 1991) has shown that there is the potential for water to diffuse through the quartz lattice and along dislocations at conditions predictable for metamorphic environments. These diffusivities seem to be high within the metamorphic realm (Blacic, 1981), but drop by orders of magnitude as one approaches diagenetic conditions. Although thorough experiments have not yet been performed for fluid inclusions in common diagenetic minerals under diagenetic conditions, there is abundant empirical evidence to suggest that diffusion of components through diagenetic minerals does not seem to occur at any significant rate under low diagenetic temperatures. Probably the best evidence against such diffusion comes from the tremendous number of published studies in which rocks were collected from outcrop or from a mine (see summary in Roedder, 1984). These studies show that fluid inclusions have not equilibrated with the low-temperature shallow groundwater in which they were bathed after uplift; the fluid inclusions still retain signatures of fluid composition and density from deep in the subsurface. Samples taken from cores in oil and gas wells also show that fluid inclusions bathed in these deeper diagenetic systems have not equilibrated with the surrounding fluid (O'Hearn, 1985; Pagel and others, 1986; McLimans, 1987; Anderson, 1989; Wojcik and others, 1992). Moreover, if diffusion were to significantly modify the composition of fluid inclusions in diagenetic systems, one would expect that fluid inclusion compositions in various petrographic generations of inclusions would tend to be more variable. Yet there are many examples of consistent and distinct data from within petrographically distinct generations of fluid inclusions. Thus, although in the quantitative sense, we still consider diffusion an unknown in diagenetic systems, it does not appear to have imposed significant problems thus far.

The potential for significant hydrogen diffusion through minerals in diagenetic settings is another problem altogether. From a simple perspective, hydrogen should be the easiest component of all to diffuse, merely due to its size. Various experiments have shown that hydrogen diffuses with relative ease in quartz at conditions predictable for some high-temperature environments (granulite facies metamorphic and intrusive igneous). Morgan and others (1993) modified natural inclusions and Hall and others (1989) modified synthetic inclusions in quartz by simulating metamorphic P-T conditions with high hydrogen gradients. Mavrogenes and Bodnar (1994) demonstrated that initially insoluble chalcopyrite daughter minerals within inclusions in quartz could be rendered soluble by subjecting them to controlled hydrogen-rich atmospheres in hydrothermal cells at 600°C for seven days. Also, hydrogen diffusion has been implicated in discrepancies between calculated fluid compositions (as constrained by mineral assemblages and thermodynamic data) and the observed natural fluid compositions (Hall and others, 1991). However, for quartz, there may be some temperature below which hydrogen diffusion would be insignificant even over geologic time: Dubessy and others (1988) reported the simultaneous occurrence of molecular hydrogen and oxygen within individual 2 Ga old natural inclusions in quartz that had been subjected to temperatures at least as high as 200°C since entrapment of the inclusions. Nevertheless, hydrogen diffusion in diagenetic minerals is possible; the truth of the matter is that the data necessary to assess the diffusion rates have not yet been collected.

Change in Volume of the Inclusion Vacuole

As many fluid inclusions are entrapped at depth at elevated temperature and pressure, it is predictable that the volume of the mineral and hence the volume of the inclusion vacuole would change as the mineral thermally contracts and as its volume expands as confining pressure is released. In the sedimentary realm thermal contraction is largely reversible through heating in the laboratory (Skinner, 1966; Bodnar and Bethke, 1984; Bodnar and Sterner, 1985; Zhang and Frantz, 1987), but the mineral is not normally recompressed in typical laboratory conditions because commercially available heating stages are not equipped to apply an external pressure. Thus, the volume of the fluid inclusion at lab conditions may be slightly different than when it was entrapped. This effect on the density of the fluid inclusion must be evaluated, because if it is significant, one of the basic requirements of much fluid inclusion research — that of maintaining a constant volume — is not met. Lacazette (unpublished) has calculated this effect and Zhang and Frantz (1987) and Bodnar and Sterner (1985) have evaluated it experimentally: in gas-poor brines in the sedimentary P-T realm, this change in volume of the mineral causes less than 1% error. However, Lacazette (unpublished) has shown that methane-rich aqueous inclusions trapped at high pressures may deviate from their true entrapment densities because of high internal pressures at Th and methane solubility relationships. These high differential pressures developed in the laboratory can cause elastic deformation to temporarily increase the volume of the fluid inclusion on approach to Th. The volume increase results in a slight increase in value of the measured Th. Lacazette evaluated this effect for rather high-pressure, high-temperature systems (1.15-1.55 kbars and 100-300°C), and calculated a Th increase of 1-8°C. The model has not been extended to more common diagenetic P-T conditions because of uncertainties in the equations of state for H_2O-NaCl in that range (Lacazette, personal communication, 1993). However, inclusion workers must keep this effect in mind as a minor source of error when evaluating Th of gas-rich fluid inclusions.

Change in Shape (Necking Down)

There is a thermodynamic drive for the crystal walls of fluid inclusions to change their shape to minimize surface free energy (Roedder, 1984). Fluid inclusions with highly irregular and tortuous shapes are more unstable relative to those that are negative crystal or spherical in shape. Fluid inclusions respond to this thermodynamic drive over time by changing their surfaces to achieve more spherical or negative crystal shapes. This process, known as necking down, is a mechanism by which secondary fluid inclusions form (see Chapter 2), and is a mechanism that affects the shapes and numbers of primary fluid inclusions. The variables that affect necking down (or shape change) include time, temperature, fluid and mineral composition, original size and shape of the fluid inclusion, and perhaps strain in crystal hosts.

Time.—

Because the process of necking down involves dissolution-reprecipitation and diffusion through the fluid, or perhaps even surface diffusion, to achieve a lower surface free energy state (Nichols and Mullins, 1965; Smith and Evans, 1984), there is a strong kinetic control on the degree to which it occurs. Thus, the longer the time available, the more likely it is that significant necking down will have occurred. Experiments have shown that necking down may take place with geologic rapidity, especially in soluble minerals and at high temperature (Pecher, 1981; Brantley and others, 1990; Brantley, 1992). Brantley (1992) suggested that microcracks in quartz may heal within 100 years or less at temperatures as low as 200°C in the presence of dilute pore fluids. Nevertheless, irregularly-shaped inclusions in diagenetic minerals are common, despite hundreds of millions of years since inclusion entrapment, so the process has obviously not progressed to the point at which all ancient fluid inclusion walls are at a minimum surface free energy state (negative crystal shape). Thus, although time is an important factor in necking down, it is but one of several, and so it is improper to assume all ancient aqueous fluid inclusions have necessarily undergone major changes in shape since they were entrapped.

Size and shape.—

The original size and shape of a fluid inclusion may also affect the degree of necking down. Primary fluid inclusions forming from the same fluid along the same growth zone in the same crystal, but with different initial cavity morphologies, may have different resulting shapes and different histories of necking down. For example, if one inclusion were to form with a negative crystal shape and another were to form with an irregular or elongate shape, the negative crystal-shaped inclusion would be relatively stable because its original shape was one of low surface free energy, whereas the other inclusion would tend to neck down to a more globular or negative crystal shape with time. Work by Nichols and Mullins (1965) showed that inclusions or cracks with sharp tips would remove surrounding material to fill in the tip and create a more blunt to spherical end. Thus, inclusions with such originally sharp angles have a large drive to change shape. In the experiments of Brantley (1992), microcrack geometry appeared to exert a significant control on rate of healing, and there was a suggestion that increases in crack aperture decreased the rate of healing.

Temperature.—

Another variable affecting the rate of necking down is temperature. Shelton and Orville (1980) produced negative crystal shaped secondary fluid inclusions in quartz in several hours to several days at 600°C and 2 kbar. Brantley (1992) observed significant decreases in rates of microcrack healing with decreasing temperature. In epithermal precious metal ore-forming environments and active geothermal systems, Bodnar and others (1985) have shown that shapes of fluid inclusions in quartz can be directly correlated with formation temperatures, the lowest temperature inclusions having the most irregular shape. Ancient secondary fluid inclusions in diagenetic quartz and calcite which have been heated to temperatures above 100°C (for geologic time) commonly preserve extremely irregular shapes and smoother shapes. Thus, although temperature definitely has an effect on the rate of necking down, it is but one of many significant variables.

Fluid composition.—

Fluid composition appears to exert a significant control on the phenomenon of necking down. For example, it is common to observe primary fluid inclusions of petroleum in fluorite (Roedder, 1972) and in calcite cements (Anderson, 1989; Peter and others, 1991) that have globular shapes or meniscus shapes suggestive of the original shapes of the oil droplets on the surfaces of crystals prior to entrapment, but in the same crystals, secondary fluid inclusions of aqueous composition have necked down to negative crystal shapes. Apparently, the low solubility of the host mineral in the petroleum has reduced the rate of necking down relative to the aqueous inclusions. Similar consistent shape differences among a single population of oil-filled and aqueous primary fluid inclusions have been observed, but it remains unclear if these differences were the result of original differences in inclusion shape or subsequent differences in necking down. Even among aqueous fluid inclusions, composition seems to have a significant effect. Roedder (1984) described secondary or pseudosecondary fluid inclusions in single quartz crystals, one plane containing spherical fluid inclusions, and another plane containing negative crystal-shaped fluid inclusions. These differences in shape may have been caused by compositional differences in the various fluids (Roedder, 1984), but other variables such as timing or temperature may also have been important. Experimental studies have demonstrated that aqueous fluid composition exerts a major control on rate of necking down. Brantley (1992) determined the rate of crack healing in quartz is slowest for CO_2-H_2O solutions, is more rapid for pure H_2O solutions, and is even more rapid for 1 *m* NaCl, 2 *m* $CaCl_2$, and 6 *m* NaCl aqueous solutions. There was also a suggestion that crack healing rate increased with increase in NaCl concentration, especially at high temperatures.

Host composition.—

The composition of the host mineral also has a significant effect on the rate of necking down. For minerals with extremely high solubilities in aqueous solutions such as various salts, necking down can take place at extremely high rates (Lemmlein and Kliya, 1952). One might also expect that compositional differences in crystals that have an effect on the mineral's solubility would differentially affect the rate of necking down.

Strain.—

Fluid inclusions have been known to change shape and even migrate through the host crystal when high differential stress had been imposed on the host crystal to create significant strain. Roedder (1971) presented evidence that fluid inclusions migrate into the more strained (most soluble) parts of quartz crystals. Gerlach and Heller (1966) suggested that fluid inclusions in halite may migrate or change shape because of strain.

Effect of necking down.—

It is predictable that necking down is a common phenomenon for many inclusions in diagenetic minerals. Thus, it is important to consider what effects such a process has had on our ability to interpret past physical and chemical fluid history from fluid inclusions. With one important exception, the effect of necking down does not appear to present major problems in our ability to use fluid inclusion data. Necking down of a monophase fluid inclusion to form multiple inclusions does not pose problems for interpreting fluid inclusion data, because the process of necking down does not change the density of the inclusion fluid significantly as material from the inclusion's wall is merely shifted from one place to another. (As discussed earlier, the process of necking down may change trace element composition and isotopic composition of fluid inclusions in some minerals, but major ion composition and bulk salinity should remain largely unaffected.) However, if an individual polyphase inclusion were to change shape to such an extent that it necked down to more than one fluid inclusion, each of the newly formed inclusions would no longer be representative of the original one. It is best to illustrate this process by considering the P-T relationships of aqueous fluid inclusions in the subsurface. Several different scenarios will be considered.

If aqueous all-liquid fluid inclusion were trapped at low temperature (near earth surface conditions, 25°C) and were then subjected to burial heating to 120°C, the conditions experienced by the fluid inclusion would approximate an isochore in the one-phase liquid field of an aqueous phase diagram (Fig. 4.1, points A to B). Of course, this is assuming that the internal pressure within the fluid inclusion does not cause its volume to expand permanently (known as *stretching*), or cause the inclusion to burst (known as *decrepitation*) — a reasonable assumption for smaller a fluid inclusion in a harder mineral. So, as warming occurs (toward point B,

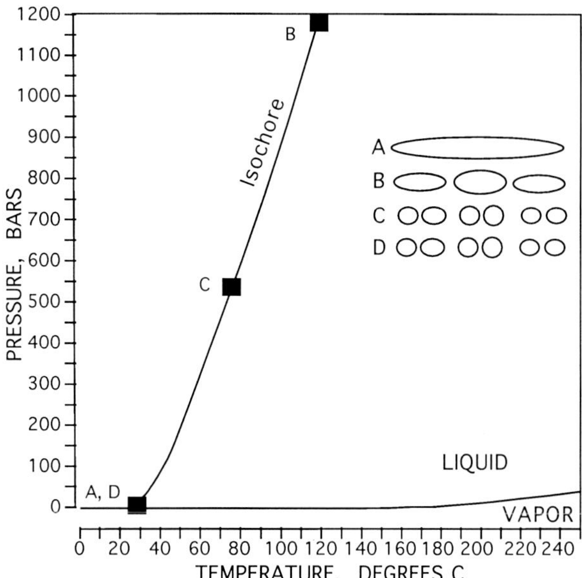

Fig. 4.1. *P-T phase diagram for pure water illustrating an isochore for pure water and the liquid-vapor field boundary. An all-liquid fluid inclusion trapped at surface conditions (point A) remains in the one-phase field as it is heated during burial to point B. If it necks down at point B, inclusions have the same density as the original inclusion. During subsequent cooling to point C, if any inclusions neck down, they will still have the same density as the original fluid inclusion. Continued cooling to surface conditions (point D) yields many necked down inclusions that all have the same density as the original single fluid inclusion.*

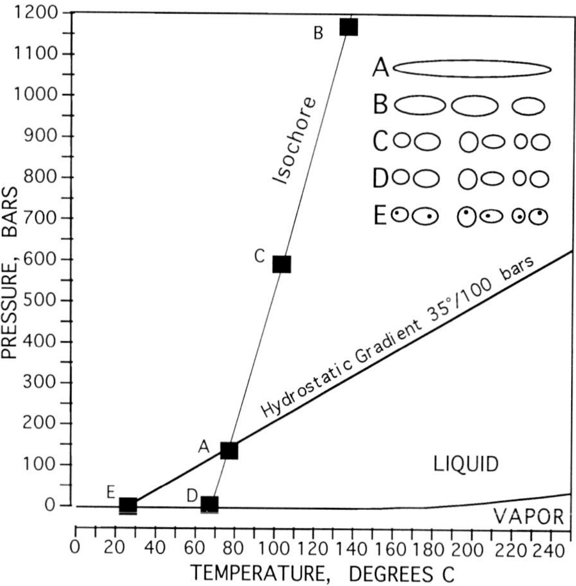

Fig. 4.2. *P-T phase diagram for pure water with an isochore for pure water and the liquid-vapor field boundary. If a fluid inclusion were trapped at 75°C (point A) and were then heated during burial to 130°C (point B), the inclusion would remain in the liquid field. If it were to neck down at point B and then were to neck down again during cooling to 100°C (point C), new inclusions would be formed that were of the same density as the original single inclusion. With continued cooling to 68°C (point D) a vapor bubble can nucleate but generally does not nucleate unless there is additional cooling. Once the inclusions have reached room temperature (point E) all have nucleated bubbles. Each inclusion now has the same ratio of liquid-to-vapor and homogenizes at point D. This is the temperature of homogenization that the single original fluid inclusion would have produced. Necking down has had no effect on homogenization temperature.*

Fig. 4.1), the inclusion remains in the liquid field (does not contain a vapor bubble). A single fluid inclusion could neck down to form several fluid inclusions anywhere on this warming path to 150°C. Even after this process had occurred, the density of the fluid in each of the inclusions would be the same as the density of the original fluid inclusion. The overall vacuole volume would also remain the same. Thus, conditions within the newly separated fluid inclusion remains on the original isochore. If cooling occurs, conditions within the fluid inclusions would remain on the isochore (toward point A, Fig. 4.1). If further necking down were to occur during this cooling (perhaps at 75°C) to create new fluid inclusions, again, density within each fluid inclusion would remain unchanged and inclusion conditions would remain on the isochore. Therefore, by the time the inclusions have cooled to surface conditions (point D), each of their densities is identical to the density of the original large fluid inclusion (point A, Fig. 4.1). So in the case of extensive necking down in the all-liquid aqueous field, all inclusions that result from necking down remain useful records of the density of the fluid inclusion that was originally entrapped.

Another scenario will serve to illustrate the effect of an individual fluid inclusion necking down to multiple fluid inclusions at a different set of conditions. If an all-liquid aqueous fluid inclusion were trapped during burial at a temperature of 75°C (Fig. 4.2, point A) and were then heated to 130°C (point B of Fig. 4.2) as burial continued (with no stretching or decrepitation of the inclusion), conditions within the fluid inclusion would approximate the isochore on which the inclusion was initially entrapped. If the original fluid inclusion were to neck down to several fluid inclusions along this heating path to 130°C (point B of Fig. 4.2) each of the newly produced fluid inclusions would have the same density as the original fluid inclusion. If the fluid inclusions were then subjected to cooling to 100°C (point C of Fig. 4.2) and necking down were to occur at that temperature to produce new fluid inclusions, again, each new inclusion would still be of the same density as the original inclusion. With cooling to the P-T condi-

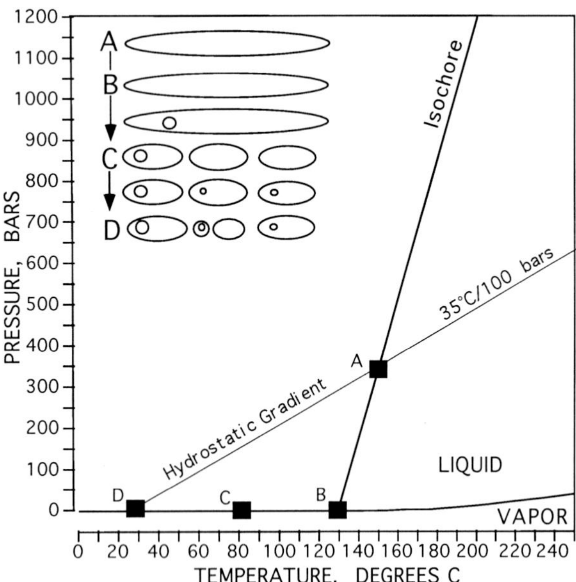

Fig. 4.3. P-T phase diagram for pure water with an isochore for pure water and the liquid-vapor field boundary shown. If a fluid inclusion were entrapped at point A and were then cooled during uplift to point B, a vapor bubble would be stable at this temperature but does not nucleate until cooling toward point C. After the bubble has nucleated and the inclusion has cooled to point C, the inclusion may neck down to form several fluid inclusions, only one of which contains the vapor bubble. With continued cooling toward point D, bubbles may nucleate in the remaining all-liquid inclusions. As the inclusions are cooled to surface temperature (point D), more necking down may occur and the bubble may be split between two inclusions. At, point D, none of the necked down inclusions have the same density as the original inclusion trapped at A. Also, none of these inclusions will homogenize at point B, they will either homogenize at lower or higher temperatures because they necked down after a phase change occurred in the fluid inclusion.

tions in which the isochore intersects the liquid-vapor field boundary (68°C, point D of Fig. 4.2), each fluid inclusion has the potential of nucleating a vapor bubble. Upon continued cooling along the liquid-vapor field boundary (to 25°C, point E of Fig. 4.2), the metastability of bubble nucleation would eventually be overcome and each fluid inclusion would contain a small bubble. If these fluid inclusions were heated in the laboratory, every bubble would homogenize at 68°C. This is the exact homogenization temperature that the original single fluid inclusion would have yielded because the necked down inclusions still contain fluids of the density representative of the fluid that was initially entrapped. Thus, for this scenario, necking down has had no effect on our ability to interpret fluid inclusion data.

A third scenario illustrates an example in which fluid inclusions that have formed from necking down are not representative of the fluid initially entrapped in the fluid inclusion. A single all-liquid fluid inclusion is initially entrapped at the maximum temperature reached (150°C, point A of Fig. 4.3). The system cools prior to completion of necking down and conditions within the inclusion travel down the isochore on which initial entrapment occurred until the liquid-vapor phase boundary is reached (130°C, point B of Fig. 4.3). At this temperature, the fluid inclusion has the potential of nucleating a bubble, but bubble nucleation commonly occurs at a lower temperature because of nucleation problems (the temperature at point B is the temperature at which a fluid inclusion of this density would normally homogenize). A bubble should nucleate as cooling continues (point C of Fig. 4.3). If the original fluid inclusion were to neck down to several fluid inclusions after nucleation of this vapor bubble, then one fluid inclusion would contain the bubble and the other inclusions produced would only contain liquid. If further cooling occurred to room temperature (point D), bubbles could nucleate in the newly formed all-liquid fluid inclusions. If one of these were to neck down to two inclusions, it would produce an all-liquid inclusion and another with the vapor bubble. Thus, each of the fluid inclusions produced by necking down after a phase change may have a different density and yield different homogenization temperatures. The homogenization temperatures produced in this way would be highly variable. For example, if a two-phase fluid inclusion were to neck down to two fluid inclusions, one inclusion containing the bubble and the other inclusion containing only liquid, and if the new inclusion containing the bubble were to end up containing just 5 volume percent bubble at the time of necking down, the inclusion containing the bubble would obviously have a lower bulk density (by about 5%) than the inclusion that necked down to contain only liquid. A 5% variance in bulk density would yield variation in homogenization temperature of at least 50°C; of course, a larger proportion of vapor trapped in one of the inclusions would drive homogenization temperatures even higher. One could imagine that the random controls on the liquid-to-vapor volumetric proportions caused by necking down after a bubble has nucleated should yield inclusions of highly variable densities and highly variable homogenization temperatures that no longer reflect conditions of inclusion entrapment. Such inclusions may be recognized by their highly variable liquid-to-vapor ratios (and bubble pressures that are inconsistent with formation in a two-phase diagenetic system such as the vadose zone) among closely associated groupings of fluid inclusions (see Chapter 6). Necking down after a phase change may also change the ratios of daughter minerals to fluid in inclusions. There is a popular misconception that good evidence for necking down exists when fluid inclusions appear highly irregular in shape and have protuberances that indicate closed off necks between adjacent fluid inclusions. But inclusions with such shapes simply indicate that necking down has not yet proceeded to the lowest surface free energy shape. Smooth-surfaced and negative crystal-shaped inclusions could also have formed from necking down. Furthermore, fluid inclusions with immature shapes may have attained those shapes prior to nucleation of another phase, in which case Th data obtained from such inclusions are valid and representative of minimum conditions of formation. Indeed, what is most important for an inclusionist to note is whether necking down has occurred *after* nucleation of another phase. This can only be recognized by high variability in phase ratios or homogenization temperatures in petrographically and spatially associated fluid inclusions of the same group. In other words, inclusions in the same small area of a growth zone, or inclusions in the same microfracture must show a high variability in phase ratios (see Chapter 6).

One may ask if necking down after a phase change is a common problem to be dealt with in the diagenetic realm. The answer to this is an emphatic yes! Because of the many variables that control the degree of necking down, such as original size, shape, fluid composition, and temperature history, and because these variables are largely unknown for most fluid inclusions investigated, it is important to consider necking down as a possibility for any fluid inclusion. Of course, if necking down has not occurred to a significant degree, or if it has occurred while only one phase was present in a fluid inclusion, the inclusion studied will still be representative of conditions that existed prior to necking. But if necking down to multiple inclusions has occurred while multiple phases were present in inclusions, the inclusions have been altered and this negates our ability to interpret conditions prior to necking. The likelihood of this occurring in the diagenetic realm is high, especially for inclusions entrapped at maximum temperature or during cooling of the system. Such can be illustrated by assuming a simple aqueous system and considering Figure 4.4. A fluid inclusion trapped at a hypothetical maximum burial temperature of 150°C would require only 20°C of cooling for the inclusion fluid to intersect the liquid-vapor phase boundary. Continued cooling past the point of metastability would allow a vapor bubble to nucleate while the inclusion is still at a relatively high temperature (at which the rate of necking down can be relatively rapid). Moreover, given a fluid inclusion

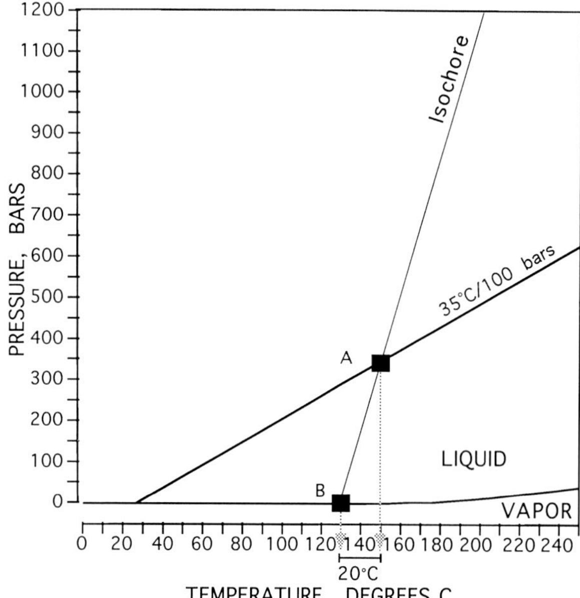

Fig. 4.4. *P-T phase diagram like that in Figure 4.3. If a fluid inclusion were trapped at its maximum burial temperature of 150°C (point A) and were then cooled during uplift, only 20°C of cooling would be needed for the inclusion to reach the liquid-vapor field boundary (point B). Much of the rest of the inclusion's uplift history would be on the liquid-vapor curve, once the bubble nucleates. Thus, necking down after a phase change should be a valid concern for aqueous fluid inclusions from the diagenetic realm.*

system in which the aqueous fluid was saturated with respect to methane, and which was also trapped at maximum temperature (trapped on the bubble point curve; Fig. 4.5), any cooling would place the inclusion in the two-phase field. Thus, a vapor bubble would be present while the inclusion remained at relatively high temperature. Therefore, given the constraints on sedimentary systems, one must always evaluate whether or not necking down has affected the phase ratios preserved in fluid inclusions.

In many diagenetic studies, necking down is not an insurmountable problem. For instance, one may find primary fluid inclusions along a growth zone in one sample in which the liquid-to-vapor ratios have been highly altered by necking down after a phase change. These inclusions would be amenable to analysis of salinity through freezing work or analysis of compositions using other techniques, but would not be useful for study of homogenization temperature, because the density of each inclusion would no longer be representative of density of the original inclusion fluid entrapped. However, one may study another sample taken from the same area, or even the same rock sample, and find that the primary fluid inclusions have

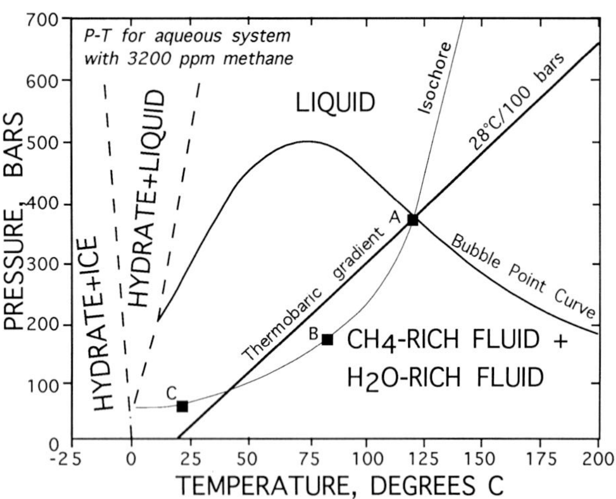

Fig. 4.5. *P-T phase diagram illustrating the bubble point curve for 3200 ppm methane in water. Notice that if a fluid inclusion is trapped on the bubble point curve (saturated with respect to methane; point A), then with cooling during uplift (points B and C) the inclusion remains entirely in the two-phase field. Thus, any necking down during cooling occurs in the presence of a gas bubble. Modified from Hanor, 1980. Also, see Chapter 3 and Figure 3.7 for a more detailed explanation.*

not undergone significant necking down after a phase change, and that their homogenization temperatures may prove valid.

Recrystallization of Mineral Host

At the low temperatures in diagenetic systems, it is apparent that recrystallization of minerals requires a fluid medium in which the minerals have significant solubility. For the most part, this constrains recrystallization to take place in the presence of an aqueous fluid. Folk (1965) inferred that a fluid film might migrate through a crystal, precipitating the new phase on one side and dissolving the unstable phase on the opposite side of the film. Reeder (personal communication, 1993) is further evaluating the mechanism by which recrystallization occurs and the possible geometries of such fluid films. It is probable that such a fluid film would be in equilibrium with the surrounding fluid pressure, and would approach compositional equilibrium with the surrounding pore fluid to varying degrees, depending on diffusion through the fluid film and local modification of the pore fluid's composition. Thus, if during recrystallization of an unstable precursor mineral, such a fluid film were to encounter fluid inclusions, it is probable that such fluid inclusions would be opened and would reequilibrate with the fluid present in the fluid film. If the inclusion cavity were still present after recrystallization, the inclusion fluid would provide a record of the conditions of recrystallization rather than conditions of precipitation of the precursor mineral. These fluid inclusions would still be considered primary, but are records of recrystallization rather than initial precipitation. Some types of recrystallization may reduce the likelihood of preserving workable fluid inclusions. Gaffey (1988) has shown through IR spectroscopy that water tends to be excluded from crystals during the aragonite-to-calcite recrystallization process. Studies of modern marine aragonite botryoids from Belize (Goldstein, 1986a) show abundant entrapment of workable-sized fluid inclusions in such material. However, ancient botryoids of aragonite that have recrystallized to calcite (Permian Capitan Limestone and Laborcita Formation) rarely contain large fluid inclusions. Nevertheless, this does not preclude their preservation in other samples of similar material.

Many diagenetic minerals are subject to recrystallization within the diagenetic realm. Careful petrography and geochemical work are helpful in identifying if such recrystallization has taken place. It is common for aragonite to recrystallize to calcite; high Mg calcite may recrystallize to low Mg calcite; low Mg calcite may recrystallize to more stable low Mg calcite; dolomite may recrystallize to more stable dolomite; ankerite may recrystallize to more stable ankerite; halite may recrystallize to halite; and gypsum and anhydrite may undergo multiple dehydration and rehydration reactions. Abegg (1990) studied dolomite that had clearly formed very early, because rhombohedra were truncated on the margins of sedimentary clasts. Yet the dolomite contained fluid inclusions that yielded temperatures and salinities consistent with the late, elevated-temperature fluids that had circulated through the system. One explanation for this is that fluid inclusions were trapped during late recrystallization of the dolomite. Wojcik and others (1992) illustrated ankerite cements in sandstone in which backscattered electron images showed that early zones had been recrystallized during later ankerite-producing events. Many of the primary fluid inclusions in that study were trapped during this late recrystallization event. Also, James and Bone (1992) studied relatively young marine calcite cements that are presently recrystallizing in shallow saline groundwater. This recrystallization has caused entrapment of the fluid responsible for recrystallization.

Alternatively, some believe that recrystallization takes place in a relatively closed system (Lohmann, 1978). If this mechanism is valid, then it may be that fluid inclusions in minerals that have recrystallized by this mechanism do not differ very much compositionally from those initially entrapped in the mineral. It has been suggested that recrystallization by this mechanism may actually employ fluid inclusions as the aqueous medium of recrystallization. Lohmann (personal communication, 1988) proposed that high-Mg calcite cements recrystallize in a closed system by migration of fluid inclusions through the crystal, accumulation of Mg in inclusions to form crystals of microdolomite within fluid inclusions, and precipitation of low-Mg calcite in the wake of the fluid inclusions. Whether this is indeed the mechanism by which low-Mg calcite recrystallizes and whether enough fluid inclusions in recrystallized calcites actually physically contain microdolomite inclusions has yet to be verified. There have also been suggestions that some halite may recrystallize in relatively closed systems. Chemical analyses of large fluid inclusions in coarsely recrystallized halite seem to contain fluids that can be interpreted as those originally present during deposition (Lazar and Holland, 1988). Thus, closed system recrystallization could have taken place, leaving the original fluid inclusion population altered only the small amount caused by its reaction with the host crystal.

Location in the Crystal

For the most part, the petrographic location in which one finds a fluid inclusion in a diagenetic mineral seems to be the location in which it was initially entrapped. In the sedimentary realm, it is rare for a fluid inclusion to migrate through a crystal a great enough distance for its origin to be unclear. Any movement of a fluid inclusion through the crystal results from a mechanism similar to necking down, but the driving force is not always the same. Thus, rates of migration may be affected by similar variables to those of necking down. Some types of movement are worthy of discussion for the diagenetic realm. Some small fluid inclusions along the same healed fracture may coalesce to make larger ones (Lemmlein, 1951). The process leaves fluid inclusions along the same healed fracture and does not alter the composition of inclusions significantly. In the metamorphic realm, Swanenberg (1980) observed that fluid inclusions in quartz migrated away from healed fractures. This can be attributed to strain in the crystal leading to greater solubility, and migration of fluid inclusions in the direction of greatest solubility (Roedder, 1984). Gerlach and Heller (1966) proposed migration of fluid inclusions through halite caused by strain in the crystals. Roedder and Belkin (1980) found that fluid inclusions in halite may migrate towards heat sources given unnaturally high thermal gradients, and that given an extremely localized heat source, such as a canister of radioactive waste, fluid inclusions in surrounding halite would migrate in the direction of the heat source. In summary, fluid inclusions in minerals from the sedimentary realm do not travel significant distances under natural conditions: secondary fluid inclusions are confined to planes, and primary fluid inclusions are confined to their original growth zones.

Deformation

Intense deformation of crystals containing fluid inclusions is likely to cause inclusions to be opened and potentially to refill with fluids present during deformation (Ypma, 1963; Kalyuzhnyi, 1971); this process is one mechanism of fluid inclusion reequilibration. If one is studying fluid inclusions in which minerals show evidence of shearing, or if deformation twins are present, it is best to assume that fluid inclusions may have reequilibrated during the deformation event. Rough sample handling may also generate fractures. Some coring operations in salt have generated fractures that breach fluid inclusions (Roedder, and Belkin, 1981) and "percussion sidewall coring" fractures most lithologies (D. Hall, personal communication, 1994).

Irreversible Phase or Chemical Changes After Entrapment

Most fluid inclusions trapped at elevated diagenetic temperatures undergo phase changes during their subsequent history and some may undergo other chemical changes. Many of the phase changes that take place during cooling are reversible in the laboratory, such as nucleation of a bubble or nucleation of a daughter crystal. However, some of these changes cannot be reversed in the laboratory. In these situations, it would appear that the fluid inclusions had changed in some way, and that it may not necessarily be representative of its composition upon initial entrapment.

One example of such a change is the formation of organic solids within petroleum fluid inclusions. It is common to observe petroleum fluid inclusions with brownish solid material adhering to the inclusion walls. This solid bitumen presumably precipitates from the hydrocarbon inclusion fluid. When the inclusion is heated, the bitumen does not dissolve. Thus the composition of the petroleum in the inclusion has changed since initial entrapment. However, one must be careful to avoid confusing these bitumens with other solid organics in petroleum fluid inclusions. For example, if a particle of solid organic material were to adhere to the surface of a growing crystal, it could cause the entrapment of a primary fluid inclusion as the crystal grew around the poisoned surface of the crystal. In this situation, no change in composition of the fluid inclusion would be indicated by the presence of solid organics that will not dissolve upon heating.

Other changes may affect some aqueous fluids in inclusions. It is well known that some subsurface brines contain significant concentrations of dissolved organic molecules, particularly, the aliphatic acid anions (Collins, 1975; Carothers and Kharaka, 1978). These anions may achieve concentrations as high as 4,900 mg/l (Hanor, 1980). There is now abundant evidence that thermal decarboxylation reactions occur in nature to break these anions down to form methane and bicarbonate at diagenetic temperatures of at least 80°C and above (Boles, 1978; Carothers and Kharaka, 1978, 1980; Milliken and others, 1981; Kharaka and others, 1983, 1986). Hanor (1980) has suggested that if these components are entrapped in fluid inclusions, they may break down to increase the methane content of aqueous fluid inclusions that reside at high temperature for some time. This increase in methane content would shift the equilibrium of the fluid inclusions so that homogenization temperature would be increased. The importance of this effect is unknown, but many studies (summarized by Barker and Goldstein, 1990; Prezbindowski and Tapp, 1991) illustrate that most aqueous fluid inclusions homogenize below their true entrapment temperature, as opposed to the higher temperatures that would result from a large amount of in situ methane production. Moreover, many diagenetic fluids are low in dissolved organic matter initially, and therefore, in situ methane production would be insignificant. However, it remains possible that in situ cracking in oil-filled fluid inclusions has altered some oil-filled inclusions subsequent to entrapment.

Nucleation Metastability

Some fluid inclusions do not contain the room temperature phase assemblage that would be expected for the inclusion's density and composition because phases that should be present according to equilibrium considerations have not nucleated. The most commonly observed example of this is a one-phase liquid fluid inclusion that should have a vapor or gas bubble, but has not yet nucleated that phase. This phenomenon can be illustrated by considering an analogous example from the laboratory. If an aqueous fluid inclusion were homogenized in the lab at 100°C (Fig 4.6, point A) and were then cooled, it is common that the vapor bubble does not immediately reappear as equilibrium conditions would dictate. In reality the fluid inclusion may need to be cooled several tens of degrees below the homogenization temperature to instigate bubble nucleation. If the inclusion in this example were cooled by several tens of degrees (Fig. 4.6, point B), its internal pressure would follow a metastable isochore below the liquid-vapor field boundary (path A-B in Figure 4.6). Thus, as the inclusion is cooled below its homogenization temperature, the liquid is under tension (referred to as a *stretched liquid*) and its internal pressure may be less than zero. As the fluid inclusion continues to cool, the drive for the stretched liquid to separate from some point on the wall of the inclusion increases until finally it does. If one were viewing an inclusion at the instant this happens, one would see a vapor bubble (or sometimes several) instantly pop into view. At this instant the inclusion conditions return to the liquid-vapor phase boundary (Fig. 4.6, point C). This procedure can be repeated on the same inclusion, but the temperature at which the bubble nucleates will vary slightly from run to run because nucleation is controlled by kinetics. For some fluid inclusions that have been heated to homogenization in the laboratory, it may take hours or weeks for the bubble to reappear. Sometimes, in fact, a bubble may never reappear during the course of the study even with tens of degrees of undercooling. Thus, two-phase fluid inclusions should not be homogenized during sample preparation or during heating runs before they are measured unless the researcher has "geologic time" to wait for bubble reappearance. Also, because of this potential nucleation barrier in nature, many fluid inclusions that should have bubbles at room temperature may have existed in a metastable state for millions of years, and are observed to be all-liquid fluid inclusions in the laboratory.

Most inclusions which have the densities that should yield homogenization between room temperature and 50°C exist as all-liquid inclusions at room temperature because metastability has prevented nucleation of a bubble. It is rare for aqueous two-phase inclusions less than 20 μm in size to yield homogenization temperatures less than about 40-50°C. This cutoff in homogenization temperature occurs not because fluid inclusions do not have the density for a bubble to be present at 39°C, but because they have not nucleated a bubble because of metastability. Sometimes with cooling below room temperature, or with prolonged cooling, bubble nucleation can be stimulated. Large aqueous inclusions (>50 μm) particularly in evaporites, can yield homogenization temperatures below 40°C. This is because the diameter of the bubble in such a large inclusion is great enough for the bubble to exist at such low temperatures. At the interface between the bubble and aqueous fluid, tensional forces on the surface of the bubble actually encourage its collapse. The

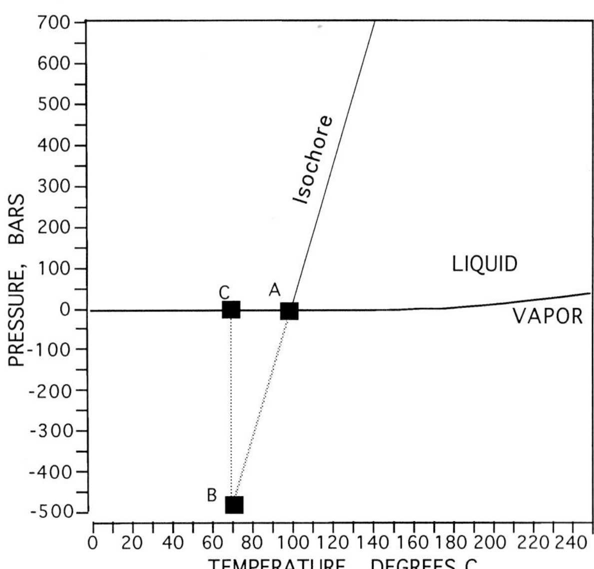

Fig. 4.6. P-T phase diagram for pure water illustrating the amount of undercooling typically required for nucleation of a vapor bubble. If a fluid inclusion had homogenized at 100°C (point A) and were then cooled, a vapor bubble would not immediately nucleate. As the inclusion is cooled toward point B, the pressure in the inclusion drops to negative values. Once the inclusion has been cooled several tens of degrees, the "stretched" liquid finally detaches from the inclusion wall to form a spherical bubble, returning the inclusion to equilibrium conditions on the liquid-vapor curve (point C).

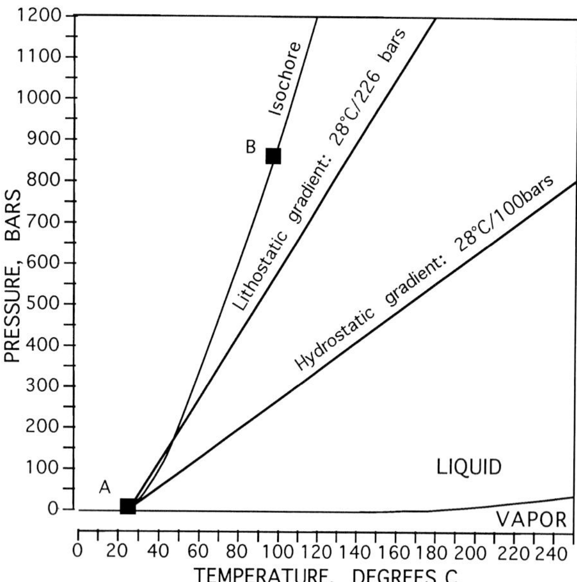

Fig. 4.7. P-T aqueous phase diagram for pure water with hydrostatic thermobaric gradient of 28°C/100 bars and lithostatic thermobaric gradient of 28°C/226 bars superimposed. If a fluid inclusion were trapped at surface conditions (point A) and were then subjected to heating during burial to point B, the fluid inclusion's internal pressure would rise significantly above the surrounding hydrostatic pressure and lithostatic pressure at that temperature.

smaller the bubble, the greater the forces and the more likely is the bubble to collapse; the larger the bubble, the less likely surface tension will cause the bubble to collapse on itself.

From the above analysis, it is apparent that for a suite of inclusions of a given density, the size can exert a significant control on whether or not a bubble can be formed and exist at room temperature. It is common to observe fluid inclusions of a single population, perhaps all trapped along the same healed fracture or in the same growth zone, in which the large fluid inclusions all contain a vapor bubble (and all homogenize at the same temperature) whereas the smaller fluid inclusions below a certain size all lack vapor bubbles. In these situations it may be inferred that the small fluid inclusions are probably metastable and exist as stretched liquid at room temperature. This effect is especially common for small (<3 µm), lower temperature (<100°C) fluid inclusions. To the contrary, in populations of inclusions that have homogenization temperatures above about 100°C, even the smallest workable fluid inclusions (about 1 µm) commonly contain bubbles.

As some fluid inclusions cool, solubility relationships would dictate precipitation of a solid daughter crystal of a mineral such as halite. In a single related population of fluid inclusions, one may observe that most inclusions have a crystal of halite and that the ratio between volume of halite and volume of aqueous fluid is the same for each. (Alternatively, if the ratio were not the same, then the inclusions with daughters necked down after a phase change, or crystals are probably not daughter minerals, but are from "accidental" enclosure during precipitation of the mineral.) For the inclusions of the same population that may lack the daughter crystal of halite, it may be inferred that these inclusions are metastable and are supersaturated with respect to halite. Commonly, the daughter crystal can be nucleated by freezing the fluid inclusion.

Overheating Effects on Density and Composition (Thermal Reequilibration)

In the diagenetic realm, when rocks are buried, they are subjected to temperatures and pressures that occur within the bounds of a hydrostatic thermobaric gradient and a lithostatic thermobaric gradient (Fig. 4.7, and also see Chapter 3). However, recall that when all-liquid aqueous fluid inclusions are subjected to heating during burial, the pressure conditions within the inclusions

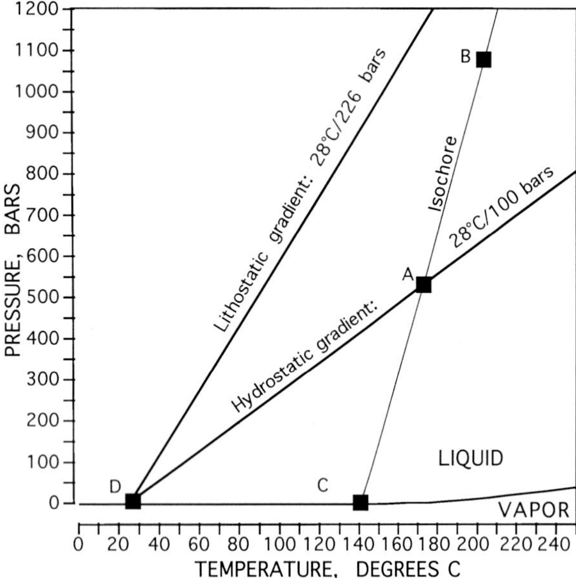

Fig. 4.8. *P-T phase diagram for pure water. If a fluid inclusion were trapped along a hydrostatic gradient at depth (point A) and were then subjected to heating during burial, the fluid inclusion's internal pressure would rise significantly above the surrounding hydrostatic pressure. During uplift and cooling from points B to A to C and to D, as soon as the inclusion has cooled below point A, the inclusion remains internally underpressured relative to the surrounding hydrostatic gradient.*

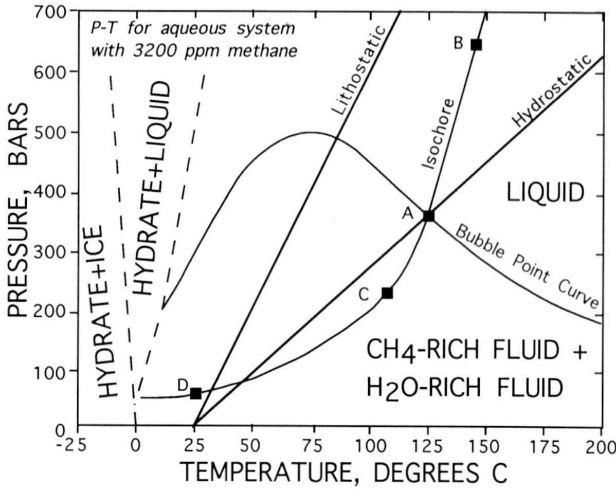

Fig. 4.9. *P-T phase diagram for 3200 ppm methane in water, with hydrostatic thermobaric gradient of 28°C/100 bars and lithostatic thermobaric gradient of 28°C/226 bars superimposed. If a fluid inclusion were trapped along a hydrostatic gradient at depth (point A) and were then subjected to heating during burial to point B, the fluid inclusion's internal pressure would rise significantly above the surrounding hydrostatic pressure at that temperature. During uplift and cooling from points B to A to C and to D, as soon as the inclusion has cooled below point A, the inclusion becomes internally underpressured relative to the surrounding hydrostatic gradient. Notice that the inclusion becomes slightly overpressured again as the temperature approaches surface conditions (point D). Modified from Hanor (1980).*

follow a different path. For example, if a fluid inclusion were trapped at surface conditions (point A), as a single freshwater liquid phase, and then heated during burial to a temperature of about 90°C (point B), the P-T path that the fluid inclusion would follow would be along the appropriate isochore (path A-B, Fig. 4.7). Note that the pore fluid surrounding the crystal (that encloses the fluid inclusion) would have P-T conditions somewhere within the wedge defined by the lithostatic and hydrostatic thermobaric gradients, but that the fluid inclusion P-T conditions must be on an isochore. Thus, at a given temperature during progressive burial heating, the internal pressure within the inclusion is greater than the external pore fluid pressure. With increasing temperature during burial this internal overpressure increases.

If a liquid fluid inclusion were trapped at high temperature (point A, Fig. 4.8) and were then subjected to additional heating, similar internal overpressures would develop as the inclusion conditions would progress up a similarly steep isochore (path A-B, Fig. 4.8). Such internal overpressures would develop either if the fluid inclusion were trapped as an aqueous-salt system (Fig. 4.8, path A-B) or if the fluid inclusion were trapped as an aqueous fluid inclusion with dissolved CH_4 (Fig. 4.9, path A-B).

Alternatively, note that during uplift (paths B-A-C-D, Fig. 4.8, Fig. 4.9), the internal pressures within the aqueous inclusions (with or without CH_4) will be below the external pressures during most of the uplift history, except for near-surface conditions in the CH_4-bearing inclusion where the internal pressure will at some point become greater than the pressure in the pore fluid (Fig. 4.9).

Natural overheating can also cause overpressuring within a petroleum fluid inclusion, but the magnitude is generally less because the slopes of the isochores for petroleum are commonly less steep than aqueous isochores (Burruss, 1987a; see also Fig. 3.8, this volume). Other types of natural heating (forest fires on outcrop, hydrothermal fluid flow, nearby intrusive) or laboratory heating during preparation or analysis may also cause internal overpressures to develop within fluid inclusions.

Rarely, internal overpressure may develop without overheating. For example, fluid inclusions that are trapped on a lithostatic thermobaric gradient and subsequently subjected to a pressure drop to hydrostatic thermobaric conditions while still at the entrapment temperature would become internally overpressured.

Evidence for this potential mechanism of developing internal overpressure has not been reported from the diagenetic realm, and therefore such a mechanism for developing overpressure is probably of far less significance as compared to internal overpressure developed during heating above entrapment temperature.

If the enclosing mineral is weak, internal overpressures developed within fluid inclusions can cause an inclusion cavity to stretch (irreversibly expand through plastic deformation) to relieve this pressure or to decrepitate (explode), sometimes allowing the inclusion fluids to leak out and new fluids to fill the inclusion. Thus, overpressuring from natural heating can cause reequilibration of fluid inclusions by either leakage and refilling, or stretching. The physical processes of reequilibration may involve stable crack growth or plastic deformation involving mobile dislocations. Unstable crack growth (B. Tapp, personal communication, 1991) is a possible mechanism of reequilibration, but there is not much evidence that such occurs in natural systems (A. Lacazette, personal communication, 1991). Unstable crack growth could produce some odd, higher than expected homogenization temperatures; however, correlations between maximum temperature and homogenization temperature (Barker and Goldstein, 1990) argue against this. Bodnar and others (1989) observed a correlation between necking down or shape change of inclusions in quartz and stretching of fluid inclusions. In general, it appears that most reequilibration results from internal overpressures developed from natural overheating of fluid inclusions beyond their original entrapment temperatures.

Some minerals are strong containers for enclosing fluid inclusions and other minerals appear to be relatively weak. As a first approximation, the ability of a mineral to contain fluids under pressure within inclusions is related to mineral hardness. Turgarinov and Vernadsky (1970) performed experiments in which fluid inclusions in a variety of minerals with a wide range of hardnesses were heated to the point of reequilibration. It was found that there was a strong positive relationship between a mineral's ability to withstand high fluid inclusion internal pressures and its hardness. Thus, one might expect that a hard mineral like quartz would be a more effective bottle for fluid inclusions than a soft mineral like calcite. Other characteristics of the enclosing mineral also appear to be important. For instance, easily cleavable minerals deform more readily than those with no cleavage. Some minerals have a tendency toward plastic deformation (calcite and quartz) and others tend toward more brittle deformation (dolomite and feldspar; Prezbindowski and Tapp, 1991). The dislocation density and location of fluid inclusions relative to lattice imperfections or lattice heterogeneity should also have an effect on the ability of an inclusion to withstand high internal pressures.

Each inclusion may behave differently during overheating. Some inclusions will reequilibrate easily and repeatedly, and other inclusions may not reequilibrate at all. Variables that control whether reequilibration takes place are the strength of the crystal, the amount of overheating, the confining pressure, other stresses, the P-V-T properties and composition of the fluid, the size, shape, and orientation of the inclusion, and its position relative to dislocations or discontinuities in the crystal. Each of these variables could exert an effect on a population of fluid inclusions altered by overheating to cause significant variability in the Th data that would otherwise have been consistent for an originally homogeneous population. This variability in data is one of the most important clues for recognizing thermally reequilibrated fluid inclusions (see Chapter 9).

Stretching results in an increase in volume of the overall inclusion cavity; once stretching has taken place the density of the inclusion fluid is lower (Fig. 4.10). There is abundant evidence that this mechanism of reequilibration (permanent plastic deformation or stretching) occurs (Bodnar and Bethke, 1984; Reeder and Ward, 1985; Ulrich and Bodnar, 1988). Figure 4.10 illustrates the result of stretching in nature. A fluid inclusion is trapped during burial at about 75°C (point A). Cooling after entrapment normally would result in an inclusion that would homogenize at 65°C. However, if the inclusion were subjected to additional burial to about 135°C, inclusion pressure would have increased greatly: as the inclusion is supposed to remain on the isochore to 135°C, the inclusion should be at point B. If the internal pressure causes volume expansion of the inclusion cavity through permanent plastic deformation, the inclusion fluid density would decrease and the conditions in the fluid inclusion would be on a lower-density isochore (point C). If this inclusion were now uplifted and cooled, it would intersect the liquid-vapor curve at 100°C (point D) and hence would yield a higher homogenization temperature than the originally entrapped fluid inclusion. Thus, the result of the volume increase of the inclusion with burial is a new and higher homogenization temperature for the stretched inclusion.

For inclusions that decrepitate due to natural overheating, the inclusion cavity may or may not become connected to the surrounding pore fluids. For those inclusions in which a fracture extends to the edge of the crystal, the inclusion cavity will come to equilibrium with the pore fluids, and the fluids within the inclusion cavity will most likely achieve a density

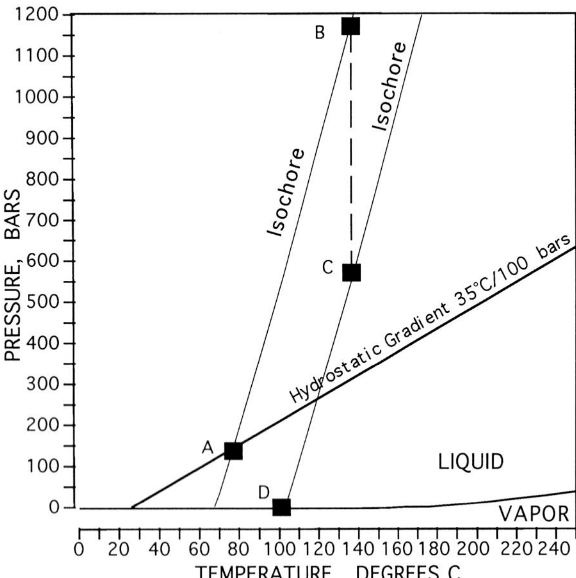

Fig. 4.10. *P-T phase diagram for water illustrating stretching of an all-liquid fluid inclusion during heating beyond temperature of entrapment. If the fluid inclusion were trapped at point A (75°C) and were then heated during burial, the inclusion's internal pressure would follow an isochore toward point B and its internal pressure would increase greatly compared to the surrounding hydrostatic pressure. If the inclusion were to stretch from this internal overpressure, its volume would increase and density and pressure would decrease. This event is represented by a drop to point C, and places the inclusion on the new isochore. Cooling of the inclusion would result in P-T conditions within the inclusion intersecting the liquid-vapor curve at about 100°C, and the inclusion now has the density that would produce Th measurements of 100°C. If the inclusion had never stretched, it would have yielded a Th at 65°C.*

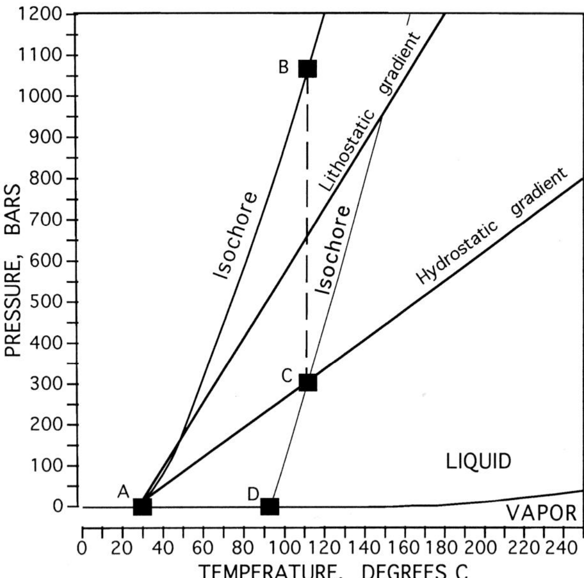

Fig. 4.11. *P-T phase diagram for pure water with hydrostatic thermobaric gradient of 28°C/100 bars and lithostatic thermobaric gradient of 28°C/226 bars superimposed. If a fluid inclusion were trapped as liquid at the surface (point A) and were then heated during burial, the fluid inclusion's internal pressure would rise along an isochore toward point B. At point B, the inclusion is internally overpressured relative to the surrounding hydrostatic thermobaric gradient. If the inclusion were to decrepitate from internal overpressure it would immediately equilibrate with the conditions in the surrounding pore fluid and jump down to point C on the hydrostatic thermobaric gradient. (If the example were for lithostatic, this would be possible also.) Because the inclusion has decrepitated it can refill with whatever pore fluids happen to be present at the time of opening of the inclusion. Rehealing of the inclusion followed by cooling would result in fluid inclusion P-T conditions following a new isochore (different density and possibly different composition) that intersects the liquid-vapor curve at a higher temperature (point D). This thermal reequilibration causes an increase in homogenization temperature of fluid inclusions and may result in their refilling with new fluids.*

and composition identical to the surrounding pore fluid before the crystal heals, resealing the inclusion cavity. By such a mechanism an early, shallow, primary fluid inclusion could be totally changed by filling with later burial fluids. This mechanism of leakage and refilling of fluid inclusions has been identified through empirical (Goldstein, 1986a, 1988, 1990) and experimental studies (Comings and Cercone, 1986). One can envision the cause and result of this mechanism of reequilibration by considering Figure 4.11. An inclusion of water is trapped at earth surface temperature and pressure (point A). It is then subjected to burial heating until a temperature of 110°C is reached (point B). Pressure within the inclusion is about 1.1 kbar, whereas the hydrostatic pressure surrounding it is about 0.3 kbar. The high internal pressure causes the hydrofracturing of the crystal around the inclusion so that the inclusion contents are open to the surrounding pore fluids (point C). At this point a new fluid may be introduced into the inclusion. If the inclusion cavity were to reseal and it were then cooled, it would follow a new isochore down to the liquid-vapor curve at point D. This inclusion would yield a Th of about 95°C and would contain a fluid other than the one originally entrapped in its cavity. The 95°C homogenization temperature is considerably higher than the original entrapment temperature, and the new fluid may bear no resemblance to the fluid originally trapped within the inclusion. Thus, if an inclusionist does not recognize that reequilibration has occurred (see Chapter 9) a completely incorrect interpretation may be formulated. However, if one can recognize reequilibration in a population of primary inclusions, a record of burial temperatures and fluid compositions might be gleaned from the inclusions.

Given that reequilibration is predicted to occur commonly in diagenetic environments, and that reequilibration commonly causes changes to inclusions which leaves them unrepresentative of original entrapment conditions, it is of crucial importance for an inclusionist to determine if a particular fluid inclusion has reequilibrated from internal overpressure. At present, there is no unequivocal technique to accomplish this through simple petrographic means, but one may assess the "likelihood" of a particular fluid inclusion having reequilibrated, and one may assess whether reequilibration has affected a petrographically related group of fluid inclusions. Techniques by which one can distinguish between inclusions that have reequilibrated and those that have not reequilibrated are covered in subsequent chapters. Assessing the likelihood of reequilibration of an individual inclusion is discussed below from theoretical, empirical, and experimental data.

Assessing the likelihood of reequilibration of an inclusion.—

One might wonder if there is some technique by which one can examine a fluid inclusion petrographically and determine if it has been subject to thermal reequilibration. Many petrographic criteria suggestive of reequilibration have been noted, especially in experimentally reequilibrated fluid inclusions, but most of these may not necessarily be preserved reliably in ancient samples. For example, Reeder and Ward (1985) used TEM imaging to identify dislocations associated with experimentally reequilibrated fluid inclusions. However, it is possible that such dislocations would be the preferential sites of dissolution-reprecipitation (necking down) and thus, some would be destroyed with time by such a process. Moreover, dislocations associated with a fluid inclusion may form from processes other than thermal reequilibration; for instance, the dislocations could have been the nuclei that initiated inclusion entrapment. Roedder (1984) observed an expansion and shape change associated with volume change of fluid inclusions in halite. Such a shape change is likely a reversible process that could ultimately revert to negative crystal shapes. Also, evidence for changes in shape could not be visually recognized in ancient samples for lack of evidence of the original shape. Prezbindowski and Tapp (1991) predicted the formation of protuberances from crack propagation associated with thermal reequilibration of fluid inclusions and illustrated some possible examples from ancient crystals. However, in working with ancient samples, it would not be clear if the inclusions originally formed with such shapes, if the shapes resulted from necking down processes, or if indeed the shapes formed from crack propagation. Furthermore, such protuberances should be preferred sites of necking down, and thus, may or may not be preserved. Swanenberg (1980) observed decrepitation clusters, or secondary fluid inclusions that emanated from reequilibrated fluid inclusions in quartz. Of all of the above examples, decrepitation clusters are the strongest petrographic evidence of fluid inclusion reequilibration, but such a petrographic observation has not previously been observed in minerals subjected to conditions below low-grade metamorphism. Clearly, observation of changes to fluid inclusion morphology presently is not the solution to assessing the likelihood of reequilibration. It thus becomes important that an inclusionist has an understanding of all parameters that could enhance reequilibration.

To a first approximation, the mineral in which the fluid inclusion is enclosed must be considered the most important control on whether a fluid inclusion has been altered by thermally induced reequilibration. As was pointed out above, some minerals simply are strong containers, and other minerals are weak ones that tend to deform when an inclusion becomes internally overpressured. This behavior has been known for some time from experimental and theoretical grounds. In 1970, Turgarinov and Vernadsky reported a positive correlation between amount of heating required for fluid inclusion decrepitation and mineral hardness (Fig. 4.12). From theoretical grounds, a mineral's ability to resist deformation from internal fluid inclusion overpressures should be related to its tensile strength and other crystallographically-controlled parameters (Prezbindowski and Tapp, 1991). Thus, soft cleavable minerals such as calcite or fluorite would be less resistant to fluid inclusion reequilibration than a strong mineral such as quartz.

Fluid inclusion size appears to have a strong influence on the likelihood of reequilibration of a fluid inclusion. With all other variables held constant, experimental data show that large fluid inclusions are more likely to reequilibrate than small ones. Experimental data from fluid inclusions in quartz illustrate that many large inclusions (approximately 100 μm in diameter) can only withstand differential pressures (internal fluid inclusion pressure minus external confining pressure) of about 500 to 1000 bars, whereas many smaller inclusions (approximately 1 μm in diameter) can withstand about ten times that internal pressure (Leroy, 1979; Bodnar and others, 1989). In contrast, Hall and others (1993) reported that a size dependence was lacking for decrepitation of synthetic inclusions in calcite. Other studies have detected a size control for reequilibration of fluid inclusions in calcite: overall, much lower internal pressures were required for reequilibration (McLimans, 1987). Thus, for an individual group of fluid inclusions, one might expect that more of the large fluid inclusions will reequilibrate than will the small inclusions.

Another important factor which could affect the likelihood of fluid inclusion reequilibration appears to be fluid inclusion shape. Bodnar and others (1989) reported that inclusions in quartz with negative crystal shapes were more resistant to reequilibration than those with amoeboid shapes. McLimans (1987) also noted shape controls for reequilibration of inclusions in calcite, but determined they were relatively minor compared to size controls. Personal experience of the authors indicates that extremely flat fluid inclusions and inclusions with sharp protuberances tend to reequilibrate more easily than spherical or negative crystal-shaped inclusions. Moreover, there may be a tendency for secondary inclusions elongated parallel to the healed fracture to "unzip" the healed fracture and easily reequilibrate (Burruss and Hollister, 1979). Intercrystalline fluid inclusions may suffer from similar weaknesses (Roedder, 1984, p.19). The same could happen to inclusions (if flat, irregular, or abundant) along the boundary between detrital quartz and an overgrowth.

Lacazette (1990) assumed linear elastic fracture mechanics theory to integrate these three important controls on the likelihood of fluid inclusion reequilibration. If his assumptions about reequilibration mechanism are correct, then his model can be applied to fluid inclusions in many minerals. The results of this fracture mechanics model (Fig. 4.13) illustrate that because of shape and size factors there is probably no single unique critical effective (differential) internal pressure required for reequilibration of a fluid inclusion

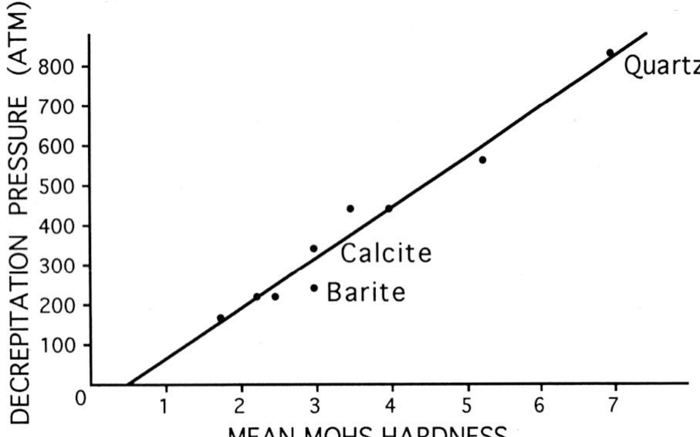

Fig. 4.12. Correlation between mineral hardness and pressure needed to initiate decrepitation of fluid inclusions. Pressure data are ±50 atmospheres. Data are from Turgarinov and Vernadsky (1970).

in a mineral. The model includes shape factors for a circular disk, an infinite flat tunnel, a sphere, and an infinite cylindrical hole. For example in quartz, the model predicts that for a population of large (10 μm), low-temperature fluid inclusions of these various shapes, reequilibration of some inclusions would begin at an effective internal pressure of 1 kbar (could correspond to burial of 3-4 km; see Fig. 4.7), and that most of the inclusions of 10 μm size or less would survive without reequilibration. Upon further burial heating, fluid inclusions with different shapes would begin reequilibrating, and all of the 10 μm fluid inclusions would not reequilibrate until an effective internal pressure of 3.5 kbars is reached (requiring metamorphic temperatures of about 290°C). If there were small fluid inclusions (<3 μm) in the same population, reequilibration would begin at an effective internal pressure of about 1.75 kbars and inclusions with particularly strong shapes would survive until inclusion effective internal pressure reached 6 kbars (requiring metamorphic temperatures at least 485°C!) Thus, reequilibration in all but the largest fluid inclusions in quartz appears to require severe overheating. Therefore, Lacazette's model predicts that in a normal population of aqueous fluid inclusions in quartz of various shapes and sizes subjected to severe overheating within the diagenetic temperature realm, some fluid inclusions may have reequilibrated, but many would survive unaltered.

Lacazette's model predicts similar behavior for fluid inclusions in a weak mineral such as calcite, but shows that much less overheating (lower differential pressure) is required to initiate reequilibration or to cause reequilibration of all the inclusions. Interestingly, even though calcite is a relatively weak mineral, it may be a surprisingly strong capsule for many fluid inclusions under certain circumstances. For 3 μm fluid inclusions the model shows that some inclusions with weaker shapes may begin reequilibration with 300-400 bars differential pressure, but it is not until 1.4 kbar differential pressure that inclusions with the strongest shapes begin reequilibration. Thus, for a low temperature aqueous inclusion population, heating to about 140°C (burial of about 4-5 km) would be required to reequilibrate all inclusions of 3 μm size in calcite. So in a soft mineral such as calcite, if fluid inclusions have been heated significantly beyond their entrapment temperature, it is probable that at least some fluid inclusions have reequilibrated; however, it would require severe overheating to reequilibrate all inclusions. Therefore, an inclusionist should expect that fluid inclusions in calcite subjected to moderate overheating would contain a mixture of both reequilibrated fluid inclusions and unaltered fluid inclusions.

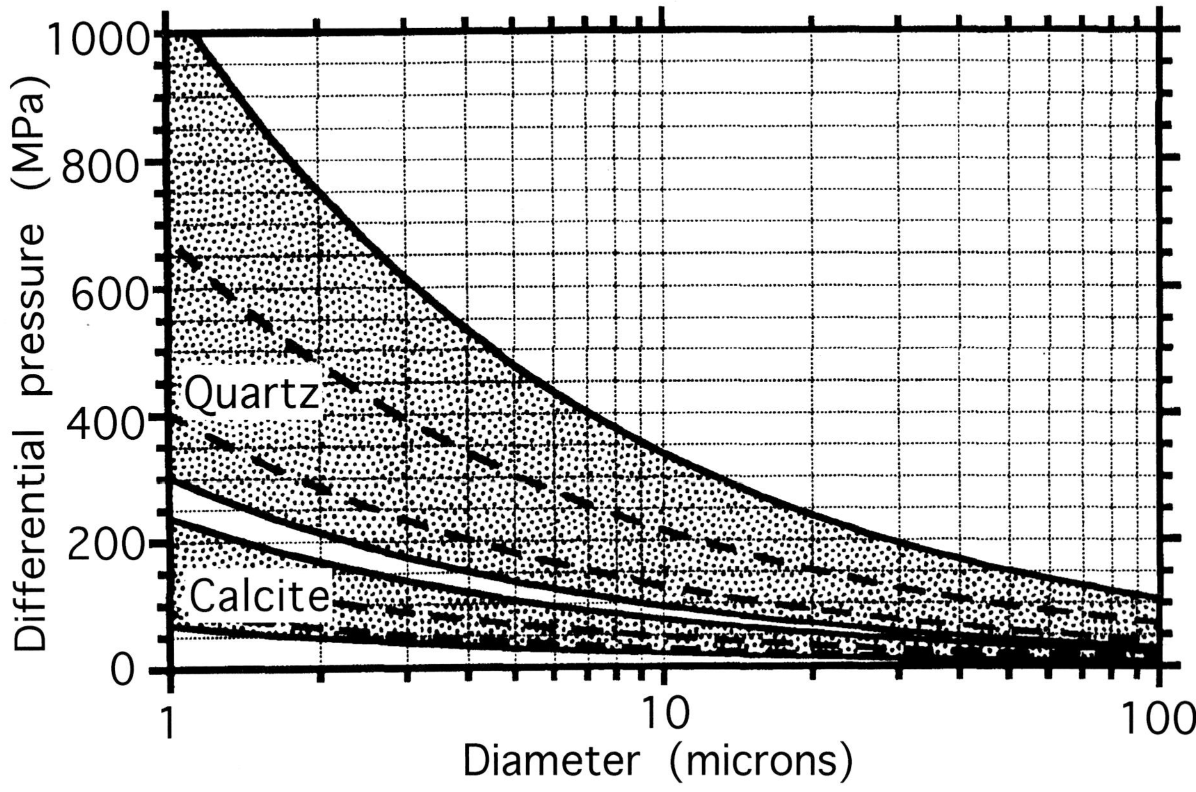

Fig. 4.13. Differential pressure (internal inclusion pressure minus external pressure) in MPa (MPa/100 = kbar) required for decrepitation of fluid inclusions of various shapes and sizes in calcite and quartz. The bottom solid line of each shaded area is the differential pressure at which the weakest shape modeled should decrepitate. The top solid line of each shaded area is the differential pressure at which the strongest modeled shape should decrepitate. Dashed lines within each shaded area define the differential pressure at which decrepitation should occur for shapes of intermediate strength. No inclusions should decrepitate at conditions below each shaded area. All inclusions should decrepitate above each shaded area. Curves are defined using shape factors in Lacazette (1990) and illustrate the relationship between size and differential pressure of decrepitation for inclusions of a given shape within each mineral. Notice the important controls of both size and shape on whether a fluid inclusion decrepitates. See Lacazette (1990) for more details. After Lacazette (1990).

The predictions of Lacazette appear to agree well with experimental data on quartz summarized in the literature (Leroy, 1979; Pecher, 1981; Bodnar and others, 1989; Hall and Wheeler, 1992), and support the empirical evidence that aqueous fluid inclusions in quartz may require severe overheating before significant reequilibration occurs. There are also empirical data that are important to consider to evaluate reequilibration of fluid inclusions in quartz in the diagenetic realm. Recently, Prezbindowski and Tapp (1991) cited published data from quartz fluid inclusions (Visser, 1982; Haszeldine and others, 1984) to note a "slightly more resistant than softer minerals" correlation between burial depth and Th; this was used to suggest the possibility that fluid inclusions in quartz may thermally reequilibrate to higher temperatures as they are heated beyond their entrapment temperatures in the diagenetic realm. Osborne and Haszeldine (1993) showed a similar relationship in which Th's of fluid inclusions in North Sea quartz overgrowths approached maximum burial temperatures. The significance of these studies is difficult to assess because the low-temperature origin of the quartz has not been unequivocally constrained, and because data were not separated and presented for individual, petrographically distinct assemblages of fluid inclusions. It remains plausible that in these studies, only unaltered inclusions formed at high temperature were encountered. In contrast, Guscott and Burley (1993) have obtained Th measurements from petrographically distinct populations of fluid inclusions of various sizes and shapes in North Sea sandstones. These data show populations with narrow ranges and exhibit characteristics suggestive of unaltered populations that have not reequilibrated from overheating. Thus, although much experimental, theoretical, and some empirical data argue against significant thermal reequilibration of fluid inclusions in quartz in the diagenetic realm, there remains a need for additional well-constrained studies. At this time, it is probably best to assume that some thermal reequilibration of fluid inclusions in quartz is possible, and that such an occurrence must be evaluated for every sample studied. However, it has been the authors' experience that fluid inclusions in quartz are relatively resistant to thermal reequilibration in the diagenetic realm.

The Lacazette model predictions for calcite are in strong agreement with experimental and empirical results. The experiments of Prezbindowski and Larese (1987) and McLimans (1987) illustrate clearly that some fluid inclusions in calcite can reequilibrate with only minor overheating, but that others survive a large amount of overheating. Hall and others (1993) illustrated survival of fluid inclusions after experimental overheating of fluid inclusions in calcite. Empirical observations on calcites of well-constrained origin indicate results in which many fluid inclusions have reequilibrated, but many have survived reequilibration altogether (Goldstein, 1986a, 1988, 1990; Barker and Goldstein, 1990; Prezbindowski and Tapp, 1991; Wojcik and others, 1994). It is clear from many of these natural examples, that some fluid inclusions in calcite survive overheating of at least 100°C beyond their entrapment temperature. Most inclusion populations in calcite that have experienced overheating show wide distributions of Th, and in some populations, Tm ice shows variability as well. It is thus possible that these populations consist of some fluid inclusions that have survived overheating and some that have reequilibrated at various stages in the overheating history. Subsequent chapters of this volume present techniques to separate the unaltered inclusion data from those that record conditions of subsequent reequilibration.

Behavior of fluid inclusions in dolomite and dolomite-like minerals such as ankerite have not been studied in as much detail as calcite. Fluid inclusions in dolomite may behave similarly to those in calcite, but dolomite probably deforms in a more brittle manner than calcite. Prezbindowski and Tapp (1991) illustrated a relationship between Th data from dolomite and maximum burial to suggest the possibility of reequilibration of fluid inclusions during overheating. Many studies of fluid inclusions in dolomite have illustrated fluid inclusion data that were highly variable within individual, petrographically constrainable, growth zones (Shelton and others, 1992; Wojcik and others, 1994); but in these studies, the variable data can either be interpreted to represent initial entrapment over a wide range of conditions or to represent partial reequilibration of the fluid inclusion populations. Some studies have assumed that fluid inclusions in dolomite reequilibrate during overheating, and have used Th data to indicate near-maximum thermal history (Tobin, 1991). Other studies have found that fluid inclusions in dolomite have not reacted to overheating and that fluid inclusions along an individual growth zone yield data that are highly consistent (R. Spencer, personal communication, 1993). However, Stephens (1988) and Goldstein and others (1991) presented data on low-temperature dolomite that had experienced heating up to about 150°C. In this work, most of the originally low-temperature, all-liquid inclusions had reequilibrated to produce two-phase inclusions yielding variable Th and Tm ice data. However, about 10% of the low temperature inclusions survived overheating and maintained faithful records of conditions of dolomite precipitation. So, when approaching a fluid inclusion study in dolomite, it is best to assume that thermal reequilibration is a reasonable possibility that must be evaluated.

Fluorite is a common minor mineral in diagenetic systems. Because fluorite is an important ore mineral, it has been subjected to extensive fluid inclusion study. Experimental work on overheating fluid inclusions in fluorite indicate that it is a very weak mineral which can experience significant thermal reequilibration of fluid inclusions with only minor overheating (Bodnar and Bethke, 1984; Rowan and others, 1985). Roedder and Howard (1988) reported fluid inclusion data from fluorite that implicated reequilibration as a potential cause for data variability. Thus, a careful inclusionist studying inclusions in fluorite must evaluate the possibility of thermal reequilibration.

Halite appears to be one of the weakest and most plastic fluid inclusion hosts in the diagenetic realm. Overheating experiments by Roedder and Belkin (1979) showed that overheating by only 20°C causes near complete resetting of the Th. Slow overheating resulted in plastic deformation around the inclusions, whereas only the most rapid (geologically unreasonable) heating rates caused brittle decrepitation. It appears that overheating in halite results in significant volume increase of fluid inclusions, the reequilibration mechanism may be one in which simple plastic deformation is favored over leakage and refilling through more brittle fracture generation (Petrichenko, 1973). Also, many workers have studied natural inclusions in salt that remain all-liquid at room temperature or yield Th data consistent with original conditions of salt formation (Lowenstein and Spencer, 1990; Casas and others, 1992; T. Lowenstein, personal communication, 1992). This suggests that the volume of some fluid inclusions in salt may be retained perhaps by absence of significant natural heating or by presence of lithostatic confining pressure in salt bodies. Thus, it appears that fluid inclusions in halite can provide useful information on the evolution of ancient brines, but the careful inclusionist must always consider thermal reequilibration, and critically evaluate any data from halite.

Although barite is not extremely common in diagenetic systems it may be found as an accessory mineral. Ulrich and Bodnar (1984, 1988) conducted experiments on the conditions of reequilibration of fluid inclusions in barite. It was found that barite is a relatively weak mineral with respect to fluid inclusion thermal reequilibration from internal overpressure. All inclusion studies in barite deserve careful analysis to determine if thermal reequilibration has altered the fluid inclusions.

Anhydrite is an extremely common mineral in the diagenetic realm. Although fluid inclusion reequilibration from recrystallization should be of great concern for this mineral, the possibility of thermal reequilibration must also be evaluated. Moore and Adams (1988) experimentally evaluated the behavior of fluid inclusions in anhydrite during overheating in the laboratory. Their work illustrated that at one atmosphere external pressure and during rapid heating runs, fluid inclusions in anhydrite would only withstand overheating of about 5 to 55°C beyond Th before they started to reequilibrate. Therefore, it is best to treat anhydrite as a relatively weak mineral in which thermal reequilibration must be evaluated for each study.

There are other common minerals that contain fluid inclusions in the diagenetic realm, but little is known of their behavior. There is much yet to learn about fluid inclusion reequilibration in zeolites, gypsum, and authigenic feldspar, for example. The best advice when faced with fluid inclusions in a poorly studied mineral is to assume thermal reequilibration may have affected the inclusions, and analyze the inclusions accordingly.

RETROGRADE FLUID INCLUSION REEQUILIBRATION

Retrograde readjustment from internal underpressure does not appear to occur in the diagenetic realm. First, such behavior in the diagenetic realm would be contrary to fracture mechanics theory (Prezbindowski and Tapp, 1991). More importantly, many studies have published Th data that record high temperatures reached by the sample (Barker and Goldstein, 1990; Prezbindowski and Tapp, 1991); clearly, in these cases inclusions did not reequilibrate to low temperatures. Perhaps the best evidence are the numerous cases of fluid inclusions sampled from outcrop, oil wells, or taken from mines, that are not filled with the near-surface fluid in which they were bathed at the time of sampling, but contain fluid from an earlier episode of inclusion entrapment when different pore fluids were present. There has never been any evidence presented yet to suggest that retrograde fluid inclusion reequilibration from internal underpressure could significantly affect fluid inclusions in minerals precipitated from the diagenetic realm.

CONCLUSIONS ABOUT REPRESENTATIVENESS OF FLUID INCLUSIONS

So, are fluid inclusions representative samples of diagenetic fluids? The answer is generally yes. However, without a grasp of the information presented in this chapter, a novice or naive inclusionist could totally misinterpret observations of fluid inclusions. The entrapment phenomenon that can cause the most confusion is the presence of more than one fluid phase in a pore. The post-entrapment processes that may cause the most difficulty in recognition and interpretation are necking down after a phase change and thermal reequilibration due to overpressure from natural heating. Thus, a careful, discerning inclusionist should always approach natural samples with healthy skepticism, and should constantly be looking for the clues that the inclusions provide for evaluating the degree to which the fluid inclusions are representative of the diagenetic system the inclusionist would like to sample. Successful approaches employed for evaluating representativeness of fluid inclusions are the subject of the remainder of this text.

Chapter 5

THE PHILOSOPHY OF CONDUCTING A FLUID INCLUSION STUDY

INTRODUCTION

To think of fluid inclusion analysis as a "black box technique" is a grave mistake. A fluid inclusion study is not something that one "just does" — it requires a more scientific methodology, first involving the formulation of *specific* questions along with conceivable hypotheses for possible outcomes. Such a philosophical approach is necessary because very commonly the fluid inclusion data that are required to answer a particular question may not be present in the rocks! So, a more structured approach may allow recognition of such a predicament at an early stage in a proposed fluid inclusion study, preventing the ordeal of wasting considerable effort in the collection of meaningless data.

The authors know of many horror stories in which a research supervisor has told a student, consultant, or subordinate to "do a fluid inclusion study" on some set of samples. Of course, the doleful soul taking the orders assumes it must be possible because the boss told him to do it, and feels compelled to come up with data no matter what the fluid inclusion population looks like. Time and time again, these researchers spend months conducting a fluid inclusion study on material that may not have the fluid inclusions appropriate for answering a particular set of questions. For example, if the research supervisor wanted to know the temperature and salinity from which some authigenic mineral precipitated, the subordinate would feel compelled to find primary fluid inclusions in that mineral, whether they were present or not. The subordinate would deceive himself (or herself) into believing that the inclusions must be primary ("my supervisor would not lead me astray") and might spend months measuring secondary inclusions or inclusions of dubious origin, producing a volume of data that does not answer the question being posed. Unfortunately, much of these data make it into the literature, and only after years of painstaking effort by careful researchers are the poor studies discredited. In general, the most common mistake that is made concerns the origin of fluid inclusions. Therefore, the most important thing to remember before conducting a study of fluid inclusions is that every suite of samples does not necessarily preserve the fluid inclusions that will answer the particular question being posed. If one keeps this in mind, and remains skeptical about the feasibility of the fluid inclusion study at the outset, one will not waste time on essentially worthless studies that may even lead to totally erroneous conclusions. Also, one should never forget the importance of conducting an inclusion study within a framework of sound geologic and petrographic observations.

PROCEDURE

A good fluid inclusion study should employ the following procedure, without skipping any of the steps. However, one need not follow the entire procedure to completion as there are several points along the progression in which the researcher can either "bail-out" because the fluid inclusions preserved are inappropriate for answering the question(s) posed, or because the question(s) posed has (have) been answered by an earlier step in the procedure. The procedure follows.

1. The first step in any fluid inclusion study is to pose a question. Any person who finds himself (or herself) in the position of having to collect fluid inclusion data must determine why the study is being conducted, and must attempt to focus on very specific aspects of the question posed. Admittedly, this is often "easier said than done." Further, the authors understand that such a search for the reasons for a proposed study may require more prodding of a supervisor, client, or professor than a novice or consultant might be comfortable with — but we plead with you to persevere in this endeavor! We assure the readers that "doing fluid inclusions in a basin" will not be a specific enough problem to be able to effectively and efficiently focus on the potential significance of data from a single (or even hundreds), of tiny fluid inclusions in the basin! In fact, with such a broad undertaking, the authors would maintain that any data collected could probably be interpreted with a myriad of hypotheses that might even

outnumber the frequency of collected data — in other words, the data in most cases would be meaningless! A more specific question might be to unravel the diagenetic history of a porous dolomite formation. We hope each of the readers can see the difference in the type of questions one may ask. This first step of asking an appropriate and specific question is fundamental to the scientific method, and we cannot stress enough that it will be crucial for the successful implementation of the fluid inclusion technique. Another point to be made here is that it should prove helpful to those considering a fluid inclusion study to develop a set of specific requirements that, if not fulfilled, will allow complete abandonment of the study if the fluid inclusions needed to answer a particular question are not present. For the reader who has only reached this point in the text, it may not be clear what such requirements might be, but by the end of this chapter, and certainly after a reading of the complete text, a better feeling for such requirements should have been acquired.

2. The next step is to select samples that have appropriate field and stratigraphic constraints that will allow one to answer the question(s) posed. This seems simple enough, but there is more to this statement than meets the eye. A randomly collected rock will rarely relinquish the secrets of its history, no matter how many thousands of dollars worth of instruments and hundreds of hours of attentive time the rock may be subjected to: a scientist must develop a frame of reference to have the "right" rock or suite of samples. This could involve considerable efforts in the field or in core sheds, where one must study the rocks to know as well as possible the geologic context from which the samples come. Without such a framework, the interpretation of much fluid inclusion data must be appropriately, and often severely, limited. Unfortunately, those who conduct fluid inclusion studies without fundamental geologic knowledge often are not aware of the limitations imposed by their own lack of basic geologic information.

3. If one is conducting a diagenetic study, one must examine the paragenetic context of the mineral of interest so that the fluid inclusion study is put into a paragenetic framework. Having a sound foundation in optical mineralogy and sedimentary petrology will be necessary to accomplish this step. A few diagenesis courses under the belt may not hurt either. But in all seriousness, diagenesis is a process, and a process occurs through time; so for data from any fluid inclusion to be able to assist in unraveling a diagenetic event, the fluid inclusion must be placed in a temporal framework. This can only be accomplished with a complete and thorough understanding of the relative timing of the diagenetic phases, and of the relative timing for the origin of any inclusion of interest. Let's not deceive ourselves: such understanding will commonly require considerable time (hours, days, months, or years depending on the scope of the project) peering down a petrographic microscope, as numerous specimens will probably have to be studied for a reasonable "confidence level" to be achieved.

4. The next step involves characterizing the fluid inclusion petrography (explained in detail in the following chapter). The more samples surveyed, the greater chance one has to find the inclusions required to answer the questions posed, and the more comfortable one becomes in the representativeness of the inclusions that potentially will be selected for microthermometric determinations. It is in this step that a careful inclusionist will determine the origin of fluid inclusions and how they fit into the paragenetic framework. Also, the inclusionist will observe and record the consistency of volumetric ratios of liquid to vapor in fluid inclusions within individual growth zones or microfractures in order to evaluate approximate temperatures of entrapment, potential number of fluid phases during entrapment, and possibilities for potential occurrence of reequilibration. Finally, simple compositional determinations can be performed in this step using the crushing stage and UV epifluorescence equipment. After completion of this step, a researcher may often be able to interpret fairly concise limits for the diagenetic conditions at which the fluid inclusions formed, so much so that the diagenetic environment might be predicted with confidence from the fluid inclusion petrography alone.

5. If the appropriate fluid inclusions are present, the next logical step is to perform microthermometric analyses of fluid inclusions (see Chapter 7). Laboratory measurements will provide density data from homogenization temperatures and compositional data from cold stage work. There is a particular methodology for data collection that will be required in order to evaluate fluid inclusion microthermometric data: data must be obtained from several individual inclusions within a group that are petrographically associated with one another, such as inclusions from a single growth zone or inclusions from a single plane of secondary inclusions. Thus, it is most important for an inclusionist *not* to focus attention on a single inclusion, but to focus on a petrographically constrained group of inclusions, in order to be able to interpret the data. The mind of a knowledgeable and careful

inclusionist must see planes and growth zones first, and individual inclusions are viewed only within the context of a plane or growth zone. Gone are the days (after this volume!) when significance is ascribed to mean values of large groups of data collected from randomly selected inclusions — each value determined for each inclusion within a petrographically constrained group of inclusions means something, and therefore must have a discernible explanation!

How much data is to be collected will be dictated by the question posed and the degree of confidence required, as well as the nature of the samples and the time and money available. Basically, once again, if one knows exactly what one is after and can determine exactly which fluid inclusions contain the answer, then the amount of data required should be minimal if the first data collected are very consistent. But if the data turn out to be more variable, or the confidence of knowing that the necessary inclusions exist is low, then the amount of data required for a believable and defensible interpretation could increase by one or two orders of magnitude!

6. Finally, one may eventually want to apply more sophisticated analytical techniques (see Chapter 12) to better evaluate the compositions of fluid inclusions. Some approaches may involve bulk analyses of many fluid inclusions, and some will be possible to accomplish on individual inclusions. Some methods will destroy the inclusions being studied, and others are nondestructive. Often several techniques may be used for a given inclusion or group of inclusions to obtain the most complete analysis. But just as explained above, under no circumstances should these special analyses be undertaken just for the sake of collecting data in hope that something will "fall out." Only after a thorough understanding has been gained from petrographic and microthermometric studies should inclusions be subjected to these sophisticated analytical instruments. Unfortunately, such rigor is too often ignored by researchers who use these instruments.

So, are there some of you still interested in becoming inclusionists? Just think of all those billions and billions of tiny fluid-filled entities just waiting:

Can you hear me?
I am trying to say something.
I know you can love me.
And when you do
I will tell it to you all. [1]

For those of you who are patient, persistent, and meticulous — read on!

---

[1] From a poem entitled "A Fluid Inclusion Speaks Its Mind" by Jayanta Guha in Roedder (1984, p. 1).

Chapter 6

FLUID INCLUSION PETROGRAPHY

INTRODUCTION

This chapter presents practical methodologies for examining the petrographic characteristics of fluid inclusions in diagenetic minerals, together with a format for interpreting the diagenetic environment and thermal history of a sample from the petrography of the fluid inclusions alone! Accomplishing a study of fluid inclusion petrography is much like any other petrographic study in that there are certain necessities, and some amount of thought and effort is required before a sample can be studied: samples must be collected with respect to the problem at hand, samples must be prepared for microscopic observation, and a properly adjusted microscope must be available. The difference between standard thin section petrography and fluid inclusion petrography is that each of these mechanical steps must be carried-out with great care; otherwise, great barriers will impede even the most persevering scientist from obtaining useful information from the fluid inclusions. Luckily, the reader will be relieved to know that these mechanics have been solved to the extent that all that is required for their successful implementation is to take note of the information presented in this chapter.

The next hurdle will be for an inclusionist to record the fluid inclusion petrography. This is not a trivial undertaking — remember what it was like when you took your first thin section petrography course! In a single thin section there can be thousands of crystals. The object is to be selective: a petrographer learns to answer a series of mental questions (e.g., minerals present, textures present, etc.) in order to be able to describe the petrography. It is the same with fluid inclusions. There can be millions and millions of inclusions in a single section; recording them all would have no meaning, and is impossible anyway. Examining fluid inclusions petrographically involves a series of steps that an inclusionist follows. Such a methodology is specified in this chapter.

Just as the standard petrography of a thin section can be very illuminating with regard to many basic aspects of a rock's history, so it is with fluid inclusion petrography within diagenetic phases. *The power of information gained from the petrography of fluid inclusions in diagenetic phases is surprisingly great.* Much of what a researcher or explorationist might want to learn from the inclusions can be gained from the petrography alone — other techniques (such as microthermometry) may not be necessary to answer many questions. The fluid inclusion petrography, when viewed in an appropriate context, provides fairly concise limits for the diagenetic conditions at which the fluid inclusions formed, so much so that the diagenetic environment can be defined from the fluid inclusion petrography, and in many cases, information regarding later thermal history can be gleaned from the fluid inclusion petrography as well. However, for an inclusionist to be able to make such interpretations requires a framework of information into which fits the observations of the petrography. Such a framework is also presented in this chapter.

Finally, even though fluid inclusion characteristics will be stressed in this chapter, please remember that for a fluid inclusion study to yield information that is *geologically* meaningful, the observed inclusions must have some geologic context. Therefore, in most instances it will be crucial that an inclusionist conducts a study within a geologic framework and to pay careful attention to field and petrographic relationships.

SAMPLING PHILOSOPHY

An appropriate sampling philosophy begins with an understanding of the scientific question that has been posed. In some cases, all that may be desired is an approximation of the maximum temperature a rock has reached; for others, more detailed information about the temperature and the salinity of mineral precipitation is desired; and for others, the entire fluid history may be of interest. Thus, how sampling is approached will be dictated, to some degree, by the scope of the proposed study.

Other factors also will affect sampling procedures: potential variability caused by time (relative temporal positions of samples), and those caused by space (relative spatial locations of samples). In the first case, enough samples must be collected in order to document the paragenesis of the diagenetic phases and the inclusions within them, and in the second case, sampling must be undertaken over a sufficient area to evaluate potential spatial variability present in the system under study.

For a sampling program to yield useful specimens for study, there is one final important factor that can exert a compelling condition on the outcome: call it what you may, but it boils down to luck. It is very difficult to predict if nature was so kind as to trap the inclusions needed to solve a problem. The more samples collected, the greater chance for success; however, success is never guaranteed by quantity. Know that some studies will require searching through tens or even hundreds of samples before the appropriate inclusions will be located, but a reality that the authors have observed is that in many proposed studies, the necessary fluid inclusions never will be found — and this means that some fluid inclusion studies have to be totally abandoned. Clearly then, the number of samples chosen will be different for every study and may also be limited by the time or funds available, or by the availability of material with measurable fluid inclusions.

In summary, when a researcher is deciding on an approach to sampling, there are several important aspects to consider. First, what is the geologic or paragenetic sampling framework that may answer the geologic question being asked. Second, given the field and petrographic framework, what would be the potential variability in time and space.

SAMPLE SELECTION CAVEATS

Samples that have been subjected to significant overheating in the laboratory or on outcrop, typically are not appropriate for fluid inclusion study. Overheating can cause thermal reequilibration (see Chapters 3 and 4), or cause two-phase fluid inclusions to be heated to the point of homogenization causing them to remain metastable at room temperature (explained in more detail later in this chapter). Therefore, avoid well cuttings in which the drying procedure is unknown. This eliminates most well cuttings from fluid inclusion analysis! Also, most cores subjected to "whole core analysis" or "full diameter analysis" have been heated to between 110°C and 230°C. As such temperatures can certainly cause thermal reequilibration (see Chapter 4), samples subjected to these types of analyses should be avoided. Porosity and permeability plugs may have been similarly altered. Outcrop scorched by forest or grass fires should be avoided. Most old thin section rock blanks, impregnated samples, and old thin sections that have been heated on a hot plate during sample preparation should not be used. Any sample that has been heated in a cold or hot cathode luminescence system (Barker, 1992), or even in an SEM may be altered. Polished sections that have been acid etched for staining may have lost their fluid inclusions altogether.

Many people wish to know if thin sections can be used for fluid inclusion studies. Generally, the answer is no, however, one may use thin sections in a preliminary search to "high grade" for the best samples before correct preparation. There are several reasons why thin sections are not recommended. First, in most cases there could have been thermal or structural damage caused by equipment and procedures used for preparing standard thin sections. Second, in most cases thin sections are too thin: many larger inclusions would have been destroyed, and the more three-dimensional perspective that is possible with a thicker section (commonly necessary for making appropriate petrographic observations) is lost. So, if one finds fluid inclusions in a thin section, then this information may be interpreted to mean that this rock sample is more likely to contain the required inclusions, as compared with another specimen. Nevertheless, it is not uncommon to have numerous useful inclusions in a thick section, and to find nothing in a thin section. Thus, finding little in a standard petrographic thin section does not necessarily mean that the sample is a poor prospect — do not give up completely until a thicker plate has been prepared and surveyed.

PREPARATION OF THICK SECTIONS FOR FLUID INCLUSION ANALYSES

There are three types of thick plates that can be prepared: a cleavage fragment, a "quick plate," and a doubly polished plate. For any of these methods, what is required is that a smooth surface results so that light is not refracted as it passes through the surface to the observer's eyes. It must always be remembered to avoid heating the sample or deforming it mechanically — so be as gentle as possible when preparing a thick section for fluid inclusion study.

Cleavage Fragments

The most gentle method is to cleave fragments with a razor-sharp blade; it works well on large crystals of calcite, dolomite, halite, fluorite, and anhydrite. Unfortunately, one will generally not have large cleavable crystals of a diagenetic phase available. Furthermore, much potentially useful paragenetic information is lost by using cleavage fragments. Thus, cleavage fragments may be useful in some cases, but in most cases, this technique will not be fruitful.

Quick Plates

A "quick plate" is so named because it is so much faster (and less expensive) to prepare than a doubly polished plate. It is prepared in the same way as a doubly polished plate, except that no polishes are applied. As long as the precautions given below for sawing and grinding specimens (in the procedures for preparation of doubly polished sections) are adhered to in the preparation of quick plates, all petrographic observations of fluid inclusions can be made with quick plates. For quick plates, instead of polishing, an immersion oil with an index of refraction which closely matches that of the mineral in which one wishes to view inclusions is simply smeared onto the upper surface of the section. This oil fills in all the pits on the surface of the mineral. As the top surface of the oil will be smooth because it is a liquid, and because the index of refraction of the oil matches that of the mineral, light is passed straight through without being refracted. The oil coating the surface acts just like a polish. Note that oil immersion objectives are not necessary. Standard petrographic lens immersion oil with a refractive index of about 1.51 is similar enough in refractive index to diagenetic phases to be sufficient for use in this procedure. Kerosene works well for fluorite. For some carbonates an oil with a refractive index of 1.60 may yield the best results.

One will find that quick plates greatly facilitate the progress of the early stages of an inclusion study; incorporation of this procedure greatly enhances the prospects for success, because so many samples can be viewed with little time and money having been expended for sample preparation.

Doubly Polished Plates

The preparation of doubly polished sections is not required for observing the petrography of fluid inclusions, but can improve optics over quick plates. Furthermore, it will be necessary to prepare samples with polished surfaces for microthermometry: oil on the surfaces of the quick plates will be rendered useless for filling in pits during microthermometry because the index of refraction will change during heating and cooling, and the physical state of the oil can change during heating and cooling as well (e.g., oil can evaporate on heating).

Procedures for preparing doubly polished plates have been described previously (Roedder, 1984; Shepherd and others, 1985), but special precautions are required for samples from the diagenetic environment, particularly for the authigenic minerals that are soft, and for low-temperature inclusions found in authigenic minerals (Barker and Reynolds, 1984; McNeil and Morris, 1992). There are five sequential steps — impregnating, cutting, grinding, polishing and mounting — that will be discussed separately.

Impregnation.—
Any impregnating technique that involves heat and/or high pressure should be avoided. Cold-curing, vacuum impregnation seems to work well. Coating a surface with superglue and letting it sit overnight to seep in and cure also has been used effectively. If the rock is very porous and permeable, multiple coats of superglue should be applied and allowed to cure until the surface is smooth (i.e., all pores are filled). Do not impregnate samples if microthermometry above 250°C is required, because most epoxies will start to degrade at temperatures >200°C.

Cutting.—
Sawing rocks in order to prepare thick plates must be accomplished with knowledge of the potential thermal and mechanical damage that could occur during the process. Harm to fluid inclusions by sawing can be minimized in three ways: (1) by taking care to cut gently and with ample coolant when using high-speed trim saws, and to leave a little extra rock to be hand ground at least 0.5 mm from sawn faces, (2) by using low-speed trim saws, or (3) by using precision high-speed trim saws that can be adjusted so that the blade will neither wobble nor bounce at the point of contact with the rock being cut. Many employ the second alternative when slicing off a piece of rock after one face is prepared and glued to a glass slide. A few professional labs find expensive high-speed precision equipment advantageous.

Grinding.—
With all of the preparation materials on the market today, a polish can be obtained relatively rapidly (less than five minutes) if the surface has been ground to a high degree of flatness. There are two different grinding media. One involves diamonds bonded to some type of substrate. The advantage to the bonded diamond products is that they last a long time and, contrary to what might be thought, are relatively inexpensive. But since they cut using sharp tips of protruding diamonds, relatively large fragments are cut from the surface being ground. Thus, when using these products, keep in mind that an important objective is to get the fragments of removed material off of the lapping surface as soon as

possible to prevent them from scratching the surface. To accomplish this, it is very important to use a hard, steady, stream of water on the lap, and it should be rotating at 700 to 1000 revolutions per minute for best results. The disadvantages of the loose grits are that they must be replenished, and it is easy for contamination to occur between steps, especially in labs with multiple users.

If grinding is done properly, polishing will be simple. One general rule to follow is to never push hard on a sample. Ample pressure can be ascertained by observing the nail at the tip of the finger: when you see the color in this area change from pink to white, enough pressure has been achieved. The first coarse grinding step should remove at least 0.5 mm from a face showing saw blade marks. Subsequent finer steps should remove all scratches from the previous grinding steps. After the final grinding step (about a 15 µm-sized abrasive), the surface should not have any scratches that can be seen with the naked eye, and it may even appear to have the beginnings of a polish.

Polishing.—

Both sides of the plate must be polished. As stated above, if the surface that results from grinding is appropriate, a polish can be obtained in short order. Again, there are two schools of approach: to use loose diamond abrasives of various sizes on a felt-like lap, or to use bonded diamond abrasives on a substrate. Both work well, but the advantages of new "three-dimensional resin-bonded" polishing products, which are just coming on the market, are that they are virtually indestructible, they last for hundreds of samples, and they are relatively inexpensive. For these products only about 1-2 minutes is required for each step. Use the same pressure and water flow as for grinding, but the speed of the lap should be faster — about 1000 to 2000 revolutions per minute for best results. Rapid polishing with loose diamond abrasives normally requires great pressure to the surface to obtain results in a relatively short time. However, this can prove fruitless when attempting to polish sandstones with carbonate cements because higher pressure results in gouging or undercutting, and thus, a poor surface on the softer authigenic phases. Carbonates in general can be polished in less than 2 minutes with a single polishing step of 3 µm or less. Silica cemented sandstones with no carbonate cement normally can be polished in 3 steps (10-6 µm, 3 µm, and 1 µm). The most difficult samples to polish are sandstones with carbonate cements, so if one is developing a personal technique, one should not begin with such a sample! One excellent method to determine the quality of polish with the naked eye is to hold the polished surface above an incandescent light bulb and tilt it in such a way that the reflection of the light bulb is visible: if one can read the label in the reflection, the polish is probably acceptable.

Mounting.—

Once a surface of a specimen is polished, it must be glued to a petrographic slide; either frosted or unfrosted glass slides can be employed. Before gluing the sample to the glass, it is very important to wash the polished surface well (with acetone or dishwashing detergent) to remove all grit and oil. (Fluorescing oil globules at the glued surface of a slide are a nuisance when one is studying a sample under UV light to find petroleum inclusions.) There are two approaches for mounting: one is to use a glue that can later be dissolved to remove the polished plate before subjecting it to microthermometry tests, and the other is to use an epoxy, and leave the plate mounted on the glass for all subsequent tests. As hard-rock geologists typically heat samples to temperatures above 200°C, the first method is used most commonly, and the many varieties of superglues present in local hardware stores worldwide are sufficient. Most dissolve away overnight in acetone. However, a little residue always will remain on the surface. Therefore, once the surface has been separated from the glass, the surface should be rubbed delicately over acetone-soaked paper a few times to remove all residue. For sedimentary rock applications, usually it will be necessary to leave the polished rock plates on the glass, because the rock will typically be so thin as to require a substrate for stability. This is normally not a problem for microthermometry, because most of the epoxies are useful to at least 200°C. The standard petrographic epoxies that cure with UV light seem to be very strong and do not require heat for curing.

If the thickness of a glass slide beneath a doubly polished plate causes thermal gradient problems or optical problems in a heating/cooling stage, a possible solution would be to epoxy the sample to a coverslip, and then glue the coverslip to the glass slide with superglue. In this way, the glass slide can be removed by dissolving the superglue, but the very thin doubly polished plate will still be held together by a substrate of coverslip glass.

Thickness of the plate.—

So far, we have shown that preparing a doubly polished plate first involves trimming a rock to an appropriate size, grinding a surface flat, polishing a surface, and mounting the polished surface to a glass substrate. Then after trimming the bulk of the rock from the glass slide (usually with a low-speed trim saw), the process is repeated for the second, opposite surface of the plate. The question that remains is how thin the final plate should be. The answer for most sedimentary rocks is very thin — less than twice the thickness of a normal thin section, or roughly 40-60 μm. Sometimes, very fine-grained, or very cloudy authigenic minerals may require no more than 35 μm thickness for an inclusionist to be able to see anything. If a microscope is readily available during grinding, treat the sample like a "quick plate" and use immersion oil to view the sample. Once light is easily transmitted, and once fluid inclusions are clearly visible, then the proper thickness has been achieved.

Polishing machines.—
A few words about polishing machines are in order: the authors have known few people who have had success with them. Labs that have been successful have a few well-trained technicians who are the only people who use the machines — the key being few, well-trained users. Thus, for labs with multiple users (i.e., students), the authors know from experience that well-designed, high-speed rotating laps, paired with manual labor, work well.

MICROSCOPE REQUIREMENTS

The final mechanical problem to overcome prior to a study of the fluid inclusion petrography is to have a petrographic microscope with the appropriate components properly adjusted. Normal petrography can be accomplished by simply plopping a sample down on a rotating stage and focusing. Such is not the case for fluid inclusion studies, particularly in diagenetic minerals: one must be intimately familiar with all of the microscope adjustments, for each must be adjusted properly to be able to see fluid inclusions optimally.

Standard Transmitted Light Petrographic Microscope

The following are some helpful instructions to assist petrographers in the use of a standard petrographic microscope for observing fluid inclusions:

1. Use a microscope with a binocular head (two eyepieces).

2. Use 12.5X, 15X, or 16X eyepieces or a 1.25X or 1.6X magnification changer setting in conjunction with 10X eyepieces.

3. Use a 10X objective to scan samples first, and use a 40X objective (or higher) to observe individual inclusions. All objectives on a petrographic microscope are corrected for the coverglass of a thin section. Typically, a thin section will not be employed for observing inclusions, as quick plates and doubly polished plates are better for the reasons described previously. These thicker plates will not have coverslips, which is not a noticeable problem up to about 25X. For 40X objectives and greater, the optical quality of the image is reduced drastically by not having the appropriate thickness of mineral sample or coverslip above the inclusion observed. Sometimes, an increase in optical quality can be obtained for inclusions near the upper surface by simply applying immersion oil and a coverglass.

4. Most inclusionists will have the analyzer in the path at all times and will flip in the polarizer only when cross-polarization is needed.

5. If one is observing inclusions in carbonates or any other mineral that shows strong double refraction, one of the images can be eliminated by either rotating the analyzer, rotating the polarizer, or rotating the stage with either polarizer in the light path. The two images are not of the same quality. The ordinary ray will produce the best image.

6. High intensity quartz halogen light sources are the best, and they must be adjusted properly for good optical results. It is not uncommon for people to complain that their microscope optics are bad, only to find that their light sources were adjusted improperly. Consult your microscope manual to learn how to focus and center the filament of the bulb.

7. Convergent illumination, focused at the correct level, greatly reduces total reflection at the inclusion walls and improves the image. The flip-in/flip-out condenser (top element on the condenser) beneath the rotating stage always must be *in* the light path, and must be correctly positioned both laterally and vertically. The necessary steps can be learned easily by consulting a microscope instruction manual; the methodology is commonly referred to as *Kohler* illumination.

8. Always begin adjustments with the condenser diaphragm open fully. Then close as a final adjustment to achieve the desired contrast.

Fluid Inclusion Microscope

A fluid inclusion microscope is similar in many respects to a standard, transmitted light petrographic microscope, with a few important differences.

Commonly, a biological-style X-Y stage is attached in place of a rotating stage to allow for more "user-friendly" manipulation of samples. Also, since a heating/cooling stage commonly will be employed on the fluid inclusion microscope, special long-working-distance objectives and a long-working-distance condenser are required to focus on a sample within the stage, and to focus a converging beam of light onto the sample within the stage (see Chapter 7). The eight instructions listed above are applicable to the fluid inclusion microscope as well.

We must stress again the importance of having the microscope optics optimized for observations of fluid inclusions (Fig. 6.1). To learn these adjustments, it may serve one well to dealign everything and then develop a series of steps through trial and error for realignment.

Important Microscope Options

The following are various microscope attachments that a researcher might consider useful for fluid inclusion studies:

Trinocular head — This would enable a closed circuit television (CCTV) or photography system to be mounted on the microscope. It is most useful to have a trinocular head equipped with an internal beam splitter that permits viewing through the oculars while at the same time light travels to the vertical tube on which the photography system and/or a CCTV system is attached.

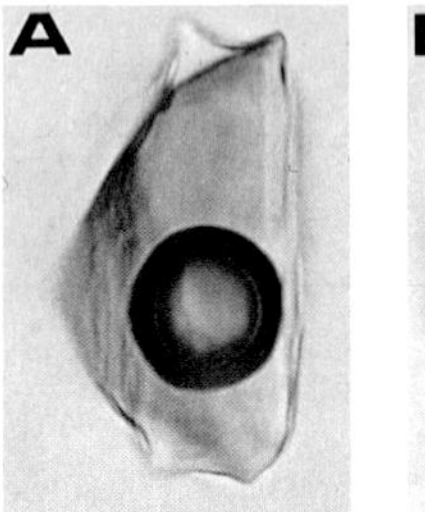

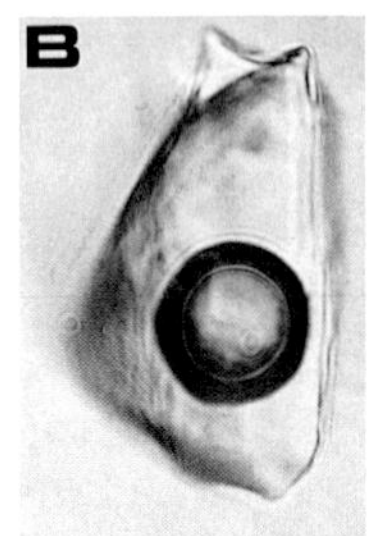

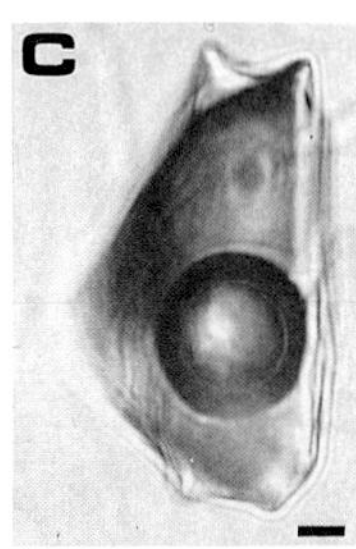

Fig. 6.1. *Photomicrographs of a large, two-phase aqueous inclusion in fluorite at room conditions to illustrate effects of incorrect microscope settings.* ***A)*** *All settings optimized. Note sharp outlines of inclusion and bubble, and absence of dark shadows.* ***B)*** *Light source of microscope out of alignment. Note that dark shadows appear on the left side of the inclusion. A good indicator that a bulb may not be aligned properly is if all inclusions have shadows on the same side.* ***C)*** *Condenser not focused. Note pronounced contrast at inclusion cavity edges and darkening of interior of inclusion. If this inclusion had been less than 5 μm, the inclusion cavity would be so dark from refraction of divergent light that vapor bubble would not be discernible — the inclusion would instead appear like a "speck of dust." Bar scale is 7 μm.*

Incident light fluorescence — Burruss (1991) reviews the principles of fluorescent light microscopy which is used to distinguish many petroleum fluid inclusions from aqueous inclusions. For inclusionists working on sedimentary rocks, such microscope attachments are a virtual necessity, especially if a cooling stage is not available (as petroleum inclusions can also be identified based on their behavior on cooling). For best results, the fluorescence system should have a 100 watt mercury bulb focused and centered properly in its lamphousing, and powered by a stabilized, direct current power supply. Also, it is appropriate to report fluorescent colors when a narrow-band ultraviolet (365±5 nm) excitation filter is employed (Burruss, 1991). However, since a wide band blue excitation filter will permit more energy to excite the sample, fluorescing inclusions will be brighter and thus easier to locate when this excitation filter is used.

Closed circuit television (CCTV) — A CCTV system is not a requirement, but high resolution systems are so affordable that their usefulness for teaching or communicating with co-workers makes them a highly recommended, invaluable aid. When equipped with a video cassette recorder (VCR) they are virtually imperative for observing crushing analyses (see below). Solid state color cameras containing "charged-couple device" (CCD) light sensing technologies and with Super-VHS (Y/C) outputs (horizontal resolution equal to 480 TV lines or better) last the longest and are the least expensive. A 13-inch monitor with similar minimum resolution specifications is recommended for use when in close proximity to the petrographic microscope; larger monitors can be selected for classroom format. A VCR should be capable of stop action and slow motion, should have Y/C input, microphone input, and should have a minimum of "three head" recording technology. Even though color capabilities are not a necessity for fluid inclusion applications, a color image is more comfortable to view, not to mention the added flexibility of being able to use the CCTV system for other petrographic applications where color capabilities would be essential. Also, the cameras with the fewest adjustments are the most "user-friendly." Another electronic device that is handy for immediate documentation is a video printer. A color printer with a good CCTV camera can produce prints that are almost as good as a photographic print, at the same price as a polaroid print. Prints from a black and white video printer are not nearly as high resolution as color prints (still, a two µm inclusion can be resolved!), but the cost of each black and white print is about an eighth that of a color print.

Photographic system — It may be necessary to photographically document various aspects of inclusions (as shown throughout this text). Microscope manufacturers supply automatic 35 mm and polaroid attachments for their microscopes. As these are typically fairly expensive, some may choose to use their personal cameras with the appropriate adapters required to mount the camera's back onto the microscope — this option yields perfectly acceptable results. For color print or slide photography, if "daylight film" is to be used, it is important to color balance the microscope light by using the correct blue filter (CB-12) in the light path with the voltage for the light bulb power supply set correctly (for halogen bulbs the correct setting is about 9.5 volts). No blue filter is necessary if "tungsten film" is employed. For black and white photography, contrast is greatly enhanced by the use of a green (545 nm) filter. Many photomicrographs in this text were produced with Polaroid 4X5 type 55P/N film with a VG-9 green filter. We like to use polaroid because we receive instant gratification, and the resolution of this film is fine; a high-quality negative is also produced with a little additional effort. A good choice for 35 mm black and white prints is Kodak type Tmax 100.

Spindle stage — Anderson and others (1992) have described a home-made, inexpensive spindle stage that is very useful for viewing fluid inclusions in three dimensions, so that more accurate volume determinations of individual inclusions can be attempted.

Image analysis system — Itard and others (1989) have described an image analysis system that digitizes geometries of inclusion cavities and/or the phases within the inclusions, and calculates the volumes. Such technology might prove very useful in conjunction with a spindle stage and in determining water/gas mole fractions and pressures within fluid inclusions (see Chapter 10).

Crushing stage — For most fluid inclusion work, a thorough petrographic study should include a mechanism for opening a fluid inclusion to determine its internal pressure at room temperature. The most readily accessible mechanism for accomplishing this is a *crushing stage*. A crushing stage is a device that will allow the inclusionist to immerse a mineral fragment containing fluid inclusions in a liquid medium, propagate a crack through the mineral to open the inclusion to one-atmosphere pressure, and observe changes in bubble size either directly through the microscope or by playback of a video recording of the action. Each lab will need to fabricate their own crushing stage until one becomes generally available. Roedder (1970) has published a useful design for a crushing stage. Also, we include a sketch of a simple crushing stage fabricated from three glass plates, glue, and using duct tape as a hinge (Fig. 6.2). Using this design, a crushing stage can be made in about 10 minutes from easily available materials. Although these crushing stages work well enough given some patience, they both suffer from one frustrating flaw in that cracks are propagated randomly during the crushing run. We have been experimenting with designing a new crushing system that uses diamond probes to direct cracks toward the fluid inclusion of interest during the crushing run. If this new design works, it should result in a substantially more user-friendly device.

PETROGRAPHIC ANALYSIS OF FLUID INCLUSIONS

Finally, the time has come to study the fluid inclusions petrographically. A specific question (or series of questions) has been devised, a thoughtful sampling program has been completed, samples have been properly prepared, the microscope has been properly set up and adjusted, and the paragenetic relationships among the diagenetic phases have been determined. Performing fluid inclusion petrographic analysis involves the following simple methodology.

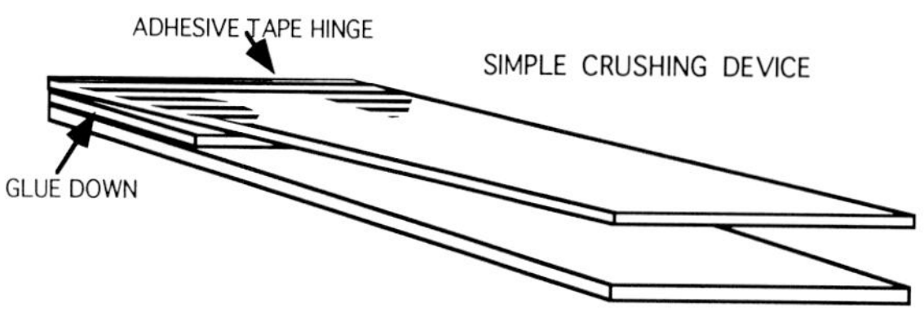

Fig. 6.2. *Sketch illustrating construction of a crushing stage device. The stage is constructed from three glass plates and duct tape.*

As a petrographer surveys samples in a search for fluid inclusions, and as inclusions are found, the petrographer must be continually cognizant of a need to answer the following questions about any given inclusion:

1. What is the inclusion origin? Subset questions are:
   - —What is the most finely discriminated, petrographically distinguishable group of inclusions? (That is, delimit a fluid inclusion assemblage.)
   - —What is the significance of the inclusion to the question being posed?

2. What is the inclusion composition?

3. What is the relative consistency of the liquid-to-vapor volumetric proportions in the fluid inclusion assemblage in which the inclusion is found?

4. What is the pressure in the inclusion?

In a way, doing fluid inclusion petrography is much simpler than standard thin section petrography because very few observations need to be made. Each of these questions is treated in detail below.

Getting Started

Upon starting, it is good practice always to check the microscope adjustments.

Begin a search for fluid inclusions appropriate to answer the question posed. Start with the idea that the entire section will be surveyed with the 10X objective, and proceed by making systematic traverses. If studying carbonate phases, the double refraction of these minerals requires that either the lower or the upper polarizer (analyzer) be in the light path and that either the stage or one of these polarizers is rotated to eliminate the double refraction as the sample is being studied. At about 125X to 150X total magnification, microfractures and growth zones (see Chapter 2) should be obvious. If something might be of particular interest at higher magnification, then rotate into position a 40X objective. If the 40X objective is a standard petrographic objective (one can tell this by viewing how close the objective is to the plate — if it is close to the plate, then it is a standard petrographic objective), then sometimes, the optics of inclusions near the surface will be improved by applying a drop of immersion oil and a coverslip. On all high power objectives, check to see if there is a rotating ring on the objective: this must be adjusted to maximize clarity of the image. If a field of view contains inclusions of possible interest, with a fine-tipped, permanent marking pen, draw an arc (open to the rear of the microscope) around the focused beam of light (roughly 3 mm in diameter) that illuminates the field of view. A 180° arc is useful, because if one is consistent, one will always know how to reorient the specimen on the microscope, which greatly facilitates navigation back to the exact area of interest. Additional documentation such as rough sketches or photographs may also be required for novices to relocate inclusions of interest (see section of this chapter "Documentation of Petrographic Relations"). As we progress into the computer age, digital storage of video images soon may become efficient and cost effective. As surveying of the sample continues, and more areas of interest are discovered, the petrographer can develop a marking code for distinguishing relative importance of the circled areas (e.g., a number of hash marks at the arc edge — the greater the number of marks, the greater the importance of the field of view).

Many novices will experience difficulty in finding fluid inclusions: pits on the surfaces of plates and solid inclusions are sources of confusion. The best way to begin is by looking at a section that certainly has fluid inclusions; for instance, thin sections of granites or even quartz clasts in a sandstone almost certainly will have abundant fluid inclusions. Once inclusions are found in the samples, then a novice should attempt to find the smallest inclusion possible to resolve at about 500X total magnification, and to ask, what characteristics of the entity are indicative of it being a fluid inclusion. Then, the novice should pretend that the inclusions contain only a single homogeneous phase (see Fig. 2.1), and make a mental note as to how it would be identified elsewhere. Finally, the novice should learn first to focus exactly on the surface of a plate, and pay attention to what the surface pits look like (they show pinkish-colored boundaries) and what dust, dirt, oil, residual glue, etc. look like (they show greenish-colored boundaries). Then the inclusionist should focus barely under the surface, and note how sharp and crisp the images are of inclusions in this position. Also, the deeper one focuses into thick plates, the fuzzier the images of inclusions will become. Since the best images will be of inclusions near the top surface, then the inclusions near the top surface will be the easiest to study.

Once these initial hurdles are overcome, surveying a thick section for fluid inclusions will become enjoyable as one always will be thinking about the four questions enumerated above. Thus, determining the petrography of fluid inclusions can become a pleasurable challenge to answer some basic geologic questions.

Determining Inclusion Origin

The first step involves determining if those fluid inclusions that should answer a particular geologic question are present. For instance, if there were interest in conditions of precipitation for a cement, one would need to find primary fluid inclusions. To identify primary fluid inclusions, remember that a relationship to growth must be shown (Chapter 2). Also, to avoid confusion with multiple diagenetic events, it is necessary to consider separately those fluid inclusions from an individual growth zone, or within a single healed fracture. In other words, analysis of fluid inclusions requires (as will be shown shortly) the petrographic capability of identifying those fluid inclusion vacuoles that formed at about the same time or during the same set of diagenetic conditions (e.g., all inclusions along the same growth zone in a single crystal or all inclusions along a single healed fracture). These finely discriminated groups of petrographically associated fluid inclusions will be termed in the remainder of this text a *fluid inclusion assemblage*, or an *FIA*, and used in favor of the less restrictive term *population*, which will have a totally different connotation. Any study of fluid inclusions to solve a diagenetic problem requires petrographic separation of fluid inclusions into fluid inclusion assemblages (FIAs).

Determining Inclusion Composition

If the appropriate fluid inclusions are present, the second step involves verifying that the fluid inclusions are aqueous, because oil-filled inclusions require drastically different interpretations. Some oil-filled inclusions may appear colored in transmitted white light, but many are clear and look identical to aqueous inclusions. A microscope equipped with an epifluorescence attachment often can be used to distinguish between aqueous and petroleum inclusions. When exposed to intense ultraviolet or blue light, most oil-filled inclusions will emit visible light (fluoresce) and aqueous inclusions will not (Burruss, 1981, 1991; McLimans, 1987). Furthermore, the color of the fluorescent light can be correlated loosely with the API gravity of the petroleum as shown in Figure 6.3.

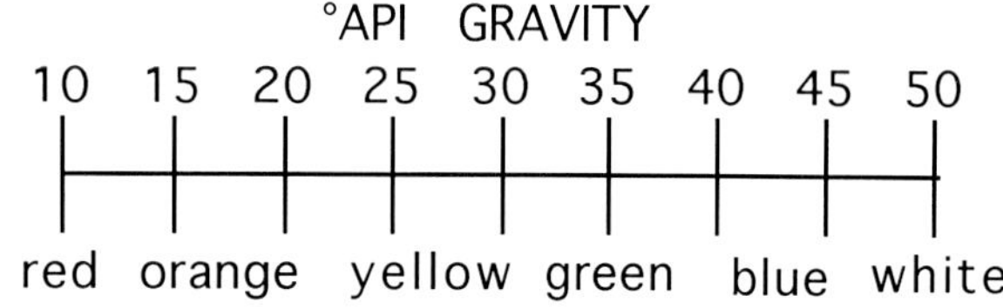

Fig. 6.3. *Illustration of comparison between API gravity of petroleum and fluorescence emission color when excitation is by narrow-band (365 nm peak) ultraviolet light. Color of fluorescence may be controlled by multiple variables, so this should be used as a gross comparison only. After Bodnar (written communication, 1991).*

Determining Ratios of Liquid to Vapor

The petrographic determination of the relative consistency of the liquid-to-vapor volumetric proportions (L:V ratios) within inclusions of a single FIA is a most useful step. If the inclusions are aqueous, the information will aid in determining the diagenetic environment of fluid inclusion entrapment (next section of this chapter), and may aid in deciphering later thermal history. Also, this information will be employed in the interpretation of microthermometric data (see Chapters 9 and 10). The resulting determinations of L:V ratios within individual FIAs should fall into one of the following categories, each of significant diagenetic value (see next section of this chapter):

—*all-liquid inclusions* that cannot be shown necessarily to result from metastability, and cannot be shown to result from necking down after a phase change, and thus, likely were trapped at low temperature

—*consistent L:V, two-phase inclusions* with small bubbles

—*somewhat consistent L:V, two-phase inclusions* with small bubbles

—*highly variable L:V*, with some vapor-rich (large bubbles) inclusions that do not result from necking down after a phase change

In order for distributions of L:V ratios to be determined, one must obviously first isolate an FIA. Again, an FIA is the most finely discriminated group of petrographically associated fluid inclusions that formed at about the same time, or during the same set of diagenetic conditions. A good example of an FIA would be a single healed microfracture defined by a plane of fluid inclusions. Isolating an FIA within a single growth zone is not quite as straightforward, because a single growth zone may form over an extended period of time. Furthermore, in minerals that commonly recrystallize (e.g., dolomite), it is conceivable that inclusions in close proximity to one another and in the same growth zone (that would be identified in transmitted light) could have been trapped at different times — perhaps even millions of years apart. In instances like this, luminescence techniques may prove valuable in isolating FIAs.

Correctly recognizing the relative consistencies of L:V ratios in FIAs comes with some experience, because the size, shape, and orientation of the fluid inclusions, and the location of the bubbles within them, can have a profound effect on the apparent L:V ratios viewed in only two dimensions with the petrographic microscope (Fig. 6.4). Thus, to compare the relative consistencies of the L:V ratios of inclusions within individual FIAs requires that similarly sized, shaped, and oriented inclusions be selected for comparison. Of course, the inclusions within an FIA could be heated for Th determination (see Chapter 7) to check the conclusion regarding relative consistencies — and this is very commonly required. Nevertheless, with a little experience, a novice inclusionist should be able to distinguish among all of the above four categories, the most difficult distinction (even for experienced inclusionists) being between the category of *two-phase inclusions with consistent L:V ratios* and the category of *two-phase inclusions with somewhat consistent L:V ratios*.

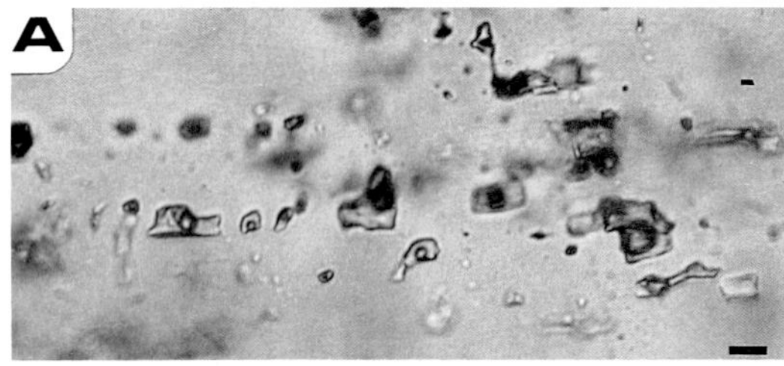

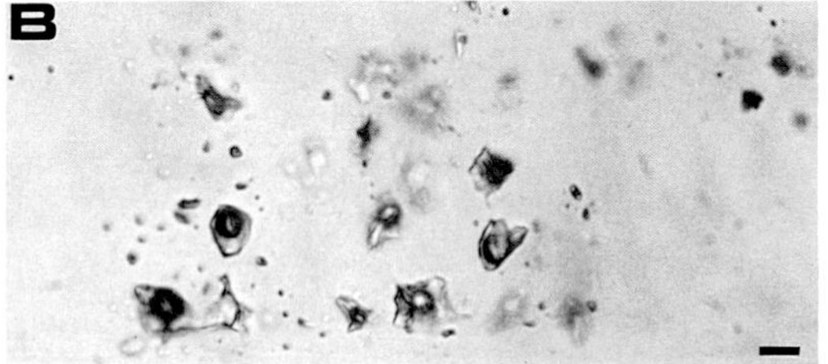

Fig. 6.4. *Photomicrographs of two separate assemblages (A and B) of fluid inclusions (FIAs) in quartz. Liquid-to-vapor ratios in the FIAs may not appear consistent to some observers; however, the Ths of the inclusions fall within a 10-15°C interval in each FIA, which means that the L:V ratios are in fact very consistent. The size, shape, and orientations of the inclusions and the positions of their respective bubbles affect the apparent liquid-to-vapor ratio when observed in two dimensions. Bar scales are 7 μm.*

Procedure.—

Initial study of an FIA should involve a search for aqueous fluid inclusions containing all liquid and lacking a vapor bubble (Fig. 2.1A). These inclusions may be difficult to distinguish from solid inclusions. Liquid-filled inclusions should yield different relief from most solid inclusions and they should appear similar in relief and color to more obvious two-phase, liquid-vapor inclusions. When polars are crossed and the enclosing crystal is rotated to extinction, many solid inclusions will be brightly birefringent whereas fluid inclusions will appear to blend in with the extinct crystal. Significant heating (the amount may vary from several tens to hundreds of degrees) of a liquid-filled inclusion should cause it to decrepitate (causing the contents to escape) or to reequilibrate through volume increase. In either case, the inclusion should appear different after it is heated and cooled to room temperature: it either will have nucleated a vapor bubble or it will have leaked and refilled with air. Most solid or empty inclusions will not have changed after the sample has been heated. Roedder (1992) provided some additional guidance for distinguishing among these various types of inclusions.

Upon identification of all-liquid aqueous fluid inclusions, one must determine their significance as records of low entrapment temperature. All-liquid inclusions can be preserved in three ways:

1. All-liquid inclusions can form from preservation of fluid inclusions trapped at temperatures below about 50°C (see Chapter 4, section "Nucleation Metastability"). These are either stable at room temperature in that they were entrapped at or below room temperature, or are "slightly" metastable in that they were trapped above room temperature, but have not nucleated a vapor bubble because they were trapped below *about* 50°C. Their usefulness as records of low temperature can only be established if other mechanisms of all-liquid inclusion formation can be ruled out.

2. All-liquid inclusions can result from necking down of two-phase fluid inclusions to produce an FIA with closely associated two-phase and all-liquid fluid inclusions (see Chapter 4, section "Effect of necking down").

3. *Significantly* metastable "stretched" liquids, trapped above about 50°C, that have a density in which a bubble should be present at room temperature but for which none has yet nucleated may also be a reason for the existence of all-liquid fluid inclusions (see Chapter 4, section "Nucleation Metastability"). These "significantly" metastable inclusions have densities that should yield Th above about 50°C and must be

distinguished from those all-liquid inclusions that formed at or below about 50°C, and that are either stable or "slightly" metastable at room temperature.

By eliminating necking down (2) and *significant* metastability (3) as possibilities, the all-liquid fluid inclusions should indicate entrapment at low temperatures. Necking down can be shown to be an unlikely mechanism if petrographic distribution of fluid inclusions in an FIA shows groups of inclusions containing abundant all-liquid inclusions but lacking the vapor-rich inclusions and the two-phase inclusions with highly variable L:V ratios that should be formed by necking down of originally two-phase inclusions. In some systems, the sparse distribution of fluid inclusions may preclude the exclusion of the necking down mechanism. As explained in Chapter 4, once heated to their temperature of homogenization, some fluid inclusions may remain metastable all liquid for significant periods of time (Roedder, 1967, 1984; Meunier, 1989). Most inclusions trapped below about 50°C remain as metastable or stable all-liquid at room temperature. Some fluid inclusions trapped above 50°C, especially the smaller ones, remain metastable all-liquid at room temperature because surface tension effects cause significant nucleation problems. Such *significantly* metastable all-liquid inclusions cannot be used to interpret precipitation below about 50°C. Thus, it is important to rule out such *significant* metastability before a low-temperature interpretation can be ascribed to all-liquid inclusions. To attempt to eliminate the possibility of all-liquid inclusions existing as a result of *significant* metastability of elevated-temperature inclusions (trapped above about 50°C), one must attempt to nucleate vapor bubbles in all-liquid inclusions. Placing samples in a kitchen freezer for several days set at a temperature that will not allow freezing of the inclusions (R. Spencer, personal communication, 1989), or prolonged cooling in a freezing stage, might encourage metastable inclusions to nucleate bubbles. We have found that inclusions in the similar size range as associated two-phase inclusions, with Th above about 50°C, tend to nucleate bubbles from this procedure. Furthermore, *significant* metastability tends to affect the smallest-sized members of FIAs and those that would homogenize at the lowest temperatures (see Chapter 4, section on "Nucleation Metastability"). If necking down after a phase change has not occurred, if attempts to nucleate vapor bubbles have been tried and have failed, and if remaining all-liquid fluid inclusions are in the same size range as liquid-vapor inclusions, it is probable that the all-liquid inclusions are not significantly out of their field of stability, and therefore, that all-liquid inclusions present were trapped at less than *about* 50°C.

As is obvious from the discussions above, the presence of all-liquid aqueous inclusions can be very significant if these inclusions can be shown to be of low-temperature (<50°C) origin. Our experience has been that most novices overlook all-liquid inclusions in favor of spotting those with dark, often bouncing bubbles. We cannot stress enough that inclusionists must attempt to find all-liquid inclusions within the FIAs.

In addition, some inclusionists also tend to ignore inclusions with low L:V ratios or vapor-rich inclusions (Fig. 2.1B, C) "writing them off" as being stretched, leaked, or damaged during polishing. To the contrary, just like the all-liquid inclusions, vapor-rich or vapor-filled inclusions might also indicate an important diagenetic condition — namely, immiscibility during inclusion entrapment. If they are identified, one must determine if they are significant as records of immiscibility, by eliminating the other ways by which vapor-rich inclusions may form: necking down after a phase change or partial leakage on outcrop or during sample preparation.

Evidence that inclusions had necked down after nucleation of bubbles in originally elevated-temperature fluid inclusions already has been discussed above. The best evidence for such necking down is the distribution of vapor and liquid in the FIA. Typically, elevated-temperature aqueous inclusions from the diagenetic realm, cooled to room temperature, should nucleate vapor bubbles that amount to no more than about 15% of the total volume of fluid inclusions. Thus, necked-down elevated-temperature inclusions of simple aqueous composition should contain no more than that approximate vol.% of the fluid inclusion assemblage. Similarly, one line of evidence that would support immiscibility would be if the overall volume of vapor within the FIA is so large that it could not be explained by necking. Furthermore, vapor-rich inclusions that have necked down from two-phase inclusions should appear to be paired petrographically with vapor-poor inclusions, if the orientation of visual observation happens to be correct. In addition, all-liquid inclusions in close proximity to vapor-rich inclusions could result from necking down after a phase change or low-temperature, near surface immiscibility. If the vapor-rich inclusions show evidence of containing an internal pressure of one atmosphere at room temperature (see next section on crushing), then this is evidence that the vapor-rich inclusions contain trapped air and formed in the vadose zone, and the necking down hypothesis is ruled out. Finally, internal pressures of bubbles from necked down vapor-rich inclusions may be above or below atmospheric pressure. But as immiscibility cannot occur for gas-free, aqueous solutions in sedimentary environments (except under extreme geothermal gradient conditions caused by a local thermal perturbation due to intrusive activity, see Chapter 3), vapor-rich inclusions with bubbles under low (nearly vacuum) pressure must not have originated from entrapment of immiscible phases but must have formed from necking down of simple, elevated-temperature aqueous inclusions.

Partial leakage of some fluid inclusions on outcrop or during sample preparation is easy to distinguish from necking down and entrapment of immiscible phases. Those inclusions that have leaked partially, seem to continue to leak in the laboratory. That means when inclusions are heated slightly, the bubble appears to grow slowly as the liquid evaporates from the inclusions. Dark inclusions that have totally leaked during sample preparation remain open, so that they can be refilled with liquid in the laboratory. Immersion of the sample in acetone (or with liquid nitrogen in a cooling stage) typically will result in refilling of the inclusion with the liquid. This liquid can be observed to evaporate and exit the inclusion under the microscope to provide evidence that the inclusion is open.

If two-phase liquid + vapor fluid inclusions are present, one must determine the relative consistency of the L:V ratios within individual FIAs. For simple aqueous inclusions trapped as all-liquid in most elevated-temperature diagenetic systems, at room temperature, inclusions would contain two phases and would consist of less than 15% vapor by volume, which corresponds to a Th of about 225°C or less (Bodnar, 1983). Fluid inclusion assemblages with such low volumetric L:V ratios should be carefully observed to attempt to identify if their L:V ratios are essentially identical or only somewhat consistent. As explained previously, the three-dimensional variability in shape and size of inclusions in a population makes petrographic determination of consistency of L:V ratios less accurate than heating the fluid inclusions to determine relative consistencies of their homogenization temperatures.

Determining Pressure

As can be understood from the previous discussion, the determination of an inclusion's internal pressure at room conditions is an important step in recording the fluid inclusion petrography. At the outset, we should warn the enthusiastic inclusionist that this step is "much easier said than done." It will require great persistence and patience, but the information obtained will be worth the effort. Because this step is such a painstaking process, it may not be necessary for all FIAs. It should be attempted if a pressure determination will assist in assessing potential reasons for variability in L:V ratios, or if it will later be used for evaluating the composition of the fluid inclusion.

Determination of the internal pressure of an inclusion at room temperature is accomplished through *crushing analysis*. Crushing analysis allows semiquantitative determination of vapor bubble pressures by opening the inclusion cavity to atmospheric pressure and observing the change in size of its bubble (Roedder, 1970). A simple crushing stage can be made from three glass plates and hinged with duct tape (Fig. 6.2), or more elaborate crushing stages can be fabricated (Roedder, 1970). A small piece of a crystal or a doubly polished section (no more than a mm on each side) is placed in the crushing stage. A drop of immersion oil is placed on the small sample fragment and the stage is closed. Under the microscope, the fluid inclusions are videotaped to preserve a record of the vapor bubble sizes before the crushing procedure. The crushing stage is then compressed to crack the crystal and open the fluid inclusions. The crushing run is videotaped so that a record is preserved of the change in size of the vapor bubbles. Boyle's law ($P_1V_1 = P_2V_2$) is used to determine the pressure. During the crush, bubbles originally near a vacuum collapse, bubbles originally at one-atmosphere pressure (trapped in the vadose zone) stay the same size, and bubbles originally under high pressure expand. For diagenetic environments, a bubble that collapses indicates elevated temperature entrapment of a gas-deficient aqueous liquid. Expansion of the bubble indicates entrapment of a dissolved or immiscible gas phase at the higher pressures developed below the water table. Even with initial entrapment at elevated diagenetic temperatures, the P-V-T relationships of such gas-charged inclusions dictate high internal pressures at room temperature (see Chapter 3).

Documentation of Petrographic Relationships

Documentation of important petrographic relationships that link the fluid inclusions to the paragenesis is an important part of fluid inclusion work. Photography and video printing with the microscope are treated in the "Important Microscope Options" section above. However, these two techniques are not very useful for documentation of spatial relationships throughout the entire thick section. To accomplish this, one simply could sketch the section by hand. But a fast and inexpensive way of mapping the thick section is to simply photocopy the section (1:1 magnification). Even better results (as the image will be magnified to fill a page) can be obtained with some color photocopy machines equipped with an attachment for printing 35 mm slides: simply mount the thick section in a 35 mm slide holder, and print it quickly with the color photocopy machine to avoid excessive heating of the slide with the intense light source. Also, a black and white enlarged copy of a slide can be produced by inserting the thick section into a microfiche printer; if the lowest magnification is used on the printer, the entire plate should be enlarged to fill one letter or legal sized page. Be careful to orient the thick section properly so that the image is not printed backwards relative to your microscope. Some inclusionists prefer the quality of a photographic print, and such can be accomplished by placing the thick section in an enlarger and first producing a negative black and white print. All of these methods can be used effectively for relating inclusions to the paragenetic framework, and for enhancing one's ability for relocating inclusions for further study.

DETERMINING DIAGENETIC ENVIRONMENTS

Using the petrography of fluid inclusions, various diagenetic environments of fluid inclusion entrapment can be determined *without performing microthermometry*! The following sections provide models of fluid inclusion petrographic characteristics, many of which are diagnostic of various diagenetic environments and subsequent thermal histories. The characteristics presented are for those inclusions in which *significant* metastability of inclusions trapped above about 50°C and necking down after a phase change have been discredited through the aforementioned techniques. For each diagenetic environment, the unaltered inclusions are considered first and are followed by discussions of the inclusion populations expected after moderate and then intense thermal reequilibration. For each scenario, the combination of L:V ratios and inclusion pressures that are diagnostic of the diagenetic history are noted.

Vadose Zone

The vadose zone is located above the water table and contains both water and air (compositionally modified) under one atmosphere of pressure. Cementation and crack healing may trap fluid inclusions representing the vadose zone's fluid heterogeneity. Fluid inclusions from the vadose zone are characterized by all-liquid fluid inclusions (diagnostic of entrapment below about 50°C) and two-phase (liquid + vapor) fluid inclusions characterized by highly variable L:V ratios (Fig. 6.5A; Goldstein, 1986a; Barker and Halley, 1988; Goldstein and others, 1990). The two-phase inclusions in some fields of view may appear somewhat consistent, so it is important to investigate the entire population to better define FIAs (Barker and Halley, 1988). Two-phase inclusions result from heterogeneous entrapment of the two fluid phases in the vadose zone. Heating of the fluid inclusions to homogenization would yield highly variable temperatures of homogenization that would produce incorrect data about temperature of entrapment (Goldstein, 1986a). Inclusions in which the bubble dominates the cavity volume are to be expected. Bubbles in two-phase inclusions are at about one-atmosphere internal pressure.

There should be no confusion between an FIA from the vadose zone and those from other environments. Superficially, an FIA from the vadose zone may appear similar to an FIA that formed in the liquid field at elevated temperature and then experienced undetected necking down after a phase change occurred (Chapter 4). However, when crushed, bubble pressures in the elevated-temperature inclusions likely would range from near a vacuum (bubbles collapse) to relatively high pressure (bubbles expand). The bubbles trapped in the vadose zone are at one-atmosphere internal pressure (bubble size does not change when sample is crushed).

After fluid inclusions from the vadose zone have experienced moderate heating during burial, subsequent reequilibration of some fluid inclusions, and cooling to room temperature, all FIAs remain petrographically diagnostic of the vadose zone (Fig. 6.5B). During heating, the all-liquid inclusions and those inclusions that trapped the smallest air bubbles would exist as homogeneous liquid inclusions that would build high internal pressures with continued heating; some of these inclusions could reequilibrate. But it is unlikely that the inclusions which initially entrapped large bubbles would experience the internal pressures needed for reequilibration. From studies of vadose-zone cements that have been heated during burial, it is readily apparent that the vadose zone signature in fluid inclusions can be preserved (Goldstein, 1990). Some of the all-liquid inclusions survive overheating and thus indicate entrapment below about 50°C. Some of the inclusions that were originally all liquid have reequilibrated at elevated temperatures and now contain vapor bubbles. From reequilibration, bubble pressures in these inclusions are either nearly a vacuum, less than one atmosphere, or at a high pressure (leaked and refilled at elevated temperature) and provide evidence for a higher temperature history postdating initial low-temperature entrapment. The vapor-rich fluid inclusions remain unchanged and still contain one-atmosphere internal pressure; they provide strong evidence for entrapment in the vadose zone.

If fluid inclusions initially entrapped in the vadose

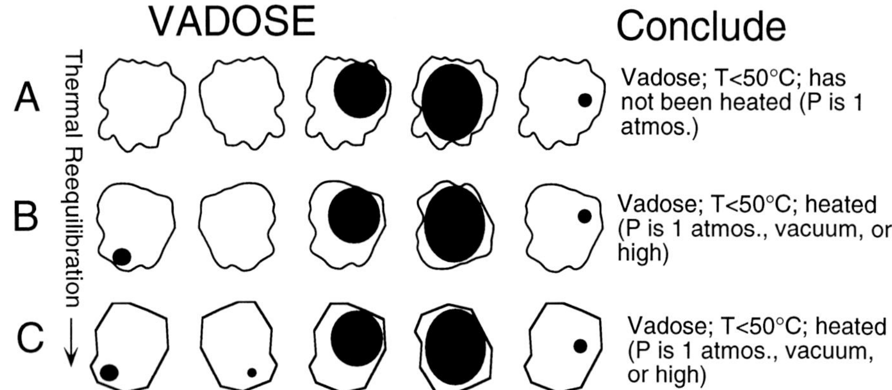

Fig. 6.5. *Schematic illustration of room-temperature distribution of liquid-to-vapor ratios within a fluid inclusion assemblage of vadose-zone origin. Sample sizes and shapes are only diagrammatic.* ***A)*** *Assemblage before any thermal reequilibration.* ***B)*** *Assemblage after moderate thermal reequilibration.* ***C)*** *Assemblage after significant thermal reequilibration. Conclusions refer to interpretations that can be made about diagenetic environment given the additional observations associated with each conclusion presented. Assume that fluid inclusions within each assemblage are of various sizes and shapes.*

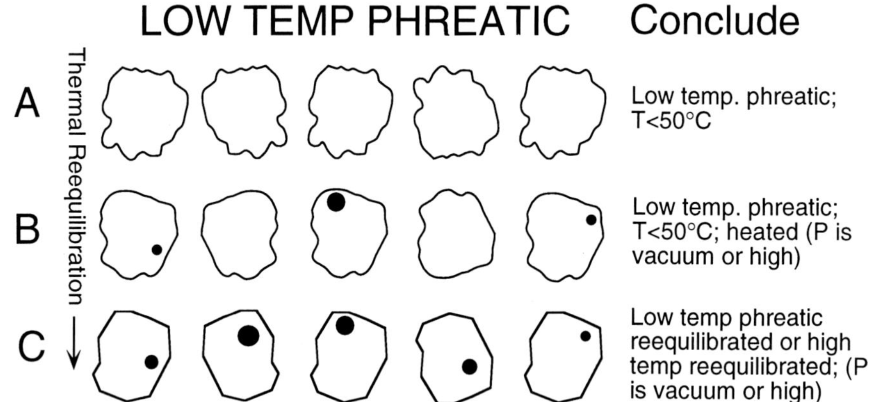

Fig. 6.6. *Schematic illustration of room-temperature distribution of liquid-to-vapor ratios within a fluid inclusion assemblage of low-temperature phreatic origin. Sample sizes and shapes are only diagrammatic. Assume that fluid inclusions within each assemblage are of various sizes and shapes.* ***A)*** *Assemblage before any thermal reequilibration.* ***B)*** *Assemblage after moderate thermal reequilibration.* ***C)*** *Assemblage after significant thermal reequilibration. Conclusions refer to interpretations that can be made about diagenetic environment given the additional observations associated with each conclusion presented.*

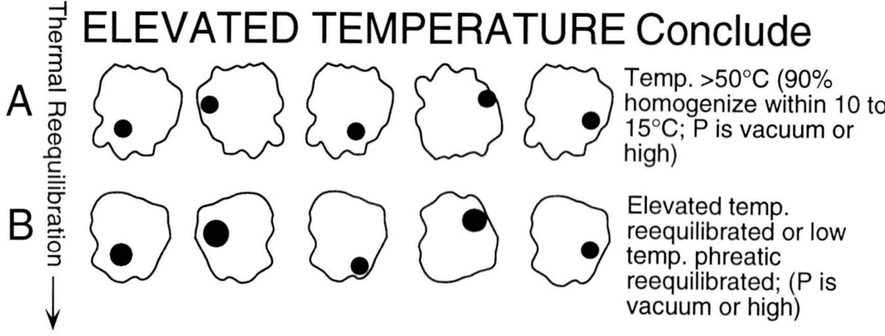

Fig. 6.7. *Schematic illustration of room-temperature distribution of liquid-to-vapor ratios within a fluid inclusion assemblage of elevated-temperature origin. Sample sizes and shapes are only diagrammatic. Assume that fluid inclusions within each assemblage are of various sizes and shapes.* ***A)*** *Assemblage before any thermal reequilibration.* ***B)*** *Assemblage after thermal reequilibration. Conclusions refer to interpretations that can be made about diagenetic environment given the additional observations associated with each conclusion presented.*

zone were subjected to heating to an even higher temperature than the example above, all of the originally all-liquid fluid inclusions could have been reequilibrated to two-phase inclusions. Although such complete reequilibration of all-liquid inclusions has not yet been described, it is a possibility that must be considered. Nevertheless, the petrographic characteristics of the fluid inclusions still would be diagnostic of the vadose zone (Fig. 6.5C). The resultant FIAs still would contain some fluid inclusions with large bubbles and some with small vapor bubbles that formed from processes of reequilibration. In them, internal pressures could vary from nearly a vacuum to high pressure, and would attest to the elevated-temperature history the rock had experienced. The overall L:V ratios would remain highly variable. Crushing studies would show that some two-phase inclusions, especially those with large bubbles, would remain under one-atmosphere pressure and would attest to the low-temperature vadose zone conditions of fluid inclusion entrapment.

Low-Temperature Phreatic Zone

The low-temperature phreatic zone exists below the water table. For the purposes of the fluid inclusion methodology described in this chapter, it will be defined to include temperatures below about 50°C. For the most part, pore fluids in this system contain a single aqueous liquid phase (although minor outgassing of gases such as CO_2 and CH_4 is known). So, entrapment of fluid inclusions in this environment mostly yields all-liquid fluid inclusions (Fig. 6.6; see Goldstein, 1986a, for examples and variability that could result from trapping separate gas bubbles at elevated pressure). An FIA containing only all-liquid fluid inclusions is diagnostic of inclusion entrapment in the low-temperature phreatic zone (Fig. 6.6A).

After some heating during burial, some of the fluid inclusions in an originally all-liquid FIA may have reequilibrated to form two-phase inclusions that have only somewhat consistent L:V ratios; the others will remain all liquid (Fig. 6.6B). The moderate variability in L:V ratios reflects the varying degrees to which reequilibration commonly progresses: for the two-phase inclusions the vapor bubbles would range from very small to a size occupying no more than about 15 vol.%. The inclusions with the larger bubbles reflect the highest diagenetic reequilibration temperatures. With a lower maximum temperature of reequilibration, the inclusions would contain smaller vapor bubbles. FIAs composed of all-liquid and liquid-rich, two-phase inclusions with only somewhat consistent L:V ratios are diagnostic of initial entrapment in the low-temperature phreatic zone followed by a later heating event (Fig. 6.6B). They cannot be confused with inclusions from other diagenetic environments, or those that have necked down. The reasons are that they lack the high variability in L:V ratios of necked and vadose zone inclusions, they retain all-liquid inclusions indicative of low temperature, and as the two-phase inclusions have formed from reequilibration with fluids at higher temperature conditions, their pressures either are near a vacuum or are high (depending on gas content of fluid that refilled the inclusion) and do not contain one-atmosphere pressure as do those inclusions from the vadose zone.

It is possible that more significant heating could result in reequilibration of all of the one-phase liquid inclusions (Fig. 6.6C). Such reequilibration would result in FIAs of two-phase liquid-rich fluid inclusions with only somewhat consistent L:V ratios. Bubble pressures would be near vacuum or high from reequilibration. These FIAs would not be recognizable as those that originated at low temperature because all-liquid inclusions would no longer be present. Although a population of FIAs that results from complete reequilibration of a low-temperature population of FIAs has not yet been identified, it might be expected in strata that have experienced very high temperatures.

Elevated Temperature

In diagenetic systems, the environment above about 50°C occurs at significant burial depths and commonly is dominated by a single aqueous phase. Here we do not cover the elevated-temperature, immiscible systems that may also exist (see Chapters 9, 10, and 11). When aqueous fluid inclusions are trapped at a single temperature in the elevated-temperature, one-phase environment, they generate small vapor bubbles (less than 15 vol.%) during uplift and cooling to surface conditions (see Chapter 3); thus, the FIAs would lack all-liquid or vapor-dominated inclusions (Fig. 6.7A). Bubble pressures either would be near that of a vacuum or high depending on the gas content of the fluid. All inclusions trapped under the same set of conditions would have the same L:V ratio and would homogenize at the same temperature if heated in the lab. For non-reequilibrated FIAs, on some individual healed microfractures we have observed all fluid inclusions to homogenize within several degrees of one another; along individual growth zones, inclusions might demonstrate the same narrow spread of homogenization temperatures and others might show a small amount of variation, with at least 90% of the inclusions in the FIA homogenizing within only 10-15°C. The lack of perfect consistency in homogenization temperature could be controlled by an inability to determine petrographically those inclusions that were sealed at the same moment. So, if liquid-rich fluid inclusions in a single growth zone or along a single healed fracture were to show extremely consistent L:V ratios (Fig. 6.7A), one would interpret inclusion entrapment to have been at an elevated temperature. As long as fluid inclusion reequilibration is controlled by variables that affect each inclusion differently to cause a wide spread in density (see Chapter 4), consistency in L:V ratio is diagnostic of elevated-temperature diagenesis and cannot be confused with other environments of diagenesis.

If an FIA that was trapped originally at elevated-temperature conditions was heated beyond those conditions, reequilibration of some of the inclusions could take place (see Chapter 4). Hence, a population of liquid-rich inclusions with only somewhat consistent L:V ratios would be produced (Fig. 6.7B). If the inclusions were heated to homogenization, they would yield variable data. Pressures in bubbles would be either near that of a vacuum or high depending on the original inclusions and the mechanism of reequilibration. The petrographic characteristics of this FIA are indistinguishable from the low-temperature phreatic example in which all the fluid inclusions had reequilibrated (Fig. 6.6C). Although this overlap is a drawback of the fluid inclusion technique, up to this point, examples of total reequilibration of low-temperature FIA populations have not been recognized. Therefore, it may be that this overlap introduces only moderate uncertainty. Moreover, other petrographic characteristics may help in distinguishing between the two. For instance, if one were to identify an early and a late FIA in a sample, and if there were overlap in fluid inclusion size, shape, and both FIAs were enclosed in the same mineral, there would be potential for distinguishing between a reequilibrated elevated-temperature and a reequilibrated low-temperature FIA: if the early FIA were to contain some all-liquid inclusions, yet the later FIA contained only liquid-rich two-phase inclusions with only somewhat consistent L:V ratios, it is unlikely that the later FIA represents total reequilibration of a low-temperature FIA given the existence of all-liquid inclusions in the earlier FIA. Elevated-temperature entrapment followed by reequilibration would be indicated for the later population of fluid inclusions in this case.

SUMMARY

The authors vividly remember being confronted with the task of "doing a fluid inclusion study" for the first time. No problem: just watch the bubbles and ice disappear for a few months. Needless to say, we now know better. Furthermore, we also appreciate that fluid inclusions are normally only a small part of a larger geologic study, and that there will always be a great temptation to begin to collect microthermometric data on the first inclusions found in order to "get the inclusion study over with." We now know that the inclusion study must be undertaken in the framework of a careful paragenetic framework and that it must attempt to answer specifically-posed questions.

This chapter describes the most appropriate methodology for tackling fluid inclusion petrography — which is a prerequisite to the undertaking of microthermometry, geothermometry, geobarometry and other analytical techniques (discussed in subsequent chapters). Without clear documentation and understanding of the petrographic relationships, the other techniques are totally unconstrained. So, just as all experienced inclusionists have said before us, take heed of our plea to allow ample time and energy for characterizing the petrography of the fluid inclusions prior to the collection of any other data from them.

As described in this chapter, the first mechanical steps are proper sample preparation and microscope set up. The authors have purposefully been rather general in the instructions because materials and equipment change all the time and there are countless products on the market. All readers should know that they are welcome to write us at any time and we will convey to them our most up-to-date methods (part numbers of equipment and materials used, as well as procedural steps).

Many readers will probably remember their first course in petrology, and how they learned to name and describe rocks: with practice and a little experience, a series of questions would eventually begin to automatically click through the brain when confronted with a rock to be described. A totally analogous situation now will exist for many readers, but rather than looking at a rock, they will now be confronted with a thick section that most probably contains millions of fluid inclusions. Assuming that samples have been collected thoughtfully and that the paragenesis of the authigenic minerals has been meticulously determined, the procedure for evaluating the petrography of fluid inclusion is summarized below:

1. Check microscope adjustments.

2. Check surface of sample; apply immersion oil if necessary.

3. Remember the specific reason why fluid inclusions are being studied.

4. Remember what sorts of fluid inclusions are required to answer questions.

5. Start with the idea in mind to survey each entire plate with 10X objective, using 40X objective when necessary.

6. Note sizes and shapes of inclusions with 40X objective.

7. Determine inclusion origins.

8. Isolate FIAs.

9. Using UV epifluorescence, check if an inclusion contains petroleum.

10. Determine relative L:V ratios of FIAs:
   —all liquid inclusions (that do not result from *significant* metastability or necking down)
   —consistent, two-phase (L+V) inclusions
   —somewhat consistent, two-phase (L+V) inclusions
   —inconsistent L:V ratio inclusions.

11. Mark areas of special interest on plate by making half circles with a felt-tipped pen.

12. Perform crushing tests if necessary, particularly on vapor-rich inclusions:
   —if bubbles collapse, P<1 atm
   —if bubbles remain the same, P=1 atm
   —if bubbles expand, P>1 atm.

13. Document petrographic relationships, if necessary.

It is that simple!

Finally, this chapter concludes by showing how powerful the fluid inclusion petrography of diagenetic phases can be in providing definitive information that can be used to ascertain the diagenetic environment of FIAs by fluid inclusion petrography alone. And even given that fluid inclusions in soft minerals are predicted to reequilibrate to some degree on burial, it was also shown that the original conditions of formation can still be recognized. One need only understand the systematics of the fluid inclusion technique to comprehend the power of the inclusion petrography.

Chapter 7

FLUID INCLUSION MICROTHERMOMETRY

INTRODUCTION

The determination of temperatures of phase changes within fluid inclusions during heating and cooling of samples is termed microthermometry. The technique is invaluable for discovering the temperatures at which minerals form, the thermal history a rock has experienced, and the compositions of the fluids that traversed a rock in its history. The fundamental principles upon which microthermometry is based are the principles of phase equilibria introduced in Chapter 3. And, as explained in Chapter 6, it is best to first spend a considerable amount of time and effort to characterize the fluid inclusion petrography prior to attempting microthermometry. This chapter elucidates useful philosophical approaches and important practical procedures that, if learned and employed, will ensure successful and efficient microthermometric studies. Also, phase changes for some simple but potentially applicable fluid systems common in the diagenetic realm will be described.

THE MENTAL PREPARATION

It is important for an inclusionist to understand early that for an accurate interpretation of microthermometric data (discussed in Chapters 9 and 10) to be possible, the data must be collected within a rather rigorous framework. When the use of the phase equilibria and their representation as phase diagrams was discussed at length in Chapter 3, in the very beginning two important assumptions had to be postulated — that an inclusion represents a chemically closed (isoplethic) system from the time of entrapment, and that the volume of an inclusion remains constant (isochoric) since entrapment. However, with the treatment of the representativeness of inclusions in Chapter 4, it can be predicted that these assumptions will not always be true in sedimentary environments: necking-down after a phase change and reequilibration due to overpressure from natural heating are phenomena that predictably occur and compromise the necessary assumptions. Furthermore, in Chapter 4 it was also pointed out that if fluid inclusions are trapped in a crystal bathed in two immiscible fluids, that the phases entrapped will most probably not be representative of the bulk composition of the diagenetic system. Thus, these natural, predictable processes either cause entrapment of phases that are not representative of the system as a whole, or cause a later change to the inclusions which destroys evidence of the conditions at the time of entrapment. So the question becomes, if one is going to collect microthermometric data from inclusions, how can one discern what the data from those inclusions represent? Are the data totally erroneous because of immiscibility? Are they totally erroneous from necking down after a phase change? Do they actually represent a later P-T condition than that of initial entrapment? One cannot assess the answers to these questions from a single fluid inclusion or even from large numbers of inclusions treated as individuals. *The only way to be able to interpret fluid inclusion microthermometric data is if data from many inclusions within a single fluid inclusion assemblage (FIA) — the most finely discriminated, petrographically associated, group of inclusions — are collected and recorded together as a group.* The goal of measuring inclusions with respect to the FIAs that they are within is to be able to discern any variability. The key point here is that variability of data within an FIA will alert the inclusionist to the plausibility of reequilibration, necking, or immiscibility having adversely affected inclusions within the FIA; whereas consistency of data among inclusions of different sizes and shapes within a single FIA will be a signal that homogeneous fluids were trapped within the inclusions, and that the assumptions of the inclusions maintaining chemically closed systems and maintaining constant volumes throughout their histories held true. Such concepts will be discussed and developed in detail in Chapters 9 and 10. At this stage though, an inclusionist should realize that it is of utmost importance that his or her attention be directed toward FIAs, not individual inclusions, as data are collected. An inclusionist must first constantly strive to isolate individual planes of secondary inclusions, or primary inclusions in individual growth zones; data must then be collected on many inclusions of different sizes and shapes within each FIA; and finally, data from all inclusions within a single FIA are denoted as being part of that single FIA when recorded. If such a philosophy to collecting data is not followed, it will be impossible to assess if the inclusions truly represent what the researcher hypothesizes; in other words, the data will be totally unconstrained, and for the most part, uninterpretable.

SELECTING FIAs FOR MICROTHERMOMETRY

Inclusions must be selected from each of the thousands and thousands of inclusions that often are found in a single plate. As explained in Chapter 6, focusing on the appropriate inclusions comes with knowing the question and knowing what inclusions would have to be found to answer the question. It should now be obvious that an inclusionist must locate the necessary fluid inclusion assemblages (FIAs) prior to microthermometry. The best examples of fluid inclusion assemblages are:

—all inclusions defining a single healed microfracture

—primary inclusions in a single, most finely discernible and petrographically distinguishable growth zone

If the petrography of fluid inclusions is first characterized through a survey of many selected samples as exhorted in Chapter 6, then often one will find FIAs of potential significance and worthy of further study by microthermometry, because of their special definable, documentable, and defensible origins within the minerals that contain them, and because of the special relationships of the enclosing mineral to the mineral paragenesis. All experienced inclusionists know that this survey step is one of the most stimulating — and therefore rewarding — aspects of fluid inclusion research, whereas microthermometry can be a very time-consuming and tedious undertaking. Thus, much of the problem of selecting an FIA is solved through the all important step of surveying samples and characterizing the fluid inclusion petrography!

Petrographically distinguishing an event of related inclusion entrapment in a single growth zone is not always as straightforward as defining a single microfracture. For example, a single growth zone may form over an extended period of geologic time, whereas a microfracture probably heals over a much shorter length of time. Thus, data from a single growth zone may vary because the petrographic resolution to define those primary inclusions that were trapped at the same time, under the exact same conditions, is not visually resolvable. Figure 7.1A provides an example of why it is sometimes difficult to define fluid inclusion assemblages. This figure illustrates the hypothetical distribution of Th data from primary fluid inclusions in a cement crystal. Note that overall, the broad growth zone contains inclusions that yield Th values from 54°C to 95°C. If the data are taken as a whole, this wide range in data and the mean or mode of the data hold little significance and may not allow one to distinguish thermally reequilibrated fluid inclusions that yield variable data (see Chapter 9) from unaltered fluid inclusions recording a wide range of temperatures during mineral growth. However, when each fluid inclusion assemblage is separated, each assemblage yields internally consistent data that cannot result from thermal reequilibration, and records a detailed history of Th starting at 54°C, progressing upward to 75°C and then 95°C, and then finally cooling to yield a Th of about 55°C again. Figure 7.1B also illustrates the care required in the selection of FIAs within growth zones. It shows that if fine zonation cannot be identified, a fluid inclusion assemblage may record a long series of events. The broad growth zone depicted in Figure 7.1B contains primary fluid inclusions that yield a wide range in Th. Taken as a whole, the population might be indistinguishable from one that had been altered by thermal reequilibration (see Chapter 9). The mean or mode would have little significance; however, further subdivision might be possible if data were gathered on transects that consistently sample outward parallel to growth direction. In this example, every outward transect yields a consistent increase in Th. Thus, one can see that a primary signal of increasing Th through time is preserved, and that even if some thermal reequilibration cannot be ruled out, useful information is preserved. So, the primary inclusions trapped in a thick growth zone could record many different conditions that existed over geologic time; the data gathered from such a thick growth zone may show extreme variability that would nevertheless be representative of conditions that existed during mineral precipitation. Therefore, when deciding on FIAs within growth zones, an inclusionist should always attempt to define the most finely discriminated petrographic context possible.

SELECTING INCLUSIONS WITHIN FIAs FOR MICROTHERMOMETRY

The quantity of inclusions within an FIA that will be selected for microthermometry, will depend on the question posed, the confidence level required for the conclusions, the consistency of the L:V ratios in the FIA, the consistency of thermometric data from the FIA, and the sizes and shapes of the inclusions in the FIA. One should never restrict measurements to the easiest fluid inclusions; that is, those inclusions most easily visible due to size, shape, depth in the plate, or positioning in a clear area. Inclusions of different sizes and shapes may even have totally different origins (Wojcik, and others, 1994). Of these variables, it is most important to measure enough inclusions to span the entire ranges of their sizes. Do not only measure the larger, easily visible inclusions, nor should one restrict data collection to the smaller inclusions. Large inclusions in an FIA may have a higher proportion of thermally reequilibrated fluid inclusions than the small inclusions because of the propensity for the large inclusions to reequilibrate thermally (see Chapter 4). Therefore, a population of data from large inclusions in an FIA could be biased. Furthermore, each population of like-sized inclusions within the FIA could yield very consistent data! Since fluid inclusions of the same size and shape may respond similarly during thermal reequilibration (see Chapter 4), in order to be able to evaluate whether or not an FIA has been subjected to reequilibration, fluid inclusions of variable sizes and shapes must be selected for microthermometry. There is no good excuse for not studying the small inclusions as certain phase changes in fluid inclusions as small as one micrometer can be determinated with the technique of cycling (discussed in detail later in this chapter).

The greater the consistency of the data between the small and large and various shaped inclusions, the less the data required. For instance, during observation of homogenization of a plane of secondary inclusions, if one first notes that there is a range of sizes and shapes of inclusions on the plane, and then sees that all inclusions homogenize between 140 and 145°C, then enough data to demonstrate consistency among the range of sizes is all that is required. On the other hand, if one notes variations in sizes and shapes in a growth zone, and as the sample is heated, if some Th's occur at about 100°C, others at 110°C, others at 120°C, and there are still bubbles left in some inclusions at 120°C, then collection of many Th's (among the entire size range) will probably be required to interpret the data from the FIA.

Novices will always be attracted to two-phase inclusions because they are easier to find and measure, but of course, as explained in detail in Chapter 6, the all-liquid, single-phase aqueous inclusions can be of immense importance, and therefore should always be sought out and never ignored. Also, other single-phase inclusions (Fig. 2.1B) and vapor-rich inclusions (Fig. 2.1C) should also be scrutinized during microthermometry, as the phase changes (or lack thereof) within them may be indicative of certain compositions (e.g., CH_4), processes (e.g., immiscibility or necking), or environments of formation (e.g., vadose zone).

In summary, enough data from various sized and shaped inclusions within a single FIA must be collected to demonstrate consistency or to represent the full range of variability.

HOW MANY FIAs SHOULD BE MEASURED?

A researcher should realize that data from each FIA will have an interpretation (Chapter 9 and 10) that follows from the possible results:

—very consistent data
—moderately consistent data
—very inconsistent data

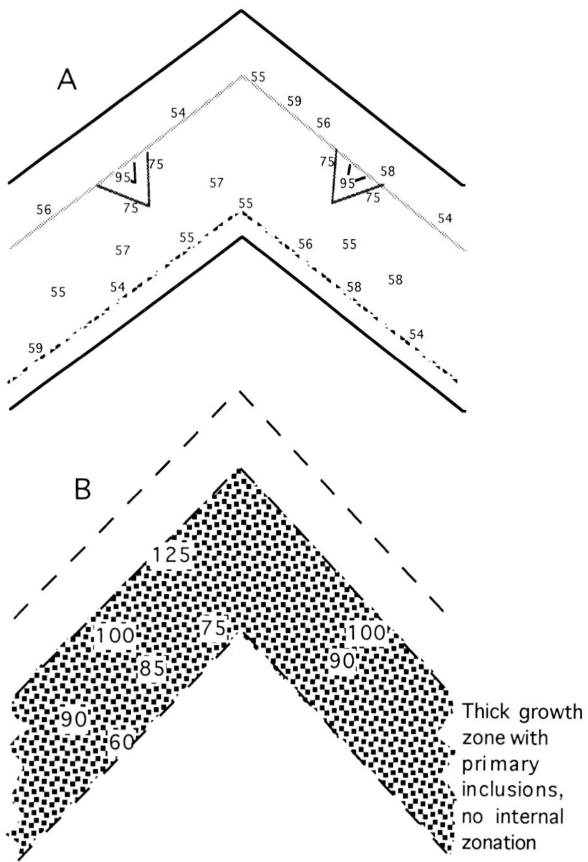

Fig. 7.1. *Hypothetical distribution of Th data from primary fluid inclusions.* ***A)*** *Distribution illustrates the need for carefully defining fluid inclusion assemblages (FIAs) to determine the degree of consistency of fluid inclusion data and for elucidating the complete history recorded by the fluid inclusion data. Note that each FIA can be defined by the growth zonation, and that inclusion data within each FIA yield consistent values that demonstrate a complex history of temperature increase, decrease, and increase.* ***B)*** *Distribution illustrates that this FIA records a series of increasing Th events which can be distinguished if the data are gathered on transects that consistently sample outward parallel to growth direction. See text for complete discussion.*

Each FIA that can be interpreted with confidence has potentially as much significance as 100 other FIAs with the same interpretation. How many FIAs are necessary will be dictated by the question posed and the confidence required, as well as the nature of the samples and the time and money available. For example, consider a case where a question might be "Is a particular calcite cement a low-temperature or an elevated-temperature cement?" Luckily the calcite has primary inclusions defining growth zones, and the first growth zone measured shows consistent Th data 125-135°C among inclusions of various sizes and shapes. What one has definitely learned from this *single* FIA is that this growth zone in this crystal formed at relatively high temperatures. But is this single growth zone indicative of the conditions from which the entire calcite crystal grew — maybe inclusions in this FIA were only entrapped during the latest phase of crystal growth. Thus, it would be appropriate that several earlier FIAs in the crystal be measured so as to cover all of the paragenesis of the crystal. Perhaps again, each yields very consistent data, and all data are >125°C. Now one is sure that this calcite crystal formed late in the diagenetic history. Furthermore, as questions posed encompass greater relative time periods (say, instead of being interested in a single pore-filling cement, one was interested in documenting the entire diagenetic fluid history that the pores experienced) or greater areas of spatial extent, again it would be necessary to study enough FIAs in the proper paragenetic context to document the full range of paragenetic events.

RESOLUTION REQUIREMENTS

For microthermometry, resolution of a temperature measurement of a phase change refers to an interval in temperature at which the operator is certain the phase change occurred (e.g., intervals as narrow as 0.1°C or 1°C, or as wide as 5°C or 10°C). These intervals are determined by the operator during the process of collecting the data (referred to as *cycling* as explained below). The resolution of the measurement, whether the temperature of the phase change is known within 0.1°C or within 10°C, is predetermined by the needs of the researcher in answering the question posed. For some studies, high resolution is needed and for others, a crude determination is satisfactory. The more crude the determination allowable, the more rapidly can data be obtained, so there are some practical trade-offs that must be addressed for each project undertaken. For example, a researcher may collect Th data to within 0.1°C, but then present the data on a histogram with 20°C class intervals, and ascribe no significance to temperature determinations in anything but crude terms. In such a case, measuring the Th data to within 0.1°C has simply been an inefficient use of time. However, an inclusionist will generally be required to obtain higher resolution data if attempting to ascertain reasons for variability of data within a single microfracture or growth zone; an example would be a case where distinguishing between thermal reequilibration and natural variability of a diagenetic system that precipitated a single growth zone was important. Thus, before a study is undertaken, it is worthwhile to evaluate the goals in terms of how finely resolved the measurements need be.

For Th determination, the authors have never had a case in which Th of aqueous inclusions had to be known with resolution greater than 1°C. But, requiring a resolution of 0.1°C is not unusual for certain cases of Tm ice measurements from diagenetic minerals. A case in point is when addressing a diagenetic question that requires information regarding the origin of near-surface fluids (e.g., fresh water, brackish water, seawater, and mixtures of these fluids). True fresh water should yield Tm ice of 0.0°C; Tm ice of just -0.2°C would indicate slightly brackish water containing about 3.5 ppt NaCl. Normal seawater around 35 ppt should yield Tm ice at -1.9°C; Tm ice of -2.1°C would indicate a fluid that is 3 ppt more saline, suggesting possible slight evaporation and providing useful information on the diagenetic or depositional environment.

Admittedly however, an inclusionist may not always know the resolution required for measurements at the beginning of a fluid inclusion study. There are two possible approaches for proceeding with the fluid inclusion study when such a dilemma occurs. One approach is to gather less precise data initially (as gross as 1-2°C intervals for freezing and 5-10°C intervals for heating) as an efficient reconnaissance tool. As the study proceeds, the researcher may find that such low resolution data are perfectly acceptable, or may decide that higher resolution data are required. Another approach is to initially collect every value to within 0.1°C, because one may need such high resolution to evaluate trends in the data. Then after doing much work, one might decide that high resolution data are not useful, and continue by collecting lower resolution data. The data previously gathered still contributes to the study.

MECHANICAL PREPARATIONS

The ability to collect microthermometric data with the greatest possible efficiency and under the best possible optical conditions requires appropriately prepared doubly polished plates, a good heating/cooling stage on an appropriately set up and adjusted microscope, and the inclusionist should be a master in the operation of the heating/cooling stage. These are all mechanical issues that have been resolved for years. To avoid potential problems (i.e., having to "rediscover the wheel") one simply needs to follow what past researchers have determined to be appropriate procedures.

Doubly Polished Plates for Microthermometry

The preparation of doubly polished sections was treated in Chapter 6; however, it does not hurt to reemphasize how exceedingly important it is to have polishes that have no microscopically visible pits remaining from the polishing procedure to facilitate the microthermometric study (Fig 7.2). In other fields of fluid inclusion research such as ore deposits and metamorphic petrology, where thickness of doubly polished plates are greater than what is required for diagenetic application (often 100 μm or more for the former versus often 50 μm or less for the latter), it is not as important to have such a perfect polish on the lower surface of a plate. For diagenetic applications, however, when the plates are so thin, it is important also to have a good polish on the lower surface, because even though the lower surface will have a glue mounting it to a glass, the glue will not have the exact index of refraction as the mineral, so light will still be refracted and dispersed by any pits present, causing degradation of the fluid inclusion images to a degree similar to that displayed in Figure 7.2B. This can cause severe impairment in an inclusionist's ability to measure phase changes in small inclusions (<2 μm). Thus, it is strongly recommended that polishes on both surfaces of a thick plate prepared for diagenetic applications have no microscopically visible pits like those shown in Figure 7.2A.

Microscope Requirements

The requirements for microscopes to be used for microthermometry are the same as those specified for fluid inclusion petrography in Chapter 6. The only additional necessity involves the fact that the specimen containing the inclusions to be studied will be inside a heating/cooling stage rather than sitting directly on the microscope stage. This requires a certain distance to exist between the sample and the objective (above), and between the sample and the condenser (below). Thus, "long-working-distance" objectives and condensers are necessary to be able to use a heating/cooling stage on the microscope. And as explained in Chapter 6, all adjustments of light sources, diaphragms, and the condenser must be optimized for each specimen for the best possible optics.

Be very careful when acquiring a microscope with the long-working-distance objectives and condensers. First pay attention to the upper and lower working distances specified by the heating/cooling stage manufacturers, and then try to match working distances specified by the microscope manufacturers for their respective objectives and condensers. Then determine the type of microscope that will be employed — type referring to the basic design of its optical system which will probably be one of the following:

—160 mm tube length corrected
—210 mm tube length corrected
—infinity (∞) corrected

For the best possible optics, it will be imperative that the long-working-distance objective be matched to microscope optical design. For instance, a 210 mm tube length corrected 100X, long-working-distance objective should be used on a microscope set-up for a 210 mm tube length corrected optical system; an 80X, infinity corrected, long-working-distance objective should only be used on a microscope set-up for an infinity corrected optical system; and a 40X, 160 mm tube length corrected, long-working-distance objective should only be used on a microscope set up for 160 mm tube length optics for the best optical results. Furthermore, for the highest possible resolving capabilities, the numerical aperture of the long-working-distance condenser should be the same as (or greater than) the numerical aperture of the long-working-distance objective selected for use; otherwise, the light will not be convergent which will result in bothersome optics along inclusion walls — making petrographic work difficult. There are some long-working-distance objectives of high magnification (40X) on the market which have adjustments that physically correct the optics for thickness of coverglass (0 up to about 2.5 mm) which may be between the sample and the objective. These objectives are highly recommended when glass windows overlie the sample in the heating/cooling stage, because the optical resolution in the stage will be improved once the objective is adjusted properly.

Heating/Cooling Stages

Since the mid-1970s there have been several commercially available heating/cooling stages that are each adequately described by Roedder (1984) and Shepherd and others (1985). The convective heating/cooling stage designed by the U.S. Geological Survey has been most widely used by researchers working in the sedimentary realm because of its accuracy and ease of use when *cycling* — a technique which is absolutely necessary for successful

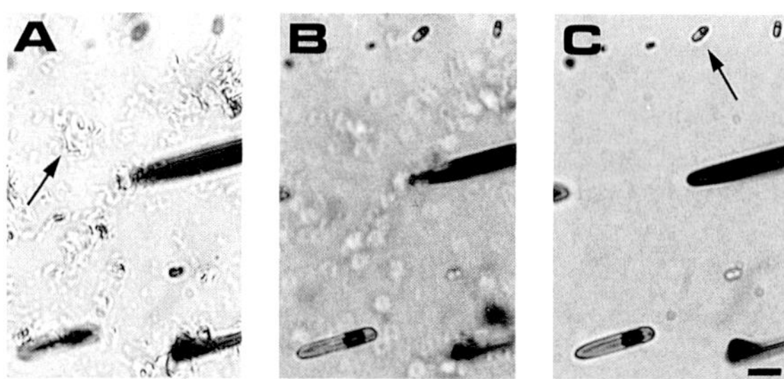

Fig. 7.2. *Photomicrographs showing the effects of a poor polish on optical quality.* ***A)*** *Poor polish on the upper surface of quartz. Arrow points to pits, which are pink in color when viewed through the microscope.* ***B)*** *Fluid inclusions just beneath the poorly polished surface shown in A. The image shown is similar to what would be obtained if the* lower *surface had a poor polish (like that shown in A), even if the upper surface had been polished well (like that shown in C).* ***C)*** *Same photomicrograph as in B, but the surface now has been polished correctly, so that there are no pits. Note that the image is very sharp and clear. This is how all inclusions should appear if polishes are appropriate and if the microscope is adjusted correctly. Arrow points to an inclusion 4 μm in longest dimension. Photomicrograph taken at room conditions. Bar scale is 7 μm.*

procurement of microthermometric data from the small (<7 µm) inclusions prevalent in authigenic minerals, explained in detail below. Perhaps the best advice the authors can provide a novice that is approaching a heating/cooling stage for the first time is to spend ample time reading the instruction manual for the stage, and to practice on samples with large inclusions with previously determined phase change temperatures before attempting either to calibrate the instrument or to study the samples of interest. Start by obtaining Th's on large inclusions (>20 µm) first, proceeding to smaller and smaller inclusions until a Th from an inclusion as small as one micrometer in longest dimension can be obtained. It will seem unbelievable at first, but the novice should understand that optically, it will be possible to obtain a Th from inclusions as small as one micrometer! Such an accomplishment will require practice, patience, and perseverance, and will require that the operator understand the instrument well enough that the *cycling technique* (explained below) can be easily and efficiently employed. It should require no more than a single day to master. The next exercise should be to determine Tm ice on large inclusions (>20 µm), and again proceed to smaller and smaller inclusions, using the procedures described in this chapter. This determination should soon become routine on low-salinity inclusions as small as 3 µm. Inclusions as small as 2 µm in longest dimension are more difficult, those as small as 1.5 µm will rarely be amenable to Tm ice determinations.

There are many temperature measuring devices on the market today that are very accurate and very stable because accurate measurements of small voltages, the electronic linearizing routines used to convert tenths of millivolts to tenths of a degree, the electronic stabilizing routines, and the precision manufacturing processes of temperature sensors are all standardized and common knowledge in the industry. As heating/cooling stages utilize these excellent new temperature measurement technologies in their electronics, the calibration of heating/cooling stages is currently fairly straightforward with the use of commercially available, polished wafers of quartz containing synthetically produced fluid inclusions. These synthetic fluid inclusions show phase changes at temperatures accurately known from experimental studies:

—pure CO_2 solid begins and ends melting at -56.6°C

—pure H_2O solid begins and ends melting at 0.0°C

—pure H_2O of the critical density homogenizes by "critical phenomenon" (disappearance of the phase boundary between liquid and vapor) at 374.1°C

Each instrument will have a particular procedure required for "standardizing," which typically involves setting three known points accurately (like the three temperatures above). It is the responsibility of the researcher to check these points whenever such is deemed necessary, and also to determine how accurately the instrument operates both between and outside these points. In addition, all researchers should assess lateral and vertical thermal gradients within the heating/cooling stage used. One method that can be used to document potential gradients is to measure many inclusions throughout the synthetic fluid inclusion wafers: by mapping and contouring the variability in the data one can readily estimate the thermal gradients at various temperatures.

In the "Resolution Requirements" section above, it was pointed out that sometimes an inclusionist may desire measurements that are accurate to ±0.1°C, particularly when documentation is required for fresh water, brackish water, seawater, or possible mixtures of these waters trapped within fluid inclusions. To ensure that such accuracies are obtained necessitates that a much more meticulous procedure be followed than the normal procedure of standardization mentioned above. Before specifying this procedure, it is important that the reader understands the meaning of the following terms as they apply to fluid inclusion microthermometry:

> *Accuracy* is a measure of the approach of the measured temperature to the true temperature.
>
> *Precision* is a measure of the reproducibility of temperature measurements collected under different experimental conditions (e.g., measurements collected on the same inclusion, but before each measurement the sample is removed from, and then remounted in, the stage).
>
> *Replication* is a repetition of temperature measurements collected under the same experimental conditions (e.g., sample is never moved).

Since in sedimentary studies ±0.1°C accuracy is desired between about 0 and -3°C, an appropriate procedure might be to determine Tm ice of an unknown inclusion multiple times, and before and after each run the Tm ice of a pure H_2O synthetic fluid inclusion in a standard wafer would be measured also. A precision could first be determined for each data set, and an analysis of error for the data set of the standard could be computed. This may seem like a considerable amount of work to obtain a single data point, but such may really be required to evaluate if an accuracy of ±0.1°C was achieved for the measurements.

PREPARING FOR MICROTHERMOMETRY

In preparing for microthermometry, it is important that fluid inclusions have not been overheated beyond their temperatures of homogenization, or frozen in the laboratory. The sample should not have been subjected to a preparation technique that involves any heating or freezing, should not have been heated during other studies such as within a cathodoluminescence stage, and cannot have been etched or stained. Any other heating or freezing of the sample that has not been part of the natural subsurface thermal history of the sample will severely jeopardize the inclusionist's ability to correctly interpret microthermometric data. As has been stated time and again, the fluid inclusions should already have been studied petrographically so that the fluid inclusion assemblage has been defined, the inclusion origin has been defined, the L:V ratios have been approximated, and the petrographic relationships of the fluid inclusions to the mineral paragenesis have been determined. For the petrographic study, efforts should have been made to generate vapor bubbles in all-liquid inclusions by cooling without freezing. Thus, the inclusionist already should know if the fluid inclusions necessary to answer the question being posed are present.

Size of Sample Used for Microthermometry

Because inclusions should never be overheated prior to the homogenization temperature measurement (see Chapter 4), several procedural precautions are required to prevent inadvertent heating of useful fluid inclusions during microthermometric measurement of other fluid inclusions. In other words, the study must be conducted in such a way that the homogenization temperature of any fluid inclusion must be measured before the inclusion has been heated above that homogenization temperature, or before an inclusion has been inadvertently homogenized during freezing. It is an unfortunate practicality that the inclusionist can only observe phase changes in one, or in at most several, fields of view during a heating or freezing run. So to minimize "wasting" potentially useful inclusions, the doubly polished plates can be cut into small pieces containing only one or several fields of view of interest. The smallest easily manipulated size is a reasonable goal, and certainly no larger than 5 mm on a side should ever be necessary. Before fragmenting the sample, be sure to map the important fields of view to maintain the relationships of the field of view with other aspects of the paragenesis (see section "Documentation of Petrographic Relations" in Chapter 6).

In the vast majority of cases, the doubly polished plates will not be removed from the glass slide (or from a cover slip that can be removed from the glass slide; see Chapter 6), as the sample glued to the slide will provide a substrate that holds the very thin plate together. The plates can be slabbed by sawing with a low speed diamond blade or wire, or particular areas of interest can be cored with a miniature diamond coring machine. The simplest method for isolating small areas of interest is to scribe a straight line on the back side of the plate (on the glass slide surface), place this line on a table edge, and then press on opposite sides of the edge to cause a clean break along the line. Continue the process until a small piece of sample to be placed in the stage results. Remember to label all pieces of the fragmented plate with a lettering or numbering system, and record all positions of cuts and fragments on the map of the thick section.

What to Do First

Over the years there has been a surprising amount of confusion regarding the relative order for conducting heating and cooling runs on a sample. Routinely, heating should precede cooling. Unfortunately, in many studies, researchers have performed freezing work first; this is very inappropriate for fluid inclusions in diagenetic phases, because volume expansion upon formation of ice could cause a bubble to disappear (and it may never renucleate, making measurements of Th impossible), or the pressures developed from ice expansion may even crack or stretch the walls of fluid inclusions making later Th measurements meaningless (Lawler and Crawford, 1983; McLimans, 1987).

MEASURING HOMOGENIZATION TEMPERATURES

The measurement of the homogenization temperature (Th) of moderately-sized (7 µm) or larger aqueous inclusions is one of the easiest tasks to accomplish in a fluid inclusion study. All that is required is for one to note the temperature at which a bubble disappears on heating (the basic principles of these phase changes were discussed in Chapter 3), as shown in Figure 7.3. In some inclusions, one will note that the bubbles will be in constant motion within the vacuole. As explained by Roedder (1984), a bubble is probably moving because of extremely rapid flow at the bubble's surface due to surface-tensional contrasts caused by infinitesimally small thermal gradients. For smaller inclusions such movement can greatly enhance one's ability to obtain Th's because humans are more adept at visually recognizing the motions of small entities rather than resolving the actual objects. Thus, a rapidly moving bubble can often be perceived until it is <1 µm in diameter, just as it disappears at Th. However, some bubbles will remain stationary through a heating run (they may be wedged between inclusion walls), or may never change position for unknown reasons, and many inclusions are so small that even bubble movement is difficult to discern. *Our experience indicates that actually being able to observe the bubble disappear occurs for less than 5% of the fluid inclusions studied in diagenetic minerals!*

However, there is a solution. Most homogenization temperatures from the sedimentary realm are measured by the technique of *cycling*, which cleverly employs the

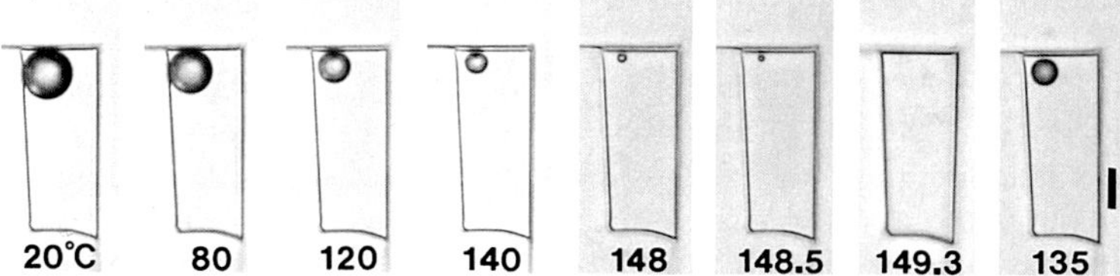

Fig. 7.3. Photomicrographs of a large inclusion in fluorite being heated to determine Th. Inclusion homogenizes at 149.3°C and must be undercooled to 135°C before vapor bubble renucleates suddenly. Bar scale is 7 µm.

inherent metastability of a homogenized liquid, that is, its failure to renucleate a bubble. After a fluid inclusion has been heated to homogenization, it must typically be cooled several degrees to several tens of degrees below the homogenization temperature (undercooled) before a bubble will reappear (see Chapter 4 for detailed explanation). For the aqueous inclusions that require undercooling of tens of degrees, the cycling technique provides an efficient and accurate method for determining Th's of inclusions as small as one micrometer in longest dimension. Figure 7.4 diagrammatically shows the procedure of cycling to obtain a bracketed Th. As the inclusion is heated from 25°C to 125°C, the bubble gradually shrinks, and in proceeding to 130°C becomes more and more difficult to perceive, so that at 130°C the observer considers that the bubble has disappeared because it can no longer be seen. At this instant the observer initiates a mechanical or electrical command to the heating stage to cool the stage to a temperature at which the bubble had just been clearly visible (125°C). In this case the bubble is indeed once again visible at 125°C, which suggests that undercooling was not required to generate the vapor bubble, meaning that the bubble was still present but simply not visible to the observer at 130°C. The inclusion is then heated to a higher temperature (135°C) and once this temperature (135°C) is reached, the stage is then immediately cooled back to the temperature at which the observer could before clearly see the bubble (125°C). The observer now checks to see if the bubble is present again at 125°C. It is present suggesting that the inclusion had not been homogenized at 135°C, so the heating stage is now taken to 140°C and then cooled to 125°C. Note that it is not necessary for the observer to strain to see the bubble during heating and cooling (125 to 140 to 125°C) because the bubble is going to appear to be gone at 130°C anyhow. So the observer now checks once again to see if a bubble is present at 125°C. But this time the observer realizes that the once clearly discernible bubble that had been seen before at 125°C does not appear to be present, so the observer now suspects that indeed the inclusion had been homogenized between 135° and 140°C. To check this hypothesis, the stage is rapidly cooled, and as it cools the observer continually directs all visual attention to the inclusion while attempting to keep the image of the inclusion in focus with the fine focus knob of the microscope. (The focus of the microscope may change rapidly because the temperature is changing rapidly; keeping the inclusion in focus at this stage is the most difficult task in cycling, and will therefore require some practice to master.) Eventually the observer sees the bubble instantly renucleate ("pop in") as a large black dot. This is then conclusive evidence that the inclusion had indeed been homogenized between 135° and 140°C, because the inclusion had to be undercooled (to 80°C) to overcome metastability in order for the bubble to renucleate, whereas when the inclusion had only been heated to 135°C, the bubble had not homogenized, so the metastable phenomenon did not occur when the inclusion was subsequently cooled — the bubble simply grew larger so that it could be easily seen by the observer at 125°C. Thus, the cycling technique has constrained, or "bracketed", the Th as having occurred

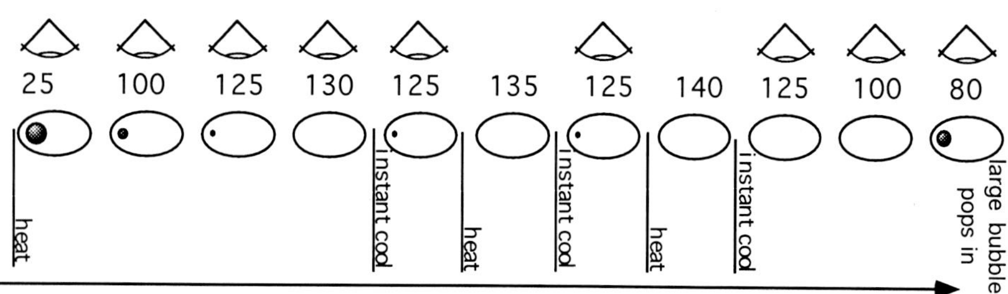

Fig. 7.4. Schematic representation of expected behavior of fluid inclusions during cycling to determine Th. Symbols above indicate when the inclusionist should be straining to see the bubble. Numbers are hypothetical temperatures in °C. Arrow indicates the progression of time during the cycling run. The inclusion had definitely homogenized by 140°C, because once that temperature had been reached, the inclusions had to be undercooled to 80°C for the bubble to renucleate. Thus, the run indicates Th took place between 135 and 140°C.

between 135° and 140°C, which for many studies is adequate resolution for such data. If higher resolution were desirable for this inclusion, the cycling technique would be applied as explained above, but smaller intervals (as small as 0.1°C are possible) would be employed.

Some highly saline inclusions require only a few degrees of undercooling for the bubble to reappear. This phenomenon may be due to the fact that more saline aqueous fluids have a more polar structure, which would mean that there is less surface tension at the contact between the fluid and the vacuole wall, permitting the fluid to separate more easily (less undercooling) from the wall (at which point surface tension forces cause the surface that separates to instantly take the form of a sphere — the bubble). For small inclusions that require <10°C of undercooling to renucleate a bubble, it is a real challenge to bracket Th by the cycling technique.

A word of caution is important: when measuring Th's on several different inclusions in the same fragment in the stage, care must be taken to avoid overheating an inclusion beyond its Th prior to the determination of the Th. For instance, of one is measuring a Th on an inclusion (say inclusion X) at 120°C, yet there is another inclusion (say inclusion Y) of interest in another field of view that was not being observed as the Th of inclusion X was determined, then inclusion Y may have been homogenized and overheated as inclusion X is being observed. If inclusion Y had homogenized at 60°C, for example, it would have developed hundreds of bars of internal pressure at 120°C (see Chapter 4), and could easily reequilibrate by stretching or decrepitation, invalidating subsequent observations. Furthermore, once a fluid inclusion has homogenized, particularly a low temperature one, it may take years before its vapor bubble reappears at room temperature because of metastability (Meunier, 1989). Thus, when performing heating runs, it is important to observe all fluid inclusions of interest as heating occurs, measure the lowest temperature Th inclusions first, and gradually work up to the higher Th inclusions. Once a Th measurement has been determined, it is commonly worthwhile (for a certain percentage of inclusions) to repeat the run to obtain a duplicate analysis to check for replicability, as gradual stretching of inclusions during measurement can occur and can be recognized by an increase in Th with each subsequent heating run.

Even though the above discussion is focused on determining Th of an individual inclusion, remember that it will be very difficult to interpret Th data unless it is collected in the context of an FIA. If an individual aqueous FIA shows very consistent Th data (90% of the inclusions fall in a 10-15°C interval), then the Th is a good measure for a minimum temperature of entrapment; if the aqueous FIA contains CH_4, then the Th is even closer to the entrapment conditions (see Chapters 3 and 9). Th's can also be collected from petroleum FIAs, and their Th's are also minimum entrapment temperatures, but potentially they are not nearly as close to the entrapment temperatures as Th's for aqueous inclusions because isochores for petroleum are much flatter, so the pressure corrections can be considerably larger (Chapter 3). To further interpret Th data, or in order to be able to apply pressure corrections, the composition of the fluid inclusion must be assessed.

MEASURING LOW-TEMPERATURE PHASE CHANGES

After inclusions have been studied for high-temperature phase behavior and all of the Th's have been recorded, the inclusions are ready to be studied at temperatures below 0°C. For aqueous inclusions, microthermometry at low temperatures provides useful information on the composition of the fluid inclusion, which may include major ions present as well as their concentration in the fluid. Not only are such compositional data useful for understanding the diagenetic environment, but it provides the basis for determining which phase diagrams to use in developing a further understanding of the Th behavior. For inclusions containing gases such as CH_4 or more complex mixtures of gases, these analyses may help to approximate gas compositions and may help define the appropriate phase equilibria, so that entrapment P-T conditions may be better interpreted (see Chapters 9 and 10).

Most sedimentary brines are so complex and fluid inclusions are so small that, in general, it is not possible to actually observe many of the phase changes that take place in aqueous inclusions upon freezing, nor can the solid phases that exist during freezing runs always be identified. One parameter that can be measured relatively accurately during freezing runs is the final melting temperature of ice (Tm ice). This measurement is used to interpret the bulk salinity of aqueous fluid inclusions, but only within certain limits because of the unknown complex composition of many aqueous fluids. In order to use Tm ice to determine bulk salinity, one must assume a model composition, as the true compositions of natural brines enclosed within inclusions are unknown. A model composition can be approximated by other low-temperature phase behavior in fluid inclusions such as the temperature of first melting (eutectic temperature, Te). Some plausible model compositional systems are discussed next, and then procedures for measuring and interpreting the low-temperature phase changes will be treated for various model composition systems.

Unquestionably, aqueous sedimentary fluids are very complex, consisting of many dissolved components in various ratios and in highly variable concentrations from 0 to greater than 30 wt.% total dissolved salts, not to mention all of the possible dissolved gases. Nevertheless, there do seem to be some general "classes" of such fluids that may be helpful in simplifying interpretations of fluid inclusion composition. Fresh meteoric ground water is the least complex, containing various components, but none in high enough concentration to have a significant effect on low-temperature phase behavior of fluid inclusions. Therefore, the most reasonable model for such fluids is the end-member, pure H_2O. However, when meteoric fluids evolve in arid, enclosed basins, their composition is anything but simple, and a huge variety of possible brine compositions could result. These continental brines will not be treated in this text. Seawater is a very complex fluid that may be concentrated, may be modified in composition as evaporite minerals precipitate, or may simply be diluted by fresh water. Seawater is dominated by the cation Na^+ with significant but decreasing amounts of Mg^{++}, Ca^{++}, and K^+, and dominated by the anion Cl^- with significant but less $SO_4^=$. But because seawater and seawater-like fluids are common in diagenetic systems, seawater may prove to be a useful model composition to assume for some systems. Subsurface or formation fluids can also be complex, but can be subdivided to allow for application of various model compositions. Subsurface brines that originate from dissolution of halite may be dominated by Na^+ and Cl^- (Carpenter, 1978) and the H_2O-NaCl model may apply to them. Most formation waters, however, are Na-Ca-Cl dominated (Dickey, 1969; Kramer, 1969; Collins, 1975) and have comparatively low amounts of Mg^{++}, K^+, and $SO_4^=$ relative to seawater. Although there is tremendous variability in formation fluids, and one cannot rule out significant (or even dominant) Mg^{++}, K^+, or $SO_4^=$, a useful simplifying model for many formation waters is the H_2O-NaCl-$CaCl_2$ system. Finally, gas components may be dissolved in any of these fluids (CH_4 probably being of dominant importance in most deeper sedimentary systems, and CO_2 being common or dominant in some), so a model that includes methane should also be considered, when appropriate.

In summary, there are two main reasons for choosing a model composition that most closely represents the inclusion fluid. First, the closer the model composition is to the inclusion composition, the closer will be the salinity interpreted from Tm ice. Second, for the interpretation of P-T conditions of entrapment (see Chapters 9 and 10), it will be important that the position of the liquid-vapor phase boundary and slopes of the isochores are constructed from a system that approximates that of the inclusion fluid. As most sedimentary fluids are more complex than the model compositions, any conclusions derived with these models are limited; one must always remember that any interpretations will only be approximations, so one should always strive to understand the possible margins of errors each time a model composition is employed. In many cases, the selection of an appropriate model composition will prove to be perfectly adequate.

Determining the Appropriate Chemical System

Selecting a model composition may be accomplished through *a priori* geological information about the likely composition of the diagenetic fluids. Commonly, such information will not be available, so attempts at determining temperatures of first melting (eutectic temperature, Te) of frozen inclusions and attempts at detecting presence of gases will be helpful.

Preparing for freezing.—
The most important preparation for studying the low-temperature phase behavior of fluid inclusions consists of the detailed petrographic study described in Chapter 6. The homogenization temperature measurements described in the foregoing section also are to be completed before freezing runs are initiated. One additional preparation may be required: purposefully stretching all-liquid aqueous fluid inclusions so that they generate vapor bubbles. The reason for generating a vapor bubble in all-liquid fluid inclusions is that interpretation of the phase equilibria for low-temperature phase behavior requires the presence of a vapor phase (Chapter 3). Thus, bubbles must be generated artificially in originally all-liquid fluid inclusions in order to apply the phase diagrams. For some all-liquid aqueous inclusions, a vapor bubble may nucleate after it has been heated during a heating run in the stage, because it stretched during the run. Typically, however, this seldom happens. More prolonged heating tends to be more effective for stretching inclusions and generating vapor bubbles. The best approach appears to be insertion of the fragment to be studied into a furnace at room temperature. Then allow the furnace to heat the sample overnight. For calcite, setting the temperature between 100 and 130°C seems to work well; dolomite commonly requires higher temperatures from 120-175°C; quartz requires even higher temperatures from 200-250°C. Such treatment commonly causes about 10% of the originally all-liquid inclusions to stretch and to generate vapor bubbles when cooled. Others may decrepitate, and still others may remain unaltered. Inclusion composition does not appear to be altered by diffusion during heating overnight. We have compared sample suites treated by short-term stretching on the stage to those heated overnight and found no differences in low-temperature phase behavior.

Freezing the inclusion.—
In the process of determining Te, one must first freeze the fluid inclusion. For the inclusion to freeze, the temperature of the stage must be lowered significantly below the temperature at which ice will melt in order to overcome the kinetics of ice nucleation. In the chemistry literature this temperature required to undercool the fluid before it will freeze is called the

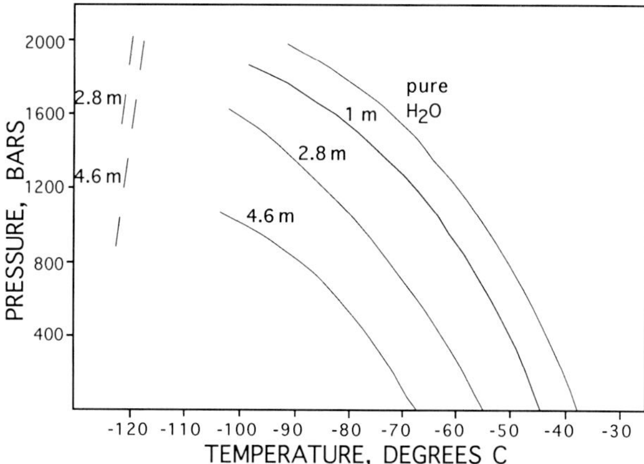

Fig. 7.5. *Homogeneous nucleation temperatures plotted as a function of pressure and composition (molality NaCl). The solid lines indicate the temperatures and pressures at which ice will nucleate upon cooling of solutions of various concentration. Dashed lines represent glass transition temperatures of aqueous solutions. Modified from Kanno and Angell (1977).*

homogeneous nucleation temperature and can be seen in Figure 7.5 to be related to the salinity and pressure: the higher the salinity and pressure, the greater the amount of undercooling required for the inclusion to nucleate a solid (heterogeneous nucleation at a higher temperature is also possible). In fluid inclusions, even more variables may control the nucleation temperature. The microscopic features that indicate when an inclusion has frozen can range from being obvious to invisible (Fig. 7.6).

On cooling an aqueous inclusion the bubble (nearly a vacuum if no gas is present, see Chapter 3) will become slightly larger as the liquid H_2O contracts, and may change position as it changes size or because of minute thermal gradients within the inclusion. Obvious indications of when freezing has occurred are:

1. Large inclusions (>10 µm) may show a sudden darkening due to refraction of light as it passes through many separate crystals in the inclusion (Fig. 7.6B, inclusion 1).

2. A sudden disappearance of the vapor bubble may occur because ice is of lower density than liquid water, so the ice may occupy all of the space of the bubble vacuum. Whether or not this happens will depend on both the liquid-to-vapor volumetric proportions in the inclusion and on the relative bulk salinity; higher L:V ratios and lower salinity both enhance the possibility of the bubble disappearing upon freezing. Higher L:V ratios correlate with less space for the ice to occupy, and lower salinity inclusions form more ice relative to other more dense phases (occupying less space) such as hydrohalite (see Chapter 3).

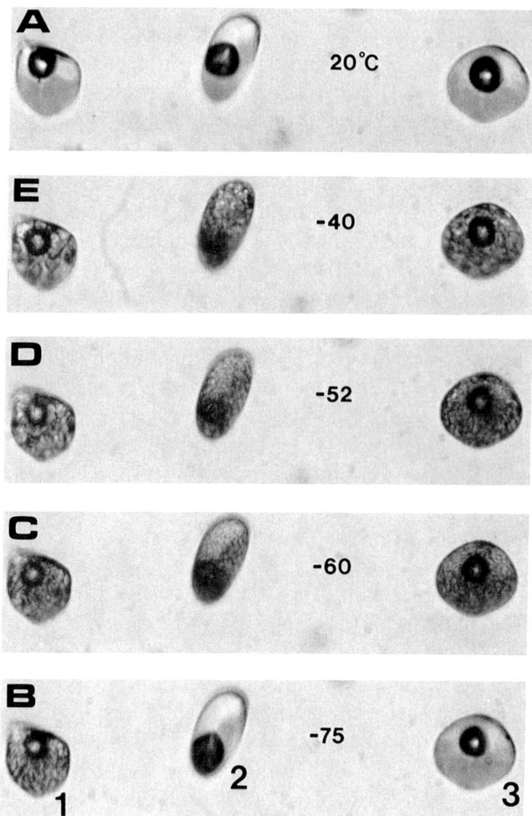

Fig. 7.6. *Photomicrographs of three synthetic H_2O-NaCl-$CaCl_2$ inclusions, in the same healed microfracture in quartz, that show different behavior on freezing and warming. Inclusion 1 froze to a distinct mosaic of crystals at about -65°C. Little change can be detected with warming to the stable Te at -52°C; however, at -40°C crystals are much larger and well defined, so melting has most definitely occurred by -40°C. Inclusion 2 froze to a clear solid at about -75°C, and this event could easily be detected by sudden bubble movement and slight deformation ("the jerk"). At -60°C a distinct mosaic of crystals is evident, but there seems to be a haze or glaziness to the image of the crystals that disappears at the stable eutectic of -52°C. At this temperature, some of the crystals become bright, in appearance and by -40°C the crystals show well-defined shapes and boundaries, indicating that melting has definitely occurred. (See also inclusion 6 in Fig. 7.17.) Inclusion 3 froze to a clear solid that had a brownish-colored tint. Upon warming it immediately darkens and the brownish tint becomes more intense, and at -65°C a mosaic of crystals is present, but again, the image of the crystals in the inclusion has a glaze or haze-like appearance. At the stable Te of -52°C, the haze disappears, the outlines of individual crystals are darker, and some of the crystals appear bright. By -40°C, crystal shapes and boundaries are more obvious. For these large inclusions (about 20-25 µm in longest dimension) most inclusionists would probably be able to say that definite melting had occurred by -40°C, meaning that Te was at a lower temperature.*

3. Rather than totally disappearing, the bubble may appear to change volume (it shrinks, but this is not always obvious) slightly (Fig. 7.6B, inclusion 3), deform slightly (Fig. 7.6B, inclusion 2), or change position (Fig. 7.6B, inclusion 2) on freezing. This may be very difficult to discern by simply comparing volume, shape, or position of the bubble before and after freezing, but it is relatively easy to observe at the instant it occurs. So it is very important to keep a watchful eye on an inclusion as one is cooling: freezing will be denoted by an instantaneous volume decrease, deformation, or jump in bubble position referred to as the *jerk* in fluid inclusion parlance. As above, the more saline the inclusion, the more difficult it will be to see the jerk because more saline inclusions will form a greater proportion of dense phases such as hydrohalite relative to ice, so that there is a smaller volume change upon freezing liquid to solid in a saline fluid inclusion compared with a dilute fluid inclusion. Also, the smaller the inclusion, the more difficult it will be to see the jerk.

4. Another possibility is that a clear solid may form on freezing (Fig. 7.6B, inclusion 2). When this occurs, one may see deformation of the bubble as described above, but this is not always observable. Sometimes an inclusion that freezes to a clear solid has a brownish tint or glazed appearance (Fig. 7.6B, inclusion 3), or may develop this appearance after slight warming. Some researchers (Roedder, 1984) call these clear solids a glass. For some of these, one normally does not have a clue that freezing has even occurred until warming the inclusion, when at some point a fine mosaic of crystals is noticeable. When this happens the inclusion typically darkens markedly, indicating that the inclusion must have been frozen (Fig. 7.6C, inclusions 2 and 3). This phenomenon is probably either an indication of a metastable reaction involving a liquid (see discussion below) or solid state recrystallization.

5. For all-liquid inclusions at room temperature, recognizing freezing can be impossible. Normally, one would undertake the necessary steps to form artificially a vapor bubble in these inclusions, as a vapor bubble is necessary to obtain a valid eutectic temperature and a final melting temperature.

6. A vapor-dominated inclusion would produce very little ice on freezing, and because the vapor-phase circumference is so very large, there will rarely be a discernible effect on the bubble when the inclusion freezes. It is always worthwhile to cool these inclusions to liquid nitrogen temperatures, for in some, certain phase changes may be seen that would indicate that an inclusion bubble contains something other than H_2O vapor (discussed later in this chapter).

7. There will be many cases in which it is difficult or impossible to observe freezing, but other phase changes may occur on heating (described below) that would indicate that freezing had indeed occurred.

First melting, eutectic temperature.—

During a cooling and warming run with the heating/cooling stage, after an inclusionist has noted that an inclusion has frozen, the temperature of first melting is the next phase change to determine, as this parameter will dictate the appropriate model composition. Eutectic melting in a fluid inclusion is the temperature at which one solid phase melts completely (Chapter 3). In practice, that phase must be present in a quantity sufficient for the melting to be observed under the microscope. Such is not always the case. For instance, an H_2O-NaCl liquid of low salinity (<5 wt.% NaCl) will contain very little hydrohalite (see Fig. 3.4 and use the lever rule to determine relative proportions of phases), so an accurate indication of first melting at Te will be difficult (commonly impossible!) to visually detect, especially in the small inclusions typically present in diagenetic phases. Thus, the observation of Te requires either that a studied inclusion be of high enough salinity that a phase can be observed to melt at Te, or that the inclusion is very large (>10 µm). Rather than attempting to determine accurately a Te, another approach is to record the lowest temperature (but higher than Te) of liquid definitely visible (e. g., definite outlines of crystals or movement of crystals; Fig. 7.6E). This information will often be sufficient for the purpose of determining an appropriate model composition. For example, detection of definite presence of melt at temperatures lower than -40°C is possible in fluid inclusions as small as 3 µm in longest dimension for some highly saline Na-Ca-Cl brines. Again, however, definite presence of melt will not always be useful for determining appropriate model composition for very low salinity fluids, as the detection of definite melt may not become obvious until just prior to final melting.

The apparent eutectic melting temperatures observable in most fluid inclusions from the diagenetic realm are controlled by the dominant ions in solution; thus, fluids of different compositions will each have a different Te (Table 7.1). Not all dissolved components will be reflected in an observed Te because theoretical Te can be controlled by unobservable minor constituents. Consider, for example, fresh water in equilibrium with gypsum: although one normally thinks of gypsum as very soluble, its solubility is actually so low that the amount of dissolved Ca^{++} and $SO_4^{=}$ has no observable effect on the initial and final melting of ice (both at 0.0°C). Obviously, the same would be true for any aqueous inclusion in equilibrium with less soluble minerals such as quartz, feldspar, calcite, or dolomite.

There is a very important phenomenon that may make the interpretation of Te very difficult: many fluid inclusions freeze to metastable assemblages, and when these assemblages are heated the metastable phases may begin melting at a lower temperature than the initial melting temperature of the stable assemblage. Thus, for most brine compositions, there is a stable eutectic melting temperature and one or several metastable eutectic melting temperatures. Recent studies have evaluated many of the metastable and stable eutectic temperatures for aqueous solutions containing common salts (Davis and others, 1990). This work has produced useful comparisons between experimentally determined Te and metastable Te, and observed Te in actual fluid inclusions; this information is summarized in Table 7.1. Other stable and metastable eutectic temperatures for common components in aqueous brines are presented in Tables 7.2 and 7.3. Even though we present these data, it is important to remember that stable as well as metastable low-temperature phase equilibria of natural fluids are complex because fluid compositions are complex. For example, a fluid inclusion containing seawater produces a minimum of six significant solid phases upon freezing (Table 7.4). To determine the detailed composition of complex fluids such as this, one must first be able to identify the solid phases that have formed, and subsequently their temperatures of melting. Then those temperatures and phases can be entered into a low-temperature model of the phase equilibria (Spencer and others, 1990) to determine the detailed composition of the inclusion. Such analyses may be possible for some highly concentrated, large inclusions with good optics, but for most inclusions in diagenetic systems, this will not be possible, and the eutectic will have to be used to assume a model composition, or more sophisticated geochemical techniques will have to be applied to extract and analyze, or nondestructively analyze the inclusion contents (see Chapter 12). For much work, the assumption of a model composition from eutectic observations works well.

H_2O-NaCl model composition.—
The selection of the H_2O-NaCl system as a representative model for the interpretation of Tm ice data is predicated on the observation of Te at -21.2°C. For many low-salinity inclusions in sedimentary environments, this simple observation is difficult to make. Furthermore, in most natural systems other components as well as NaCl exist in the fluid, so the model of the H_2O-NaCl system is useful only as an approximation for determining the bulk salinity, and does not provide the true salinity, nor is it an accurate representation of the fluid's composition. If one were to freeze a fluid inclusion and observe eutectic melting at -21.2°C, then the H_2O-NaCl model should be applied. Phases melting at temperatures as low as -30°C strongly suggest that an inclusion is dominated by monovalent cations, and may also warrant the selection of the H_2O-NaCl system (Table 7.1), because NaCl fluids are more common than KCl fluids. Generally, interpretation of the eutectic temperature assumes presence of a bubble to constrain the pressure, because first melting phenomena are pressure dependent as are final melting phenomena (Fig. 7.7).

H_2O-NaCl-$CaCl_2$ model composition.—
If one were to freeze an inclusion and observe eutectic melting at -52±5°C, one should consider adopting an H_2O-NaCl-$CaCl_2$ model, a common composition in terms of main components in sedimentary brines. Possible metastable eutectic

***Table 1**. Summary of predicted stable eutectic, predicted metastable eutectic, and observed eutectic temperatures (Te) from various aqueous systems**

SYSTEM	First Melting - T_e	T °C ----------> lower
NaCl-H_2O	stable	-21.2
	metastable	-28
	observed	(-21.1 to -21.2) (-28 to -35)
NaCl-$CaCl_2$-H_2O	stable	-52
	metastable	-70
	observed	(-47 to -53) (-70 to -85) -90
NaCl-$MgCl_2$-H_2O	stable	-35
	metastable	(-37 to -55) -80
	observed	(-33 to -40) (-45 to -50) (-70 to -80)
NaCl-KCl-H_2O	stable	-22.9
	metastable	-28
	observed	(-23.0 to -23.4)
NaCl-$CaCl_2$-$MgCl_2$-H_2O	stable	-57
	metastable	
	observed	

**data from Davis and others (1990)*

Table 7.2. *Phase data for aqueous solutions of common chloride species**

Dissolved species	Eutectic temperature (°C)	Eutectic composition (wt. %)	Solid phases	Mineral name	Optical character	Solid melting relations
			H_2O	ice	hexagonal colorless RI ε 1.313 ω 1.309	0°, congruent
NaCl	-21.2	23.3% NaCl	$NaCl\cdot 2H_2O$	hydrohalite	monoclinic colorless RI 1.416	+0.1°, incongruent
			NaCl ice	halite ice	cubic colorless RI 1.544	
KCl	-10.7	19.6% KCl	KCl ice	sylvite ice	cubic colorless yellowish RI 1.490	
$CaCl_2$	-49.8	30.2% $CaCl_2$	$CaCl_2\cdot 6H_2O$ ice	antarcticite ice	hexagonal colorless RI ε 1.393 ω 1.417	+30.08°, incongruent
$MgCl_2$	-33.6	21.0% $MgCl_2$	$MgCl_2\cdot 12H_2O$			-16.4°, incongruent
NaCl-KCl	-22.9	20.17% NaCl 5.81% KCl				
NaCl-$CaCl_2$	-52.0	1.8% NaCl 29.4% $CaCl_2$				
NaCl-$MgCl_2$	-35	1.56% NaCl 22.75% $MgCl_2$				

**modified from Crawford (1981)*

Table 7.3. *Phase data for aqueous solutions of non-chloride species**

Dissolved species	Eutectic temperature (°C)	Eutectic compositions (wt. %)	Solid phases	Mineral name	Optical character	Solid melting relations
NaBr	-28	40.5% NaBr				
KBr	-11	31.5% KBr				
Na_2CO_3	-2.1	5.75% Na_2CO_3	$Na_2CO_3\cdot 10H_2O$	natron	monoclinic white RI α 1.405 γ 1.440	+32°, incongruent
$NaHCO_3$	-2.3	5.8% $NaHCO_3$	$NaHCO_3$	nahcolite	monoclinic white RI α 1.377 γ 1.583	
Na_2SO_4	-1.2	3.85% Na_2SO_4	$Na_2SO_4\cdot 10H_2O$	mirabilite	monoclinic colorless RI α 1.394 γ 1.398	

**modified from Crawford (1981)*

melting events at lower temperatures are given in Table 7.1. If Te at -52°C is not observed, but definite melting occurs at any temperatures below about -40°C (Fig. 7.6), then an inclusion is likely to contain divalent cations (Ca^{++} and/or Mg^{++}). Since the H_2O-NaCl-$CaCl_2$ composition is most likely for sedimentary brines, this model is probably appropriate to assume. Eutectics between -30 and -40°C are difficult to interpret.

Table 7.4. *Temperatures at which various solid phases appear upon chilling of seawater: experimental data compared to model calculation* *

SOLID	EXPERIMENT	MODEL
ice	-1.921 [a]	-1.924
mirabilite	-8.2 [b]	-5.90
hydrohalite	-22.9 [b]	-22.84
sylvite	-36. [b]	-34.25
$MgCl_2 \cdot 12H_2O$	-36. [b]	-36.82
antarcticite	-54. [b]	-53.64 [c]

[*] after Spencer and others (1990)
[a] data from Fujino and others (1974)
[b] data from Nelson and Thompson (1954)
[c] calculated in sulfate-free system

The H_2O-NaCl-$CaCl_2$ model for determining salinity of the fluid inclusion can be used to its fullest extent if a temperature of intermediate melting of a phase can be determined with confidence, in addition to the final melting temperature and eutectic melting temperature. To be interpretable, however, the intermediate phase that is melting must be positively identified, which is extremely difficult without equipment more sophisticated than a heating/cooling stage (e.g., Raman spectroscopy). Furthermore, metastable assemblages commonly form on freezing which complicate matters. In our combined careers, we have yet to identify the exact intermediate melting event with 100% confidence, even in large (>20 μm) synthetic inclusions of known composition! Therefore, use of intermediate melting phenomena should be approached with skepticism.

Seawater model composition.—

In the cases in which one knows from the geologic circumstances that the fluids in the inclusions are derived from seawater, then a seawater model would be appropriate. Almost any near-surface diagenetic phase that forms in association with rocks of marine origin has the potential to have been precipitated from a fluid containing salts of marine origin. This could include true marine cements precipitating directly from seawater, minerals precipitating from evaporated seawater, minerals precipitating from mixtures between seawater and fresh water, and minerals precipitating from fluids containing sea spray or marine aerosol salts. Application of the seawater model to fluid inclusion data can be justified by identifying marine cements from petrographic and geologic criteria. For example, a marine cement could be identified using cross-cutting relationships. If a cement in a marine sediment has been truncated by erosion (hardgrounds, marine truncation surfaces, intraclast margins) and is overlain by more marine deposits, the cement is likely derived from marine fluids. In some carbonate boundstones, pore reducing cements are truncated by fissures and neptunian dikes that cut across the cements. These dikes and fissures are filled with marine sediment that dates the cements as having formed in the marine environment. If pores in a marine sediment are reduced by a cement, and marine internal sediment fills the pores and lies on top of the cement terminations, the cement is likely marine in origin. Or, if a cement zone or termination in pores in a marine sediment has been encrusted or bored by a marine organism, then the cement likely formed in a marine environment. If a cement can be shown petrographically to have a marine origin and if the cement has primary fluid inclusions, then it is probable that the inclusions contain a seawater-derived fluid, as long as the mineral did not recrystallize at any time in its history. In such cases, it would be appropriate to select the seawater model for interpretation of Tm ice data in the fluid inclusions.

An "evolved" seawater model (not covered in this text) would also be appropriate if evaporites with primary fluid inclusions are found in close association

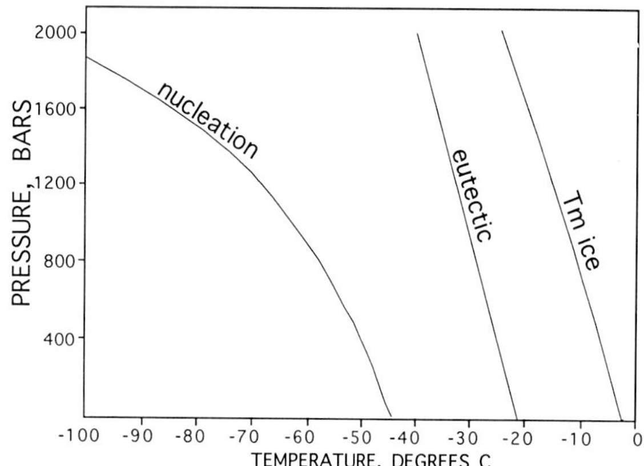

Fig. 7.7. *Illustration of the effect of pressure on the final melting temperature of ice (Tm ice), the eutectic, and the homogeneous nucleation temperature for 1 m NaCl solution. Modified from Kanno and Angell (1977).*

with marine sediments or had geochemical signatures indicative of marine origin.

In the examples discussed above, there is solid petrographic evidence for selection of the seawater model, but such definitive evidence may not always be present. Often there may only be stratigraphic or sedimentological evidence that a marine environment existed nearby at the time of cement formation and other circumstantial geologic evidence that the cement formed before significant burial. Such a situation might occur if a fluid inclusion in a mineral were trapped in a freshwater/seawater mixing zone. Here, an assumption of a seawater derived fluid might be critical in correctly modeling the diagenetic system. The difference in salinity determination between choosing the H_2O-NaCl model versus the seawater model (Tm ice of seawater is -1.9°C) is about 2.7 ppt, which is about an 8% error. This kind of error can be geologically significant. For instance, if one were to measure the final melting of ice at -1.9°C, and were to adopt the H_2O-NaCl model system, the salinity would be interpreted as 32.3 ppt. This would be interpreted as a salinity more dilute than seawater (about 8% more dilute), and might be incorrectly interpreted to represent a mixing zone or restricted marine waters. If the seawater model were adopted, one would simply interpret the salinity as being 35 ppt, the salinity of normal seawater. These potential errors may seem small to some researchers, but may have enormous implications for deciphering geologic processes in specific environments. So, if definitive geologic information constrains the system to the seawater model, then tenths of a degree centigrade differences in the temperature of melting of ice can have geological significance that can be interpreted, and how it is interpreted can depend on the selected model.

The freezing point depression of seawater at various concentrations has been calibrated from 0-60 ppt and Tm ice between 0 and -3.1°C and is presented in Figure 7.8 and in the following equation:

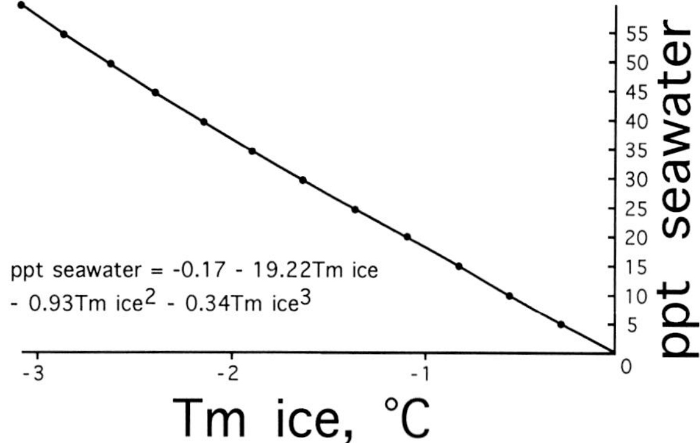

Fig. 7.8. Seawater freezing point depression constructed from data of Lyman and Fleming (1940). This model should only be applied for inclusions that range from brackish to less than twice the salinity of seawater.

$$\text{ppt seawater} = -0.17 + 19.22\,\theta - 0.93\,\theta^2 + 0.34\,\theta^3$$

where θ is the depression of the freezing point (magnitude Tm ice is depressed below 0.0°C) in °C. This model should only be applied for inclusions that range from brackish to less than twice the salinity of seawater.

H_2O-NaCl-CH_4 model composition.—

It is well known that brines in the subsurface contain significant dissolved CH_4, and therefore, some aqueous fluid inclusions should contain CH_4 mainly partitioned into the bubble. The H_2O-NaCl-CH_4 model system may be adopted if there is evidence for significant CH_4 in the fluid inclusions. The presence of CH_4 can be identified by crushing runs that produce expanding bubbles of organic gas (Chapters 3 and 6) or through cooling runs. A solid that melts at -182.5°C is most probably CH_4 because this is its triple point temperature. Sometimes cooling runs produce solid methane hydrate (clathrate) which is also an indicator of CH_4. Application of this model system is of utmost importance when interpreting Tt (trapping temperature) and Pt (trapping pressure), as the liquid-vapor phase boundary for an aqueous fluid containing CH_4 can be at much higher pressures than for a fluid without CH_4 (see Chapter 3). However, it is not important to apply this model system to determine salinity from Tm ice because the effects of CH_4 on Tm ice are relatively minor. Hanor (1980) evaluated the effect of methane on the Tm ice in aqueous fluid inclusions. There are two important effects to be considered: the inclusion's internal pressure at Tm ice and the partitioning of components into clathrate that may be present at Tm ice. For the first, methane-rich aqueous fluid inclusions are at high internal pressures at room temperature (Fig. 3.7), but as a methane-rich fluid inclusion is cooled, clathrate forms and incorporates a significant amount of the CH_4 from the bubble and some of the H_2O from the solution to make a solid. This causes the inclusion internal pressure to drop somewhat, but internal pressure should still remain significantly higher than that of a simple aqueous inclusion at this temperature. So as melting of ice takes place at a significantly high pressure (because of the CH_4), Tm ice will be depressed (Fig. 7.7), but the magnitude of this depression should be considerably less than 1°C (Hanor, 1980). Collins (1979) reported that salinity of an inclusion brine is increased by sequestering water in clathrate, but using Collins' relationship (his Fig. 5) and substituting common CH_4 contents, this effect is likely to be of secondary importance (Hanor, 1980). For an inclusion with a very high CH_4 content (about 1 mole % CH_4 for natural systems) and a salinity of 20 wt.% NaCl, the effect of forming clathrate would be to increase the salinity of the brine by about 1 wt.%. For an inclusion of seawater salinity with such a high content of CH_4, the effect would be to increase the salinity only 0.2 wt.%. Thus, even a great amount of CH_4 in aqueous inclusions has only a minor effect on Tm ice. Nevertheless, this minor effect could be significant if one were attempting to employ the seawater model composition. In such a case, the salinity determination could be in error by up to about 15 ppt — perhaps unacceptably large for some studies. So, for aqueous inclusions containing detectable CH_4, it will probably be inappropriate to apply the pure H_2O model composition or the seawater model composition for interpreting salinities from Tm ice.

Punting.—
The Tm ice for inclusions as small as 2 µm can often be obtained by use of the cycling technique as described below. However, it is rare to be able to detect any clue that would indicate the presence of gases, or to be able to identify any sign of first melting for the vast majority of inclusions typically found in diagenetic minerals, as they are generally <5 µm in longest dimension. So, for the cases where it is difficult to identify eutectic melting with a high degree of confidence, or even if there simply is not enough information to constrain the composition of more complex systems (either by use of other analytical techniques or from the difficulty of the inclusion observations), an H_2O-NaCl model composition is commonly used to evaluate Tm ice; the salinity is reported in terms of *NaCl salinity equivalent* reflecting a concentration of salts dissolved in the solution if it were pure H_2O + NaCl. As it turns out, this is a very reasonable model because most formation fluids are NaCl dominant or NaCl + $CaCl_2$ dominant, and comparisons of the cotectic surfaces where ice melts for various systems shows relatively small variations (Fig. 7.9). The NaCl, NaCl + $CaCl_2$, and $CaCl_2$ curves essentially lie on top of one another, with a maximum deviation of about 2 wt.%. The system containing $MgCl_2$ and KCl have maximum deviations from the H_2O-NaCl curve around 6 and 4 wt.% respectively, but these decrease at lower salinities. Furthermore, as explained above, the possible effects of undetected CH_4 are relatively minor as well. Thus, interpreting Tm ice from small fluid inclusions in terms of wt.% NaCl equivalent when other compositional information is lacking will, in many cases, be perfectly acceptable. Nevertheless, if the composition of major ions can be well constrained through chemical analyses or low-temperature phase behavior in the fluid inclusion, much more accurate determinations of bulk salinity are possible.

Effect of additional undetected components.—
The selection of any model composition system is only an approximation. The effects of other components that are most certainly present in any natural fluid, but not detectable, are assessed by the same arguments presented in the section above: the effects will be minimal for any models except the seawater model. If one selects the seawater model, the interpretation of Tm ice may be significantly affected by the presence of CH_4 or by significant deviations of the ionic species' proportions in the inclusion fluid.

Low-Temperature Phase Behavior for the H_2O-NaCl Model System

The theoretical description of low-temperature phase behavior from Chapter 3 (Fig. 3.4) can be expanded with practical observations and tricks for obtaining measurements. The H_2O-NaCl system phase diagram is reproduced in Figure 7.10 and practical procedures are presented below.

Recognizing eutectic melting.—
The initial low-temperature observation that indicates the applicability of the H_2O-NaCl model for aqueous fluid inclusions is the first melting temperature (eutectic temperature, Te) observed with a bubble present. At Te, a liquid forms from the breakdown of hydrohalite and the melting of ice (Fig. 7.10). The inclusion will remain at Te until one of the phases is completely gone. First melting is often difficult to discern and is thus a source of much consternation. The level of difficulty increases dramatically with smaller sizes of inclusions (<10 µm) and with lower salinities (<5 wt.%). The best way to learn how to detect first

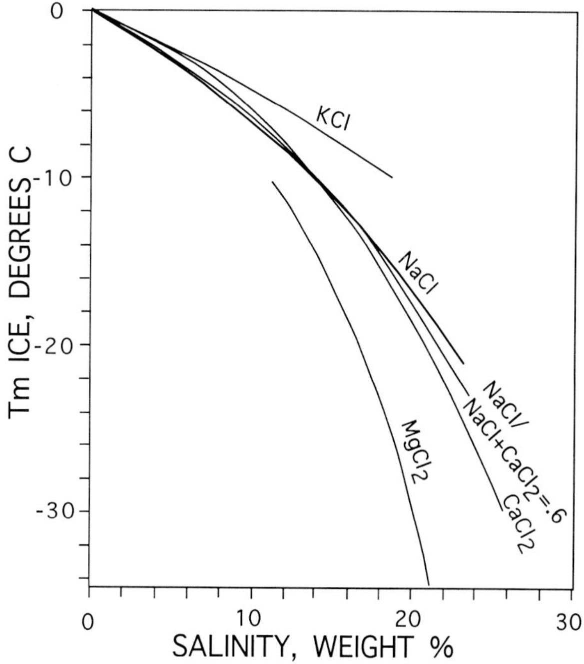

Fig. 7.9. *Final melting temperature of ice (Tm ice) versus wt.% of various salts in solution. Modified from Crawford (1981) with data from Oakes and others (1990).*

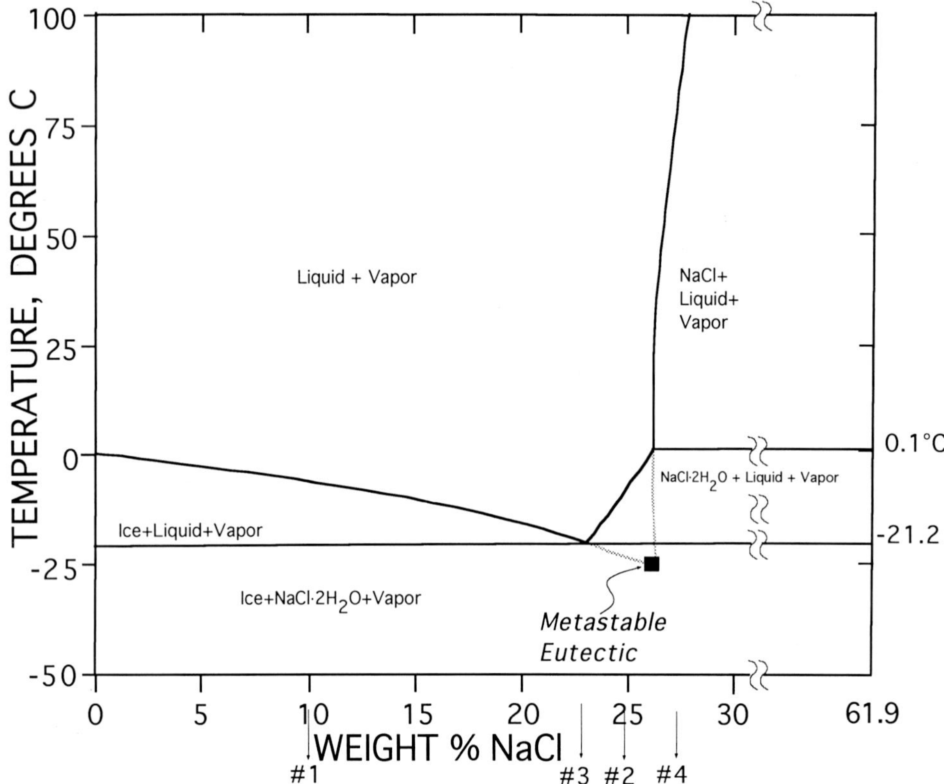

Fig. 7.10. T-X plot for the lower temperature, lower salinity portions of the system H_2O-NaCl. Each point on this diagram is at equilibrium vapor pressure. Produced from Crawford (1981), Roedder (1984), and Hall and others (1988). See text for complete discussion of phase behavior of 4 different fluid inclusion compositions: #1 yields final melting of ice; #2 yields final melting of hydrohalite; #3 yields final melting of ice and hydrohalite at the eutectic; and #4 yields final dissolution of halite.

melting is to practice on the NaCl eutectic composition standard made by SYN FLINC (commercially available from FLUID INC., Denver, Colorado, USA) which shows complete melting of all solid phases at -21.2°C (Fig. 7.12). Figure 7.11 illustrates a cool/thaw run for a synthetic aqueous inclusion of 10 wt.% NaCl composition. Figure 7.12 is a similar series of photomicrographs for a synthetic aqueous inclusion of the eutectic composition. Both inclusions freeze to a clear solid. With warming, at about -25°C a mosaic of crystals (from recrystallization?) is faintly evident (Fig. 7.11 and Fig. 7.12). Within about 1°C below Te, many of the crystals within the inclusions show a "brightness" which is a diagnostic characteristic of Te occurring or at least being very close, and is commonly referred to as the *orange peel texture* (Fig. 7.11D and E; Fig 7.12E and F). It is probable that these bright crystals are hydrohalite. The instant at which one of the phases (ice or hydrohalite) completely disappears can be rather dramatic (if the inclusion is large enough and if there is enough of the disappearing phase present), and can be best described as a "marked clearing" of an inclusion. (For the case where hydrohalite breaks down completely at Te, compare Figure 7.11E with Figure 7.11F.) But for many of the small inclusions common in the sedimentary realm, neither the orange peel texture nor a marked clearing will be possible to observe. About the best one can do in these cases is to note the temperature at which one is sure there is only a single solid phase left in the inclusion: at least one would have confidence that Te would be lower than this temperature. Not even this observation will be possible for many cases, as the inclusions will be so small (<5 µm) and the compositions too dilute (<5 wt.% NaCl).

Salinities less than the eutectic composition.—
For fluid inclusion composition 1 (Fig. 7.10), and for other H_2O-NaCl inclusions with salinity less than the eutectic composition (23.2 wt.% NaCl), the inclusion will remain at the eutectic temperature until all of the hydrohalite has broken down, at which point the inclusion contains vapor, ice. and liquid of the eutectic composition (Fig. 7.11F).
Alternatively, first melting could occur at a metastable eutectic. If a disequilibrium assemblage of halite, ice, and vapor forms upon freezing, then as the inclusion is warmed, liquid H_2O solution will form in metastable equilibrium with halite, ice, and vapor at the metastable eutectic temperature (-28°C predicted, but possibly observed as low as -35°C; Davis and others, 1990). It

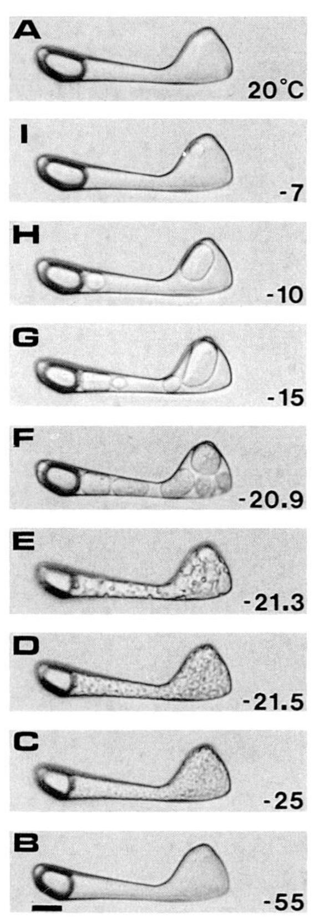

Fig. 7.11. Cool/thaw run for a synthetic H_2O-NaCl inclusion of 10 wt.% NaCl. ***A)*** *Room temperature.* ***B)*** *This inclusion freezes to a clear solid at about -55°C. Note that the bubble is smaller than in A.* ***C)*** *The first indication of any crystals (recrystallization ?) is at -25°C.* ***D)*** *At -21.5°C crystal outlines are obvious and some are brighter than others.* ***E)*** *Photographed at 0.1°C below the eutectic. At the eutectic, the brighter crystals (hydrohalite) instantly begin to disappear, and concomitantly the ice crystals change size (larger ones grow at the expense of smaller ones).* ***F)*** *By just 0.3°C above the eutectic there is an obvious "clearing" because all of the hydrohalite has broken down.* ***G)*** *Four ice crystals are left at -15°C.* ***H)*** *Two ice crystals are left at -10°C.* ***I)*** *One ice crystal is left at -7°C, and it melts completely at -6.6°C. Bar scale is 7 μm.*

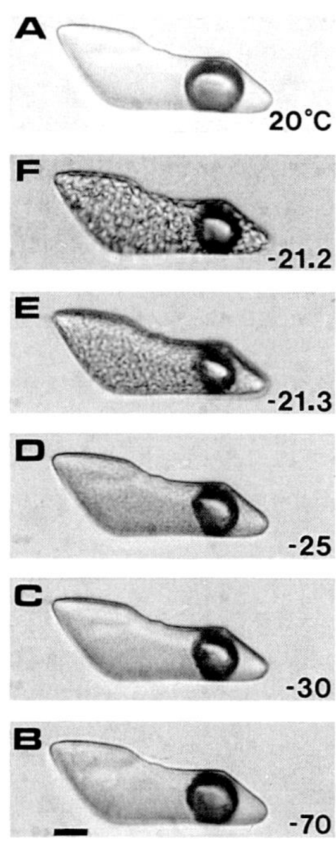

Fig. 7.12. Cool/thaw run for a synthetic H_2O-NaCl inclusion of the eutectic composition (23.2 wt.% NaCl). ***A)*** *Room temperature.* ***B)*** *This inclusion freezes to a clear solid at about -70°C. Note deformation of the bubble.* ***C)*** *Nothing has changed by warming to -30°C except that the bubble is a little more deformed.* ***D)*** *Very diffuse crystal boundaries beginning to show at -25°C.* ***E)*** *At -21.3°C crystal outlines easily noticed, and some of the crystals are bright.* ***F)*** *At the eutectic temperature, the "orange peel" texture is best displayed and followed by an immediate "clearing." In this case, everything clears! Bar scale is 7 μm.*

is possible that all halite will dissolve with addition of heat at the metastable eutectic, and that only ice will be left in the inclusion at -28°C. If this is the case, then ice will melt along the metastable cotectic between -28 and -21.2°C (Fig. 7.10). Alternatively, once liquid H_2O forms at the metastable eutectic, it will react with halite to form hydrohalite if warming proceeds slowly enough (Crawford, 1981; Davis and others, 1990). If all of the halite is consumed to form hydrohalite, then the inclusion will experience equilibrium melting at the stable eutectic temperature. The behavior of such metastable melting may not be repeatable 100% of the time. This lack of repeatability can be used as evidence of metastable behavior, and that the higher temperature melting event is in fact occurring at the stable Te rather than the metastable Te.

After the inclusion is warmed just above the eutectic, it contains ice, brine, and vapor (Fig. 7.10 and Fig. 7.11F). Ice crystals will generally appear to have low relief, are typically rounded, and lack well-developed crystal faces (Fig. 7.11F-I). In many small fluid inclusions, the ice may not be visible at all. As temperature rises, the fluid follows the cotectic surface and more ice melts, diluting the brine (Fig. 7.10). As the ice melts, the bubble may be seen to grow larger, occupying a greater volume as the saline liquid is denser than the pure ice. Also during warming, ice crystals readily recrystallize to one or a few larger crystals: larger crystals will grow at the expense of smaller ones that can be seen to shrink and disappear. This ease of crystal growth is a characteristic of ice that will enable one to distinguish it from hydrohalite. Ice crystals continue to melt as the temperature of the stage rises, continuing to dilute the brine, until finally the last ice crystal melts completely (Fig. 7.11I to A) leaving the brine of the original bulk composition.

As explained in Chapter 3, the temperature at which the last crystal of ice melts is referred to as Tm ice, also sometimes called the final melting temperature or the freezing point depression. In general, these all refer to the same value, but Tm ice is more specific as it refers to the temperature at which the last crystal of ice melts as opposed to other phases. For most inclusions in the sedimentary realm, one will not actually be able to see the last crystal of ice as it melts. In these cases, a cycling technique will have to be employed to determine Tm ice.

Cycling technique explained for freezing.— As ice melts on warming, bubbles grow in size because liquid H_2O occupies less volume than solid ice. On cooling, a growing crystal of ice will commonly push a bubble against the walls of an inclusion and deform it. Such behavior observed during cooling can be used as evidence that unobservable ice is present: if a nucleus of ice is present during cooling, then more ice will gradually overgrow the nucleus during gradual cooling and eventually deform the bubble. In the absence of a nucleus, the inclusion must be supercooled to extremely low temperatures until it eventually freezes, at which instant the bubble "jerks." This instant movement of the bubble is a completely different phenomenon than the slow shrinkage, movement or deformation of the bubble that is observable when ice is slowly growing from a nucleus, and it can be used as an indication that the ice nucleus was not present at the beginning of the cooling run. A more specific procedure of repeatedly causing a particular reaction of the bubble is typically used to indicate the presence or absence of ice. This procedure is called *cycling*, and is routinely employed to determine efficiently and effectively Tm ice in fluid inclusions found in authigenic minerals.

To determine Tm ice in inclusions in which final melting of the last ice crystal cannot be seen, the cycling technique is the only method to employ. After the inclusion has been frozen, it is often worthwhile to perform a preliminary, relatively quick, warming run to identify the approximate temperature at which the bubble radically changes in size, shape, or position in inclusion. This phenomenon may or may not be indicative of the final crystal of ice melting; however, it does provide an approximate temperature at which something may be happening in the fluid inclusion, and is a temperature near which cycling is initiated. For instance, if on a quick warming run the bubble flickers and/or gets larger and/or becomes easier to see at around -5 to 0°C, then the cycling procedure could begin at -5°C to test a hypothesis that Tm ice occurs at a higher temperature. After the initial quick warming run, be sure to warm the inclusion above 0.0°C to be sure all ice is melted prior to initiating a new cooling run.

With the quick warming run, the inclusionist has produced an hypothesis as to the approximate value of Tm ice. Refreeze the inclusion and begin the cycling run as shown schematically in Figure 7.13A and as described below:

1. The frozen inclusion is warmed rapidly to a temperature below the hypothesized Tm ice. Take note of shape, size, position of bubble, and temperature of stage, before warming to a higher temperature.

2. Warm slowly until the bubble looks in any way different than it did in step #1.

3. When the bubble shows the slightest discernible difference, immediately lower the temperature of the stage and again watch the bubble: if the shape becomes more deformed, if the bubble size becomes smaller, or if the bubble gradually moves (beware; this can take place without growth of ice, due to a thermal gradient or size change), then hypothesize that ice was still present at the highest temperature achieved in the run and begin to warm again.

4. Warm the inclusion to a higher temperature (0.1, 0.5 or 1°C depending on desired resolution) than that reached in step #2, and then reduce temperature again (step #3).

5a. If the observation of bubble change determined in step #3 is repeated, repeat step #4 to an even higher temperature. Continue repeating this procedure until step #5b behavior is observed.

5b. If during cooling, the bubble does not appear to change by shrinkage or deformation (or possibly migration), the ice probably melted within the temperature interval just preceding the highest temperature achieved in step #4.

6. One way to be sure that the ice is completely melted is to continue cooling the inclusion and to observe that the inclusion refreezes suddenly (bubble jerk) when the inclusion has been supercooled sufficiently to freeze in the absence of an ice nucleus. Once this has been observed, then the inclusionist is assured that the ice melted during step #4. If supercooling is not required for freezing and the jerk is absent, then ice must have still been present, and step #4 should be repeated to a higher temperature until step #5b and #6 are completed successfully.

The procedure above has determined the temperature interval in which ice must have totally melted (Tm ice). The whole process could be repeated but at smaller temperature intervals to determine Tm ice to a higher resolution. Figure 7.13A illustrates cycling for an example in which Tm ice is between -3.5 and -3.0°C.

In some inclusions, changes in the vapor bubble due to slow growth of ice cannot be observed. In these inclusions, the only phenomenon observable may be the rapid jerk of the vapor bubble as the supercooled fluid inclusion instantly freezes. A second cycling technique utilizes this phenomenon to determine Tm ice for these difficult but commonly encountered fluid inclusions. The procedure is summarized in Figure 7.13B and described below:

1. Cool the inclusion and notice the temperature at which the "jerk" is observed from rapid freezing of the inclusion (nucleation temperature). During subsequent runs, this will provide an important reference temperature.

2. Warm the inclusion rapidly and look for any rapid migration or change in shape or size of the vapor bubble. If one is seen, then this temperature should be noted as a hypothesis for an approximate position of Tm ice. If no change is observed during the rapid warming, hypothesize Tm ice at -10°C. Heat the inclusion above 0.0°C to be sure all ice has melted, then rapidly refreeze to the supercooled nucleation temperature determined in step #1. Watch the inclusion jerk again.

3. Warm the inclusion to just above the preselected temperature determined in step #2, and instantly start cooling it again to the nucleation temperature, taking note of the warmest temperature reached.

4a. If the bubble does *not* jerk at this temperature, then ice must have remained in the inclusion and Tm ice must be at a temperature above that attained in step #3. Repeat step #3 but warm to higher temperature (0.1 to 1°C depending on desired resolution) than the previous nucleation temperature: if the bubble does not jerk, repeat step #3 at a successively higher temperature until a jerk is observed.

4b. If the bubble jerks at the previously determined (in step 1) nucleation temperature, then the ice must have been totally melted by the warmest temperature achieved in step #3. If this is the first cycle (step #4a

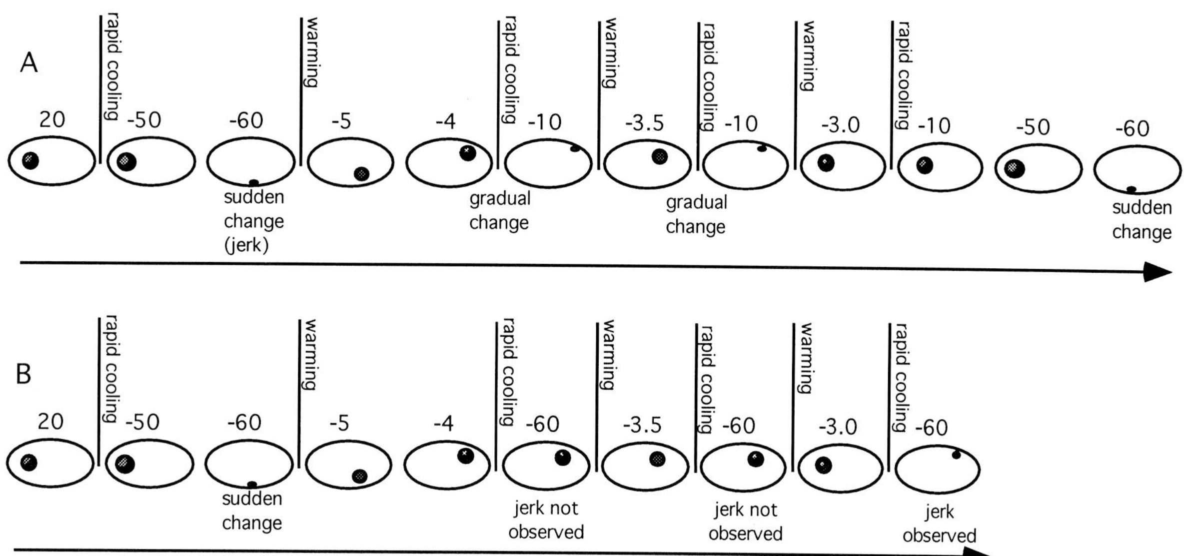

Fig. 7.13. *Schematic representation of the "cycling" technique used for determining Tm ice for small inclusions prevalent in diagenetic minerals. Arrow is direction of time. Prior to initiating cycling, a researcher should first do a "quick run" by freezing an inclusion (watch for the "jerk" in the bubble when the liquid freezes) and then warming it very rapidly (at a minimum rate of 3°C/second) until a rapid change to the bubble is noted. For the examples here, say that something happened at about -1°C, so the inclusionist begins the cycling technique at -5°C.* ***A)*** *After this "quick run," the inclusion is rapidly cooled and refrozen. Then the inclusion is rapidly (as fast as possible for equipment being used) warmed to -5°C without exceeding this temperature. The bubble characteristics at -5°C are noted. Then the inclusion is warmed more slowly. At -4°C, a slight "darkening" of the bubble may be noted. Perhaps this darkening is due to the clearing of the optics because ice in the inclusion is melting. Other common observations might be that the bubble gets ever so slightly larger, or that it moves. At -4°C the temperature of the stage is immediately lowered, and as the inclusion is cooled to -10°C one may note that the bubble has gradually become smaller and has gradually deformed. So, one reasons that ice may still have been present in the inclusion at -4°C and that it gradually grew as the temperature dropped. The growing ice deformed the bubble, or ice that had grown upon cooling impeded the image quality of the bubble. (If the ice had completely melted, then the inclusion would have had to be cooled to -60°C in order to renucleate ice.) Next the stage is warmed to a higher temperature than -4°C. A 0.5°C interval for cycling was chosen for this case. Once -3.5°C is attained during warming, the stage is rapidly cooled. Again, one notes that the bubble changes similarly to the previous cooling run. Each time the cycle is repeated to higher temperatures, more of the invisible ice crystals will melt, and if there are fewer ice crystals present when the temperature is lowered, then remaining crystals have a greater chance to push on and deform the bubble as they grow back, rather than pushing on one another. Thus, as cycling proceeds, it is very likely that one will see greater changes in the bubble during the cooling phase of the cycling run. In this example, upon cooling, after reaching -3.0°C, no change in bubble characteristics are observed as the stage is cooled down to -10°C. This suggests, that ice has melted completely between -3.5 and -3.0°C. This is verified by supercooling the inclusion: observation of the jerk at -60 is definitive proof that the ice had indeed melted completely by -3.0°C. Thus, Tm ice is between -3.5 and -3.0°C.* ***B)*** *Sometimes, perhaps because of the shape of an inclusion or the position of its bubble, it is not possible to discern any gradual changes in the bubble as ice melts or gradually regrows when using the technique explained for A. For this scenario, each cooling cycle is dropped to the temperature at which the "jerk" occurred upon initial freezing. If no jerk is observed, then ice must still have been present, and grew back to fill the inclusion at a temperature higher than -60°C. The next warming cycle is taken to a higher temperature. After achieving -3.5°C the stage is cooled to -60°C and no jerk is seen (ice must not have melted completely by -3.5°C). Then the stage is warmed to -3.0°C and again cooled to -60°C at which point one notices a jerk, which means ice must have been completely melted by -3.0°C. Since ice must have been present at -3.5°C (based on the result of the previous cycle), then the last ice crystal must have melted in the interval -3.5 to -3.0°C.*

has not been attempted), then the temperature has not been bracketed, so repeat step #3 at 5°C or lower than the predicted temperature of step #2. If this is not the first cycle (step #4a has been accomplished two or more times), then the temperature interval in which ice totally melted is that interval just below the highest temperature achieved in step #4a.

Once again, a cycling technique has been used to bracket a temperature interval in which the last crystal of ice must have melted, but ice was never actually observed! In both cycling techniques described, the necessity of supercooling for overcoming the kinetic barriers of crystal nucleation was used to the advantage of the inclusionist.

Freezing runs on inclusions lacking bubbles.— Upon cooling, temperatures of phase changes that occur in aqueous inclusions that do not contain vapor bubbles cannot be used for compositional interpretations, because the phase equilibria that are employed for interpreting the phase changes require the presence of a vapor phase (see Chapter 3). As explained in Chapter 6, bubbles may not be present due to low temperatures of formation (<50°C), metastability, or necking. But for low-salinity inclusions found in diagenetic phases, a bubble may be completely filled by expansion of H_2O on freezing to solid. Upon warming a frozen inclusion that has no bubble, the eutectic will be encountered. However, at Te the inclusion contains significant ice that may have internally pressurized the fluid inclusion, and high internal pressures may drive the eutectic to lower temperatures (Fig. 7.7), but because the internal pressure is an unknown, the magnitude of Te depression is unpredictable. Therefore, interpretation of the eutectic temperature requires the presence of a vapor bubble. As an inclusion without a vapor bubble is warmed above Te, ice melts, diluting the brine and causing a decrease in the pressure due to the volume decrease that occurs with melting of ice. At some point, enough ice has melted and pressure has dropped low enough for a vapor bubble to nucleate, but such does not always happen. Two paths are possible: either a bubble nucleates, or a bubble does not nucleate initially because of kinetic problems in bubble nucleation. In the first case, if a vapor bubble nucleates, the phase equilibria become interpretable, so if the vapor bubble returns below the temperature at which Tm ice is observed, then Tm ice can be interpreted to yield an estimate of the inclusion's bulk salinity. In the second case, if the bubble has not reappeared, yet ice continues to melt upon warming, as more and more ice melts, the pressure drops lower and lower, yet the bubble that should be occupying space has still not appeared, at which point the liquid is actually under tension (pressure is negative) because the liquid is filling the space that should be occupied by the bubble. The lower the pressure, the more Tm ice is

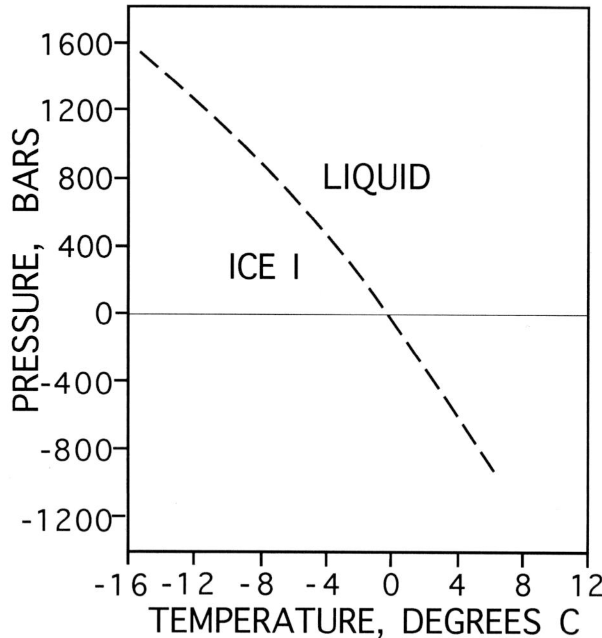

Fig. 7.14. *Ice-water equilibrium for the pure water system. Notice the strong pressure control on temperature of ice melting. Modified from Roedder, (1984).*

brought to higher temperatures, and under negative pressure ice will even become stable above 0.0°C (Fig. 7.14). The ice that exists above its stable melting temperature, with negative pressure, can be thought of as superheated ice, a phenomenon about which Roedder (1967) published observations and warned inclusionists. Superheating of 2-6°C is not uncommon. When the superheated ice finally melts, it does so by a sizable bubble renucleating with concomitant instant melting of any ice. The temperature at which this occurs cannot be used to interpret salinity. To obtain a valid salinity from Tm ice, a bubble must be present at a temperature below Tm ice.

A problematic fluid inclusion, lacking a vapor bubble during important phase transitions, may sometimes be coaxed to generate a vapor bubble for the phase change so that a meaningful Te or Tm ice determination can be made. For those inclusions that are all-liquid at room temperature, a bubble may nucleate during freezing, either because cooling the fluid inclusion causes thermal contraction of liquid that may encourage nucleation of a vapor bubble, or because once the fluid inclusion freezes, the volume expansion of ice formation is enough to cause the volume of the inclusion to increase through deformation of the crystal host. The compressibility of ice is so low, that originally all-liquid inclusions in quartz (a relatively strong mineral) have been known to increase in volume after the inclusion is frozen. Another way to provoke nucleation of a vapor bubble in fluid inclusions that are all-liquid at room temperature is to heat the inclusion in the laboratory to increase its volume through artificial thermal reequilibration (stretching). The methods useful for carrying out this procedure are covered in this chapter in the initial statements on "Measuring Low-Temperature Phase Behavior." The most difficult inclusions are those that are two-phase at room temperature and in which the bubble disappears upon freezing. There are several tricks that can be used to encourage the bubble to be present. One option is to warm very slowly after the inclusion has frozen: hold the temperature of the inclusion just below the approximate Tm ice for several minutes, and with luck, a bubble may appear so that the valid Tm ice can be measured during subsequent warming. Trying multiple freezing and warming runs is also worthwhile, as it may be that one out of ten freezing runs retains the bubble to allow for a valid measurement. If these tricks do not work, then heat the inclusion beyond its Th to cause the inclusion's volume to increase, as described above.

If leakage of a fluid inclusion occurs during any artificial procedure, then it is possible that the inclusion salinity will be altered. Therefore, if any of the above procedures are ever employed to cause bubbles to nucleate, it is most important to perform repeated freezing and warming runs and observe the liquid-to-vapor volumetric ratio: if it is changing during subsequent runs, this is a clue that an inclusion might be leaking, and thus its salinity could be changing and may not provide a valid representation of the inclusion's original composition. To test if salinity is changing, repeat freezing runs: if Tm ice is changing, then salinity must be changing during leaking, so do not record the invalid Tm ice data and proceed to another inclusion.

Interpreting Tm ice for H_2O-NaCl.— Once the system has been identified as H_2O-NaCl from the eutectic, and Tm ice data have been collected, the salinity of the fluid inclusions can be interpreted from the Tm ice data. The system H_2O-NaCl has been calibrated well and Tm ice data can be related to salinity with high reliability. The most commonly applied calibration has been that of Potter and others (1978). Using data of Hall and others (1988), Bodnar (1992a, b) produced a revised equation for the relationship between salinity and Tm ice that reproduces the Hall and others (1988) data to better than ±0.05 wt.% NaCl:

$$\text{Salinity(wt.\%)} = 0.00 + 1.78\,\theta - 0.0442\,\theta^2 + 0.000557\,\theta^3$$

where θ is the depression of the freezing point (magnitude Tm ice is depressed below 0.0°C) in degrees C. Solutions for this equation are reproduced in Table 7.5, after Bodnar (1992a).

Table 7.5. *Table of salinities (wt.% NaCl) that correspond to freezing point depressions for fluid inclusions in the presence of a vapor bubble**

FPD	.0	.1	.2	.3	.4	.5	.6	.7	.8	.9
0.	0.00	0.18	0.35	0.53	0.71	0.88	1.05	1.23	1.40	1.57
1.	1.74	1.91	2.07	2.24	2.41	2.57	2.74	2.90	3.06	3.23
2.	3.39	3.55	3.71	3.87	4.03	4.18	4.34	4.49	4.65	4.80
3.	4.96	5.11	5.26	5.41	5.56	5.71	5.86	6.01	6.16	6.30
4.	6.45	6.59	6.74	6.88	7.02	7.17	7.31	7.45	7.59	7.73
5.	7.86	8.00	8.14	8.28	8.41	8.55	8.68	8.81	8.95	9.08
6.	9.21	9.34	9.47	9.60	9.73	9.86	9.98	10.11	10.24	10.36
7.	10.49	10.61	10.73	10.86	10.98	11.10	11.22	11.34	11.46	11.58
8.	11.70	11.81	11.93	12.05	12.16	12.28	12.39	12.51	12.62	12.73
9.	12.85	12.96	13.07	13.18	13.29	13.40	13.51	13.62	13.72	13.83
10.	13.94	14.04	14.15	14.25	14.36	14.46	14.57	14.67	14.77	14.87
11.	14.97	15.07	15.17	15.27	15.37	15.47	15.57	15.67	15.76	15.86
12.	15.96	16.05	16.15	16.24	16.34	16.43	16.53	16.62	16.71	16.80
13.	16.89	16.99	17.08	17.17	17.26	17.34	17.43	17.52	17.61	17.70
14.	17.79	17.87	17.96	18.04	18.13	18.22	18.30	18.38	18.47	18.55
15.	18.63	18.72	18.80	18.88	18.96	19.05	19.13	19.21	19.29	19.37
16.	19.45	19.53	19.60	19.68	19.76	19.84	19.92	19.99	20.07	20.15
17.	20.22	20.30	20.37	20.45	20.52	20.60	20.67	20.75	20.82	20.89
18.	20.97	21.04	21.11	21.19	21.26	21.33	21.40	21.47	21.54	21.61
19.	21.68	21.75	21.82	21.89	21.96	22.03	22.10	22.17	22.24	22.31
20.	22.38	22.44	22.51	22.58	22.65	22.71	22.78	22.85	22.91	22.98
21.	23.05	23.11	23.18							

**FPD is absolute value of Tm ice in °C; after Bodnar (1992)*

Higher salinity compositions.—

With final breakdown of hydrohalite.— The low-temperature phase behavior of more saline inclusions that are in the salinity range above the eutectic composition, but below saturation with halite (23.2-26.3 wt.% NaCl; Fig. 7.10, inclusion #2), is similar to the behavior exhibited by the 10 wt.% NaCl inclusion #1 with one major exception: for these more highly saline inclusions the final event is hydrohalite breaking down rather than ice melting. It is of crucial importance that a reader note that final stable breakdown of hydrohalite can occur over the same temperature range as that of ice; namely, from the eutectic to about 0°C. Therefore, for an inclusionist to properly interpret the temperature of the disappearance of the final phase, it is absolutely imperative that he or she be able to identify whether the last phase present is hydrohalite or ice. Hydrohalite has a refractive index of 1.416, ice is 1.313, and liquid H_2O is generally 1.33-1.38 (Table 7.2); thus, hydrohalite is a higher relief mineral compared to ice. However, this difference in relief is not always an obvious characteristic. A good method for learning the distinction between hydrohalite and ice is to practice on the NaCl eutectic composition standard made by SYN FLINC (commercially available from FLUID INC., Denver, Colorado, USA). Figures 7.15A, B, and C show a synthetic H_2O-NaCl inclusion of eutectic composition after cycling between -20 and -24°C to grow larger crystals. In 7.15A, the larger "dull" crystals are ice and the smaller "bright" crystals are hydrohalite. Viewing in color down the microscope, ice will have a pinkish tint and hydrohalite will be greenish. Also, during cycling, the ice crystals grow much more rapidly than the hydrohalite (which is a very important distinguishing trait), and rapidly take much of the space, leaving no room for the hydrohalite crystals to grow. But with continued cycling, as more ice crystals are eliminated, hydrohalite crystals can be grown large enough to study easily, like those displayed in Figure 7.15B and C. Note that the ice crystals are rounded, whereas the hydrohalite displays faces, and the hydrohalite is of higher relief in the solution as compared to ice. Figure 7.15D shows a synthetic H_2O-NaCl inclusion saturated with NaCl, as is indicated by the cube of halite within the inclusion. Note that the boundary at the upper corner of the halite cube with the inclusion wall is invisible. This is because the index of refraction of halite is virtually the same as that of one of the indices for quartz. Contrast this boundary with the boundaries of several hydrohalite crystals touching the walls of quartz in Figures 7.15B and C: there is a fine line at the inclusion wall that separates quartz from hydrohalite. Once one learns the distinguishing characteristics of hydrohalite, there will be no mistaking it, even in inclusions as small as 3 µm. In small inclusions (<7 µm), hydrohalite's sluggish reaction to warming and cooling will be the best diagnostic features.

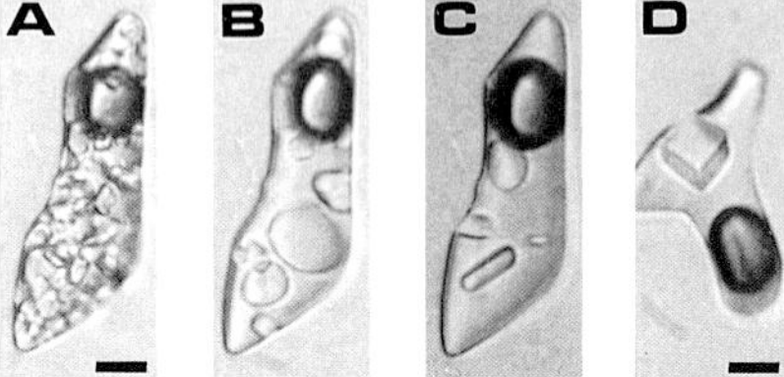

Fig. 7.15. *Photomicrographs of synthetic H_2O-NaCl inclusions in quartz illustrating the distinguishing characteristics of ice, hydrohalite, and halite.* ***A)*** *An inclusion of eutectic composition just below Te after a few short cooling and warming cycles on the heating and cooling stage. Ice crystals are rounded, large, and dull; hydrohalite crystals are small and bright.* ***B)*** *Same inclusion as in A, but after more cycling, the hydrohalite crystals are large enough to see crystal faces and differences in relief compared to ice and quartz.* ***C)*** *Same inclusion as B, showing one ice crystal and three hydrohalite crystals.* ***D)*** *A halite crystal inside an inclusion at 20°C showing no relief between quartz and halite where the two minerals touch. This will not always be the case as only one of the refractive indices of quartz is the same as the index for halite, and the apparent contact commonly has a liquid film. Bar scales are 7 µm.*

As in earlier examples, the inclusion must be supercooled to overcome the nucleation problem. As the inclusion is cooled, the bubble may gradually get larger. When the inclusion finally freezes it does so rapidly and exhibits similar characteristics to those described above. One exceptional difference for these more highly saline inclusions is that a greater volume of hydrohalite relative to the ice will exist in an inclusion upon freezing (see Fig. 3.4 and use the lever rule). Thus, it is unlikely that the bubble will disappear upon freezing, and furthermore, the deformation of the bubble (the "jerk") will be much more subtle than for the lower salinity example discussed above. On warming, the eutectic event is recognized by similar observations to those discussed above, except that it is normally easier to observe (particularly in larger inclusions) as there is rapid melting of all ice at once. Also at this eutectic, some hydrohalite breaks down to produce residual brine of the eutectic composition surrounding crystals of hydrohalite. After eutectic melting is complete, addition of heat results in an increase in temperature. During the temperature increase, the inclusion contains only hydrohalite, brine and vapor, and the hydrohalite crystals gradually break down to increase the concentration of the residual brine until the last crystal of hydrohalite has disappeared. Determining the final breakdown temperature of hydrohalite with accuracy in the small (<7 µm) inclusions typically found in diagenetic phases will usually not be possible. This is due to the fact that hydrohalite breaks down and grows sluggishly, so that rapid cycling procedures that are useful for determining the Tm ice would have to be accomplished very slowly, under very controlled thermal conditions. Such is not really a problem, because in practice, all that an inclusionist really needs to determine is which phase is the last to disappear (ice or hydrohalite). If hydrohalite is the last phase to disappear, then the salinity is known to be between 23 and 26 wt.% NaCl — the authors know of no cases in which a greater resolution in compositional determination for such a highly saline brine would be necessary. It may also be comforting to know that recognizing the differences between hydrohalite and ice are fairly straightforward, once one has had a little experience with each case, and as long as one is always cognizant of the importance of identifying the final phase that disappears.

Another possible problem in determining accurate salinities of fluids of these compositions is that some inclusions do not form the hydrohalite that would be predicted from the equilibrium phase diagrams, but form halite instead. In these cases, rather than showing final disappearance of hydrohalite between -21.2 and 0.1°C, they exhibit final melting of ice along the metastable extension of the ice melting curve between -21.2 and the metastable eutectic of -28°C (Fig. 7.10). Again, in practice, if one can determine that ice is the phase that is melting at such low temperatures, then a salinity of 23 to 26 wt.% NaCl can still be interpreted.

Just as with the lower salinity system, useful interpretation of the temperature of final disappearance of hydrohalite requires the presence of a vapor bubble to constrain the system to a low pressure. If during final melting, the inclusion is under extreme negative pressure because a vapor bubble has not nucleated, the final melting temperature of hydrohalite will be affected (Adams and Gibson, 1930). Similarly, the lack of a bubble at the eutectic causes variation in the eutectic temperature.

For the rare cases in which the final disappearance temperature of hydrohalite has been determined with precision, then this temperature can be used to determine salinity of the fluid inclusion. An adequate calibration between salinity and final disappearance of hydrohalite (Tm hydrohalite) can be extracted from the low-temperature phase equilibrium study of the H_2O-NaCl-KCl system by Sterner and others (1988) for salinities above the eutectic composition and with Tm below or equal to 0.1°C. By using an NaCl/NaCl+KCl ratio of "1" the relationship is given by:

$$\text{wt.\% NaCl} = 26.28708872 + 14.80771966\Psi$$

where Ψ = Tm hydrohalite °C/100. This equation approximates the experimental data with an average deviation of ±0.1 wt.% and a maximum deviation of ±0.23 wt.%. This equation can be applied to final melting of hydrohalite in the H_2O-NaCl system, in the presence of vapor, for Tm hydrohalite between -21.2 and 0.1°C.

Final melting at the eutectic.— If a fluid inclusion in the H_2O-NaCl system were to have the eutectic composition, in the presence of vapor, all melting would take place at -21.2°C (composition #3; Fig. 7.10). Behavior during initial cooling should be the same as for those discussed above and is shown in Figure 7.12. The bubble should enlarge slightly during cooling and the inclusion must be supercooled to achieve the nucleation temperature. At the nucleation temperature, the inclusion will freeze suddenly as indicated by the "jerk" of the bubble. Also, the bubble could disappear upon freezing, in which case it must renucleate before subsequent phase changes are recorded at warmer temperatures. During warming, metastable eutectic phenomena discussed earlier could also be observed, but by the time the inclusion has reached the stable eutectic temperature, all of the enclosed ice will melt and hydrohalite should break down to liquid at -21.2°C (the eutectic). As hydrohalite breakdown may be sluggish, it will be important either that the inclusion is held at the eutectic temperature long enough to allow for melting of any phases present, or that no cycling is employed. If large ice and/or hydrohalite crystals are grown, the system can be sluggish to respond, and the apparent Te value will be too high.

Disappearance of halite daughter crystals.— Inclusions that are more saline than 26.3 wt.% (for example, composition #4, Fig 7.10) should contain halite as a daughter mineral in inclusions above 0.1°C (Fig 7.15D). The temperature at which the halite dissolves can be used to determine the salinity of the fluid inclusion. (In theory, the phase equilibria require a vapor bubble to be present at the instant when halite totally dissolves, but the absence of a vapor phase in this case only causes a very small underestimation of salinity.) A definitive identification of halite as a daughter mineral will be required; it must be distinguished from other daughter minerals such as sylvite. Halite is a cubic, isotropic mineral (Table 7.2) that has the same refractive index as one of those of quartz; thus a useful identifying characteristic is that the quartz-halite interface will often be invisible (Fig. 7.15D). Halite can be distinguished from sylvite by its tendency to react to form hydrohalite during cooling, and its small change in solubility with temperature compared to sylvite. Upon rapid supercooling, freezing, and warming, a fluid inclusion containing a crystal of halite will produce first melting at the metastable eutectic -28°C. Roedder (1984) shows that with slow cooling, or with inclusions rapidly frozen but then warmed to temperatures above -28°C, hydrohalite may crystallize sluggishly from reaction between the liquid and the solid halite crystal to precipitate hydrohalite below 0.1°C. The hydrohalite may even form an armor around a halite crystal causing the reaction to cease. As an inclusion is slowly warmed, hydrohalite crystals will melt incongruently to form new halite as the inclusion temperature climbs towards 0.1°C. Aqueous fluid inclusions in halite are an interesting variant on this case. For these inclusions, cooling causes the aqueous fluid to react with the inclusion walls to form hydrohalite, corroding the walls of the inclusion (Davis and others, 1990).

If halite crystals are present in the inclusion at room temperature, and if the solution is pure NaCl-H_2O, then the inclusion must contain a minimum of 26.3 wt.% NaCl. Some inclusions (not in halite) may be supersaturated with respect to halite at room temperature, but have not nucleated the halite because of nucleation problems. Cooling inclusions may encourage nucleation of a halite crystal. If a halite crystal is present, to determine the composition more specifically, an inclusion is heated using the same heating methodologies discussed for determining vapor homogenization. Solubility data for halite in H_2O is used to interpret salinity based on Tm NaCl. The water in the vapor bubble is so minor that no correction for it is needed (Hall and Sterner, 1992). The equation defining salinity of a fluid inclusion from halite dissolution temperature (Sterner and others, 1988) is:

$$\text{wt.\% NaCl} = 26.242 + 0.4928\Psi + 1.42\Psi^2 - 0.223\Psi^3 + 0.04129\Psi^4 + 0.006295\Psi^5 - 0.001967\Psi^6 + 0.0001112\Psi^7$$

where Ψ is the halite dissolution temperature in °C/100. This calibration reproduces experimental data with an average deviation of ±0.11 wt.% and with a maximum deviation of ±0.31 wt.%.

Low-Temperature Phase Behavior for the H_2O-NaCl-$CaCl_2$ Model System

Use of H_2O-NaCl-$CaCl_2$ as a model for the chemical system in a fluid inclusion requires identification of the eutectic temperature indicative of this system. Remember, however, that adoption of this system composition is only a model, and that other species are likely to be present in the solution. Procedures for freezing and melting fluid inclusions have been described in foregoing sections of this chapter and will not be repeated here. However, there are particular metastable and stable phases that form upon freezing and melting in the H_2O-NaCl-$CaCl_2$ system that are summarized here. As with other systems, inclusions in this system must be supercooled to freeze. Again, the main evidence of nucleation is the "jerk" of the bubble. Examples of three types of freezing phenomena exhibited by synthetic H_2O-NaCl-$CaCl_2$ inclusions are shown in Figure 7.6.

Identifying the system.—

Once a fluid inclusion has been frozen, careful observations of the temperature of first melting must be made to identify the H_2O-NaCl-$CaCl_2$ system. The degree of difficulty in recognizing first melting will depend on the size of the fluid inclusion and the way it freezes. If an inclusion freezes to a clear solid or one with a brownish tint (Fig. 7.6A and B, inclusions 2 and 3), as the inclusion is warmed, very commonly, a faint mosaic of crystals will become discernible 10°C or more below the stable Te (Fig. 7.6C, inclusions 2 and 3). The appearance of this faint texture is difficult to interpret. It may indicate recrystallization in the solid state, or it may indicate that liquid is already present. The important distinguishing characteristic which is diagnostic of the texture that tends to appear at the stable Te, is that at the stable Te, some of the crystals in the mosaic are bright in appearance (Fig. 7.6D, inclusions 2 and 3); this is commonly referred to as the *orange peel* texture. For many inclusions, the orange peel texture will be obvious, but for those inclusions that freeze to a mosaic of crystals, it will be next to impossible to see (Fig. 7.6, inclusion 1). However, with warming above Te, many inclusions will show well-defined crystal shapes and boundaries (Fig. 7.6E): the temperature at which such characteristics are noted should always be recorded as they are indicative that definite melting is present, and that the Te must be at some temperature below this. In fact, from a practical standpoint, the only observation that an inclusionist needs to make in order to be justified in assuming the H_2O-NaCl-$CaCl_2$ model system is that well-defined crystal shapes and boundaries are noted at some temperature less than -40°C. Remember that this is only a model composition and that other ions (especially divalent cations) likely are present.

Inclusions of the H_2O-NaCl-$CaCl_2$ are notorious for producing metastable assemblages prior to Te (Davis and others, 1990; Spencer and others, 1990) such as:

$$\text{vapor} + \text{ice} + \text{halite} + CaCl_2{\cdot}4H_2O \qquad (1)$$

$$\text{vapor} + \text{ice} + \text{halite} + \text{antarcticite} \qquad (2)$$

Upon warming inclusions with metastable assemblage #1, a metastable eutectic is predicted to occur at -70°C (see table 7.1 for actual observations) in which some ice melts and reacts with $CaCl_2{\cdot}4H_2O$ to form $CaCl_2{\cdot}6H_2O$ (antarcticite). As warming continues, the antarcticite may break down at -52°C. With further warming, any halite present may react with liquid to form hydrohalite and return the fluid inclusion to a stable assemblage. For assemblage #2 antarcticite may break down at a metastable eutectic around -52°C. Other details and actual observations of the metastable phase behavior in this system (when saturated with halite) are summarized in Table 7.6.

Stable phase equilibria.—

If a fluid inclusion freezes and produces the stable assemblage vapor + ice + hydrohalite + antarcticite, stable eutectic melting in which antarcticite breaks

Table 7.6. *Interpretation of observed melting behavior in halite-saturated aqueous fluid inclusions in the system H_2O-NaCl-$CaCl_2$**

OBSERVED MELTING BEHAVIOR	$CaCl_2$ MOLALITY 0.12	0.75	1.8	2.9	INTERPRETATION
Darkening clear glassy solid	--	-85 to -75		-90 to -80	Metastable $CaCl_2$-hydrate eutectic
Granular solids coarsening	--	-70 to -60		-70 to -50	Melting & recrystallization ice
Opaque inclusions lighten. Vapor appears. Rim development	-53 to -47	-56 to -50		--	Stable $CaCl_2$-hydrate eutectic reaction
Final melt clear globular crystals	-35 to -23	-50 to -35		-36.7 to -36.0	Final melt ice
Final melt birefringent crystals	-0.1 to -0.0	-1.4 to -1.2	-5.0 to -4.4	--	Final melt hydrohalite

**temperatures (°C) of melting events for each bulk composition are presented*
**after Davis and others, 1990*

down is predicted at -52°C. The path followed by heating such an inclusion is most easily visualized using the phase diagram presented in Fig. 7.16. At -52°C, (point A, Fig. 7.16) the eutectic reaction occurs for the system and all of the antarcticite breaks down to produce an inclusion containing vapor, hydrohalite, liquid, and ice at that temperature. The inclusion can now rise above this temperature. As warming proceeds, the composition of the liquid develops along the cotectic curve (A-B) separating the ice + liquid and hydrohalite + liquid fields as hydrohalite breaks down and ice melts. If the inclusion has a low enough salinity so that ice (rather than hydrohalite or halite) is the last phase to melt, the next event is the temperature at which the last crystal of hydrohalite breaks down (e.g., point B, Fig. 7.16A, B). This *intermediate melting* temperature is important because it defines the relative amounts of $CaCl_2$ and NaCl in the fluid inclusion. After this has occurred, the inclusion contains only ice, liquid, and vapor. As warming continues, ice continues to melt, diluting the concentration of the residual brine, but leaving its NaCl/$CaCl_2$ ratio unchanged. Finally, the last crystal of ice melts (point C, Fig. 7.16B). The temperature at which Tm ice occurs must be combined with the intermediate melting temperature to determine the salinity of the fluid inclusion: the salinity (18 wt.% NaCl + $CaCl_2$ in this case) is determined by finding the point of intersection of the line defining the NaCl/$CaCl_2$ ratio (determined by the hydrohalite intermediate melting temperature) and the isotherm of Tm ice melting in the ice+liquid field (point C, Fig. 7.16B).

If the H_2O-NaCl-$CaCl_2$ model system is chosen because of melting events occurring at temperatures below -40°C (see above), and if an intermediate melting temperature of hydrohalite can be observed at a temperature below final melting of ice, the three observations can be used to determine the NaCl/NaCl+$CaCl_2$ wt. ratio and the bulk salinity of the fluid inclusion using the phase diagram presented in Figure 7.16B. It is worthwhile questioning the accuracy of any determination of NaCl/NaCl+$CaCl_2$ wt. ratio using this diagram, because the phase diagram is only as good as the experimental data on which it is based, and the statistical fit of those data. Oakes and others (1990) constructed these phase diagrams by using data from Yanatieva (1946) to construct the ice/hydrohalite cotectic curve, and by supplying new data (Table 7.7) in the ice+liquid field to construct isotherms in this field. Oakes and others (1990) pointed out that for the example illustrated in Figure 7.16, in which final hydrohalite breakdown is observed at -25°C, their data yields a NaCl/NaCl+$CaCl_2$ wt. ratio of 0.7, but when the Yanatieva (1946) data are used, a NaCl/NaCl+$CaCl_2$ wt. ratio of 0.53 results. Thus, one must remember that, a great potential source of error in determining composition of fluid inclusions from low-temperature phase changes in this system is the degree to which the phase equilibria of the system are known.

Some inclusions are saline enough that hydrohalite is the last phase to break down rather than ice. The antarcticite will still break down first at the eutectic (-52°C), hydrohalite and ice will still "melt" along the ice+liquid/hydrohalite+liquid cotectic curve (Fig. 7.16), but all of the ice will melt before all of the hydrohalite. After the ice has melted, such an inclusion will move into the hydrohalite+liquid field where hydrohalite will finally break down completely. If the inclusion is even more saline, halite will be the final phase that disappears (dissolves). The first melting is still at the eutectic and antarcticite then breaks down completely in the presence of hydrohalite, ice, liquid, and vapor. Hydrohalite and ice will then "melt" along the ice+liquid/hydrohalite+liquid cotectic curve, when all of the ice has melted, the liquid composition can move into the hydrohalite+liquid field and finally to the hydrohalite + liquid/halite + liquid peritectic as hydrohalite breaks down and halite forms. Finally, the

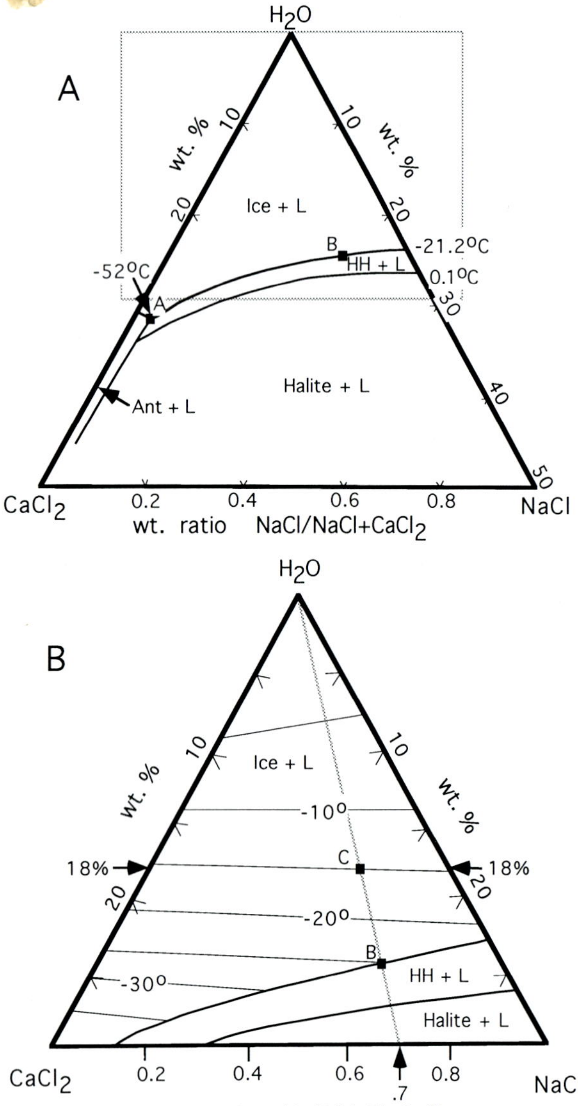

Fig. 7.16. Phase equilibria for the H_2O-NaCl-$CaCl_2$ system. L = liquid; Ant = antarcticite; HH = hydrohalite. ***A)*** *Phase diagram modified from Crawford (1981) using data of Yanatieva (1946).* ***B)*** *Expanded version of top portion of A, modified from Oakes and others (1990) using isotherms of Oakes and others (1990) and cotectic curves of Yanatieva (1946). See text for detailed discussion.*

liquid composition continues into the halite+liquid field and halite eventually melts (see data of Vanko and others, 1988; Oakes and others, 1992).

Practical observations.—

An important question about the feasibility of identifying and determining the composition of this system is whether the predicted phase changes can actually be seen and measured. In evaluating this question, the authors acquired (from M. Sterner) synthetically-produced inclusions of various compositions in the system H_2O-NaCl-$CaCl_2$ trapped in quartz. The inclusions were produced in the same experimental runs as those previously studied by Haynes and others (1988) in their evaluation of the SEM/EDA technique. The following descriptions should serve as a guide for the phase changes that might be noted in unknown natural samples. The precision with which the phase changes can be identified, the limitations of their interpretation, and tricks for making effective observations also are included

Upon supercooling inclusions of total salinity about 23 wt.% NaCl + $CaCl_2$, two significantly different styles of freezing were observed: either the inclusion froze by a jerk of the bubble and the inclusion remained relatively clear, or there was a jerk of the bubble and the inclusion suddenly looked dark, containing a fine polycrystalline mass (Fig. 7.6, Fig. 7.17). For the most part, inclusions froze between -52 and -85°C, but one sample of lower total salinity (about 11 wt.% NaCl + $CaCl_2$) showed freezing behavior above the eutectic temperature.

After freezing, a mosaic texture of some sort would appear at temperatures far below the stable Te in some fluid inclusions (Fig. 7.17). This appearance could be due to recrystallization of crystals or of glass (Roedder, 1984), or crystallization of metastable phases in the presence of a liquid (Davis and others, 1990; Spencer and others, 1990), and is a source of much confusion when attempting to identify the stable Te. Remember, the stable eutectic is best identified by the orange peel texture (around -52°C) followed by the observation of definite liquid, and both are clearly observable in some fluid inclusions. In many inclusions we have studied, melt becomes obvious close to the eutectic, commonly just several degrees above it. As explained above, all that one must recognize is some evidence for definite melt at any temperature below -40°C to justify selection of the H_2O-NaCl-$CaCl_2$ system as a model when working on inclusions from diagenetic environments. As can be seen in Figure 7.17E, melt is obvious by -40°C as evidenced by crystals showing well-defined shapes and boundaries.

The next phase predicted to break down completely for moderate and low salinity fluid inclusions is hydrohalite; once hydrohalite is gone, only ice is present. The temperature at which the last hydrohalite melts can be observed as the end of a period of clearing

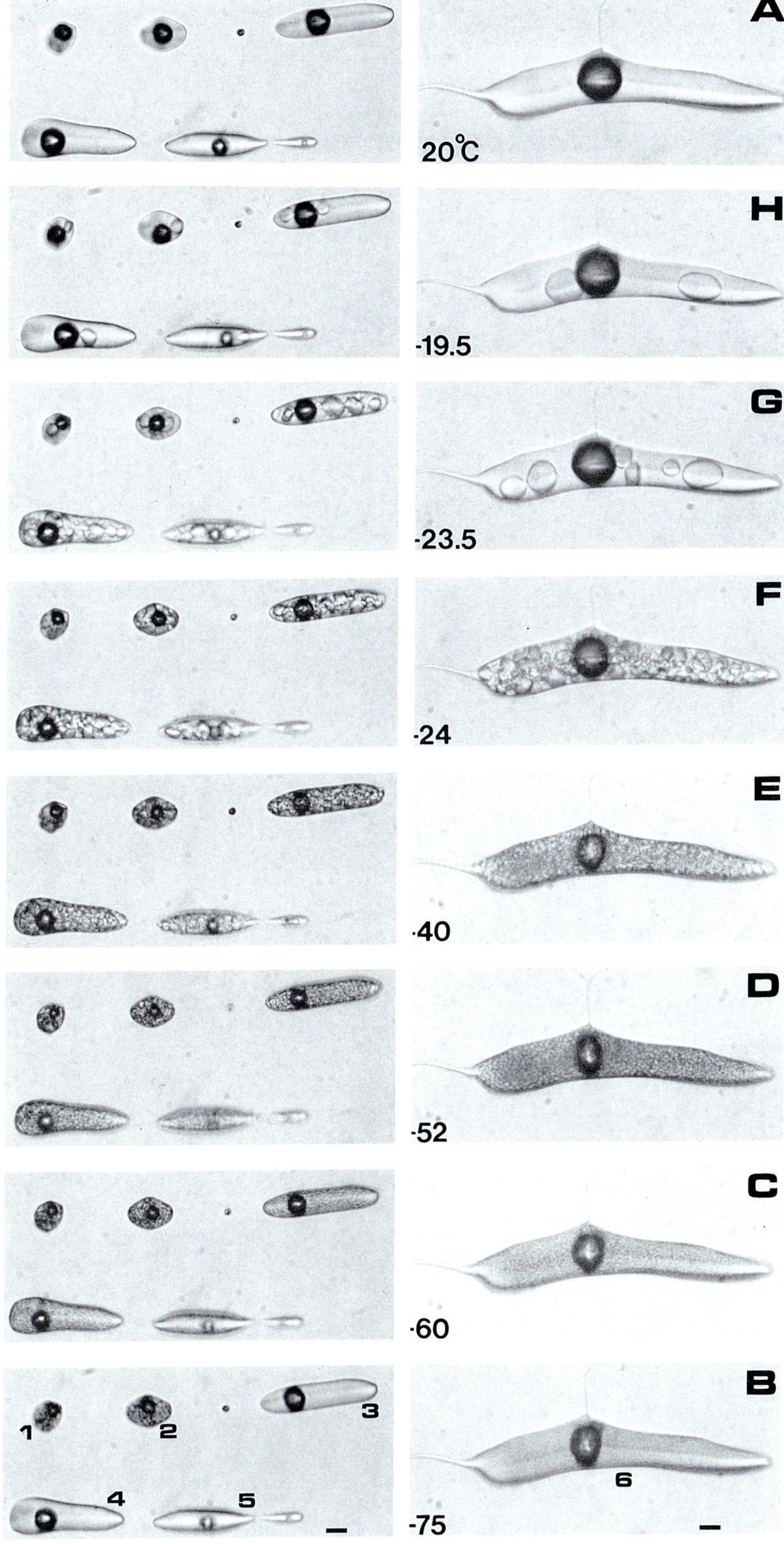

Fig. 7.17. *Photomicrographs of a cool/thaw run for six synthetic H_2O-NaCl-$CaCl_2$ inclusions from the same healed microfracture as those shown in Figure 7.6.* ***A)*** *Room temperature conditions.* ***B)*** *Inclusions 1 and 2 freeze first to a mosaic of crystals at about -65°C. All other inclusions freeze at about -75°C to a "clear" solid. Inclusion 6 shows a very diffuse mosaic texture upon freezing.* ***C)*** *With slow warming (5 minutes) to -60°C, inclusions 3-6 develop a mosaic texture, but each has a diffuse, hazy appearance. No detectable change was noted in inclusions 1 and 2.* ***D)*** *At the stable Te (-52°C), the mosaic of crystals in all inclusions is "crisper" — the haziness has disappeared — and some of the crystals in each inclusion have a brightness not evident at colder temperatures. This is the characteristic "orange peel" texture indicative of the eutectic.* ***E)*** *By -40°C, crystals have well-defined shapes and boundaries, so if an inclusionist was in doubt as to whether or not melting had occurred at -52°C, by -40°C he or she would most certainly be sure.* ***F)*** *More melting and recrystallization has occurred by -24°C: the rounded, more visible crystals are ice and they have grown larger at the expense of smaller ice crystals. Other smaller crystals (the brighter ones are probably hydrohalite) still occupy spaces between the larger ice crystals.* ***G)*** *At -23.5°C a noticeable clearing occurs that seems especially rapid in inclusion 6: in this inclusion, the small crystals that were interstitial to the large ice crystals disappear in a very narrow temperature interval, and as they dissolve, they jiggle around as if they were all in a hurry to get somewhere. Inclusions 3 and 4 still show a few of these small crystals remaining. Note that some of them appear to have very low relief, so they must not be hydrohalite! The authors have not been able to identify this phase; furthermore, this intermediate melting event (indicated by the rapid clearing) is lower than what is predicted to occur for inclusions of this NaCl/NaCl + $CaCl_2$ ratio! This example highlights how important it is for an inclusionist to be able to identify the phase that is disappearing before blindly applying the phase diagram in Figure 7.16.* ***H)*** *At -20.5°C, only a few crystals of ice remain, and they finally melt completely at -19.5°C. Bar scales are 7 μm.*

Table 7.7. *Data on freezing point depressions and total salinity for various NaCl/(NaCl+$CaCl_2$) weight ratios**

(1.000) θ	(1.000) wt%	(0.593) θ	(0.593) wt%	(0.195) θ	(0.195) wt%	(0.000) θ	(0.000) wt%
0.00	0.00	0.00	0.00	0.00	0.00	0.00	0.00
1.84	3.08	1.70	3.13	1.55	3.18	1.42	3.13
2.03	3.40	2.06	3.77	2.03	4.09	2.08	4.42
5.42	8.45	3.01	5.32	3.37	6.34	3.68	7.09
6.98	10.48	4.05	6.92	4.40	7.87	4.06	7.64
9.22	13.11	5.19	8.50	5.05	8.75	5.21	9.16
10.27	14.25	5.21	8.54	6.00	9.95	6.02	10.15
12.49	16.46	6.02	9.58	7.02	11.13	6.03	10.20
12.86	16.78	7.08	10.88	8.09	12.27	7.06	11.35
14.91	18.57	8.10	12.05	9.02	13.18	8.08	12.39
16.91	20.18	9.10	13.10	10.04	14.12	8.42	12.64
19.17	21.80	10.03	14.03	11.13	15.06	9.08	13.34
20.64	22.80	11.02	14.96	11.97	15.75	10.98	14.83
21.48	23.34	12.19	16.00	13.96	17.22	11.62	15.46
		13.24	16.89	16.09	18.65	12.03	15.72
		14.08	17.55	18.07	19.82	13.50	16.77
		15.14	18.36	20.36	21.08	14.21	17.31
		17.04	19.69	22.01	21.94	15.30	18.04
		19.06	21.02			16.02	18.42
		21.02	22.19			17.80	19.46
		23.23	23.41			19.32	20.19
		23.28	23.43			19.64	20.40
						20.78	20.97
						21.12	21.16
						21.97	21.56
						23.08	22.04
						24.02	22.41
						25.23	23.01
						25.25	22.95
						26.08	23.37
						27.06	23.73
						27.08	23.73
						27.21	23.78
						27.22	23.82
						28.08	24.16
						28.98	24.44
						30.16	24.89
						32.09	25.52
						32.13	25.56
						34.07	26.17
						35.72	26.69
						36.03	26.78
						37.07	27.08
						38.08	27.39
						39.03	27.66
						39.83	27.88
						40.80	28.14
						42.16	28.50
						42.89	28.68
						43.15	28.71
						44.00	28.97
						44.77	29.16
						45.64	29.37
						46.55	29.58
						46.66	29.61
						48.28	29.94
						49.37	30.23
						50.14	30.39
						50.51	30.46
						51.20	30.59

(0.796) θ	(0.796) wt%	(0.393) θ	(0.393) wt%	(0.169) θ	(0.169) wt%
0.00	0.00	0.00	0.00	19.99	20.86
1.71	3.03	1.59	3.12	22.11	21.94
2.01	3.54	2.04	3.88	24.05	22.85
3.05	5.20	2.10	4.02	25.98	23.70
4.01	6.68	3.18	5.84	28.02	24.54
4.55	7.43	4.22	7.40	30.08	25.34
4.94	7.97	5.46	9.08	30.21	25.39
4.98	8.09	6.94	10.88	32.11	26.09
6.12	9.57	8.06	12.10	34.03	26.75
6.13	9.56	9.16	13.24		
7.03	10.69	10.06	14.09		
8.84	12.78	11.11	15.04		
10.12	14.12	12.06	15.84		
11.00	14.98	13.00	16.60		
11.91	15.81	14.02	17.37		
13.32	17.05	15.32	18.30		
14.19	17.79	16.95	19.39		
15.19	18.58	17.03	19.41		
16.05	19.25	19.36	20.85		
17.03	19.96	21.22	21.87		
18.01	20.65	23.21	22.96		
18.99	21.30	25.07	23.85		
19.00	21.31	26.93	24.71		
20.05	21.99				
20.96	22.56				
22.13	23.25				
22.48	23.63				

**weight ratios appear above each column in parentheses*

**θ is absolute value of Tm ice*

**after Oakes and others (1990)*

during warming of the inclusion. As shown in Figure 7.17, a visible indication that such has happened rarely can be dramatic (Fig. 7.17G, inclusion 6), even for inclusions 10 μm in size (compare Fig. 7.17F and G, inclusion 1), and can be described as sudden activity (movement) within an inclusion concomitant with a "clearing," as shown in Figures 7.17F and G. The authors cannot predict if such an impressive intermediate melting phenomenon will be detectable in very many fluid inclusions smaller than 10 μm. It is important to point out that it is not always easy to identify that the phase that is melting is hydrohalite. For the inclusions illustrated in Figure 7.17G, the authors were unable to identify the phase that breaks down during the intermediate melting event. Cycling can be used to isolate or enlarge the intermediate phase to facilitate petrographic observations. The temperature just at the end of the clearing event (the temperature at which only one phase is left floating in the inclusion liquid) is worth noting if it is observed, as different compositions of inclusions may be indicated by such data. In analyzing the synthetic fluid inclusions, we have found it difficult to be able to reproduce the predicted hydrohalite melting temperature in some inclusions. Recall that several degrees of error could lead to significant misinterpretation of the NaCl/NaCl + $CaCl_2$ wt. ratios (Fig. 7.16). Also, because natural brines are very complex chemically, one must avoid applying Figure 7.16 unless the phases present at melting can be identified. Other analytical techniques could be helpful for constraining NaCl/NaCl + $CaCl_2$ weight ratios of natural fluid inclusions from diagenetic minerals.

The final melting temperature of ice can either be readily observed, or cycling can be employed as described above. All of the tricks and pitfalls described earlier for making this measurement in the H_2O-NaCl system can also be applied here. For the cases in which the NaCl/NaCl+$CaCl_2$ wt. ratio had been determined from the Tm hydrohalite or from some other method of extracting and analyzing the inclusion fluid, the Tm ice can be converted to salinity using the isotherms presented in Figure 7.16B or by using the equation and diagrams published by Oakes and others (1990, 1992).

Low-Temperature Phase Behavior for the H_2O-NaCl-CH_4 Model System

Most aqueous fluids in the diagenetic realm are in equilibrium with other components normally thought of as gases. These include helium, argon, krypton, nitrogen, carbon dioxide, hydrogen sulfide, hydrogen, butane, propane, ethane, and methane. Natural gases are highly variable in composition; typically, they are dominated by methane (greater than 80%) and contain lesser amounts of ethane, propane, butane, pentane, nitrogen, hydrogen sulfide, and carbon dioxide (Selly, 1985), but any of these gases may be present in significant concentrations in natural gases. For most natural gas systems in the burial diagenetic realm, the assumption of methane dominance is probably realistic. Ignoring the presence of other gas phases that may be present in gas-rich fluid inclusions should be viewed as an assumption of a model composition that may lead to potentially significant error in interpretation of fluid inclusion data (see Chapters 9 and 10).

As was shown in Chapter 3, immiscibility is possible in the sedimentary environment in the system H_2O-NaCl-CH_4. Therefore, the behavior one views upon freezing inclusions trapped in this system can be separated into two general categories: inclusions containing dominantly water, and inclusions containing dominantly methane. For those containing dominantly methane, the behavior of the inclusions on cooling differs depending on whether inclusions trapped are higher or lower than the critical density for methane (Fig. 3.2).

H_2O-dominant inclusions with CH_4.— Because CH_4 is present as a dissolved component in many subsurface brines, a careful inclusionist will constantly be searching for evidence indicating the presence of CH_4 or other dissolved gases in aqueous fluid inclusions from the burial diagenetic realm. The presence of CH_4 can be identified by crushing tests that produce expanding bubbles of organic gas (Chapters 3 and 6), by sophisticated instrumentation (Chapter 12), or through cooling runs. Below are described the microthermometric observations that help in identification and interpretation of fluid inclusions containing CH_4.

Aqueous inclusions containing high-pressure bubbles of CH_4 or methane-rich gas should be expected to form solid gas-hydrates, termed clathrates, as they are cooled. The observation of clathrates within an inclusion is evidence of the presence of gas in the inclusion. Clathrate stability is predicted as inclusions drop below room temperature. They can exist above 0°C in equilibrium with water or can exist at lower temperatures. Gas and liquid composition have an important effect on the P-T stability fields of clathrates (Fig. 7.18). During cooling, the point of formation of clathrate in a fluid inclusion can be detected when the bubble suddenly "jerks" or deforms at a temperature above that which is necessary for inclusions to freeze. Thus, during cooling the cautious inclusionist should be watching for two jerks of the bubble, the first caused by formation of clathrate and the second by freezing of the fluid inclusion. Figure 7.19 shows evidence of clathrate formation that was manifested only after the inclusion had been frozen, thawed, and instantly cooled after final melting of ice. In this case, clathrate crystals probably had not nucleated until everything froze at once when the inclusion was supercooled. At the time of final melting of ice, the clathrate crystals were so small that they were not detected. But with cooling and

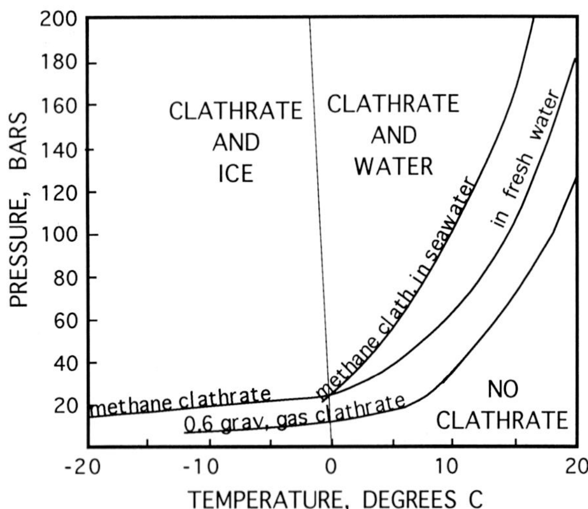

Fig. 7.18. Stability fields for clathrates. Modified from Hunt (1979) and Seitz and Pasteris (1990). Curves represent boundaries of stabilities of clathrates. No clathrate exists to the right of and below the curves. Clathrate coexists with water at higher temperatures and coexists with ice at lower temperatures. Ice-liquid water phase boundary is for fresh water. Clathrate stabilities are represented for methane clathrate in equilibrium with fresh water, methane clathrate in equilibrium with seawater, and a 0.6 gravity natural gas in equilibrium with water.

in the absence of other solid phases, they could grow without interference. Growth occurring at the bubble/liquid interface can be rapid as both compounds (CH_4 and H_2O) are readily available for formation of CH_4 clathrate; the observable result is a contorted, ragged-edged bubble (Fig. 7.19E and F). These characteristics have been most easily observed in inclusions with hundreds of bars of internal pressure at room temperature, but have also been observed in large inclusions (>7 μm) with as little as 50 bars internal pressure at room temperature. Sometimes more than one cycle is required (do not proceed to a temperature so low that ice and other phases renucleate), as too many growing crystals of clathrate may form a supportive framework that prevents any detectable deformation of the bubble. Each successive cycle should achieve a slightly (0.5°C) higher temperature, because the clathrate crystals decrease in number as they melt during warming, and the decrease in number of nuclei at progressively warmer cycling runs should allow for formation of larger clathrate crystals during subsequent cooling. Commonly, this facilitates easier observation of the crystals (Burruss and Reynolds, 1993). Gradual movement of the bubble during cooling is not indicative of the presence of gases in an aqueous inclusion. Also, in many cases one may not be able to observe the above phenomena even if gases are present,

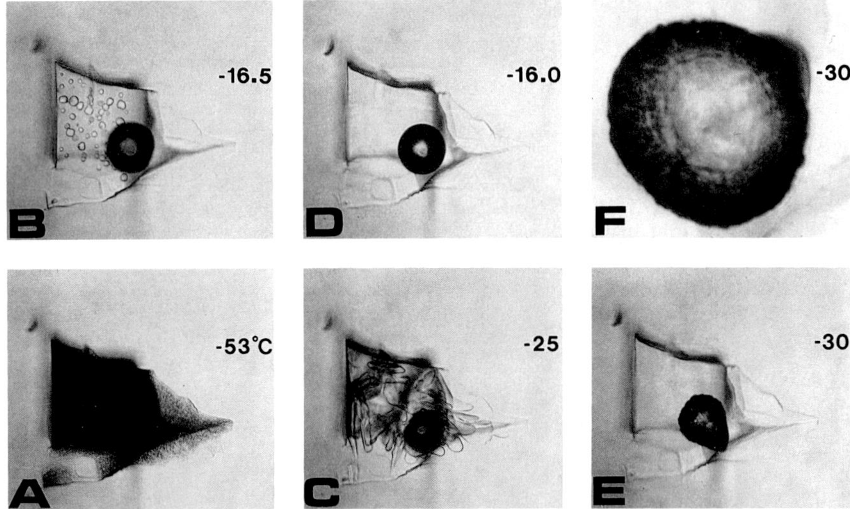

Fig. 7.19. Photomicrographs of a large aqueous fluid inclusion in fluorite that shows evidence of clathrate formation. ***A)*** *Upon freezing of the liquid, at about -53°C, the inclusion turns dark because the thick mosaic of crystals have prevented light from passing.* ***B)*** *After warming to about -16.5°C, a few ice crystals remain.* ***C)*** *With cooling to about -25°C, the ice crystals grow and compress the bubble.* ***D)*** *The inclusion is again warmed to -16.0°C and all ice crystals have melted completely. Just as the last visible ice crystal disappears the stage temperature is rapidly lowered.* ***E)*** *At about -30°C the bubble deforms, but not due to the growth of ice or hydrohalite, as no crystals are evident in the fluid as shown in C.* ***F)*** *A close-up view of the deformed bubble (on a different run). The interface between the liquid and vapor is coated with CH_4 clathrate, as was confirmed by Raman spectroscopy. A video tape of the phase changes this inclusion experiences has been published by Burruss and Reynolds (1993). On this video tape clathrate crystals can be seen to grow, and when a crystal grows outward from the liquid to eventually touch the bubble, newly-formed clathrate crystals can be seen to then rapidly coat the interface. The authors would be remiss if we did not divulge to the reader the size of the inclusion: it is embarrassing for us to admit that the diameter of the bubble in this inclusion at room conditions is about 50 μm! Such a wonderful inclusion is definitely an anomaly in a text on inclusions from diagenetic minerals.*

especially in small inclusions or those that contain low amounts of gas. A good method for learning these techniques is to practice on CO_2 clathrate found in three-phase (liquid H_2O, liquid CO_2, vapor CO_2) inclusions. This type of inclusion is common in mesothermal, greenstone-hosted, gold vein deposits and in many medium-to-high grade metamorphic environments, so should be easy to find. If not, then the synthetic H_2O-CO_2 inclusion standard made by SYN FLINC and distributed by FLUID INC. (Denver, Colorado, USA) has suitable three-phase CO_2 inclusions also.

The temperature of final melting of clathrate is best determined by the cycling technique described above: eventually, the bubble will not deform on cooling, indicating that the clathrate must have melted in the last interval of the cycle. Tm clathrate depends on gas content and salt content of the inclusion, and ranges from about -20°C to +25°C. The higher the gas content, the more clathrate will form, and therefore, the chances of making the above observations will be better in fluid inclusions rich in gas. Conversely, the lower the gas content, and the smaller the inclusions, the more difficult it will be to observe the formation or melting of clathrates.

Methane-dominant inclusions.—

At room temperature, gas-rich inclusions may look "empty" to the inexperienced eye because many of them have a "dark" appearance and no meniscus can be observed (Fig. 2.1C and Fig. 7.20A). Alternatively, some gas-rich inclusions may appear to be all-liquid H_2O inclusions (Fig. 7.21A). A wary inclusionist will never ignore features that do not contain an obvious bubble, and will attempt to determine what may be in them. Such inclusions could be crushed to identify gases (Chapters 3 and 6) or other analytical techniques could be applied to determine composition (Chapter 12). Some inclusions that are "empty" (perhaps they leaked during sample preparation and contain air) may be identified under the microscope. Either acetone can be placed on the sample, or liquid nitrogen can be allowed to fill the heating and freezing stage: if an inclusion is open and filled with air, the dark, "empty" inclusion will turn clear as the liquid fills it, and will darken again as the liquid nitrogen or acetone evaporates. Dark or clear single-phase inclusions should always be cooled to liquid nitrogen temperature (-196°C), and should be carefully observed throughout the entire cooling process to see if any phase changes occur. Some which may occur are described below and shown in Figures 7.20 and 7.21.

For methane inclusions, liquid and vapor phases can coexist only at temperatures below the critical temperature (-82.1°C for pure CH_4 — unlikely for natural gas found in nature). For methane inclusions of densities below the critical density (see Chapter 3), one should look for a "flicker" (movement) along the edge

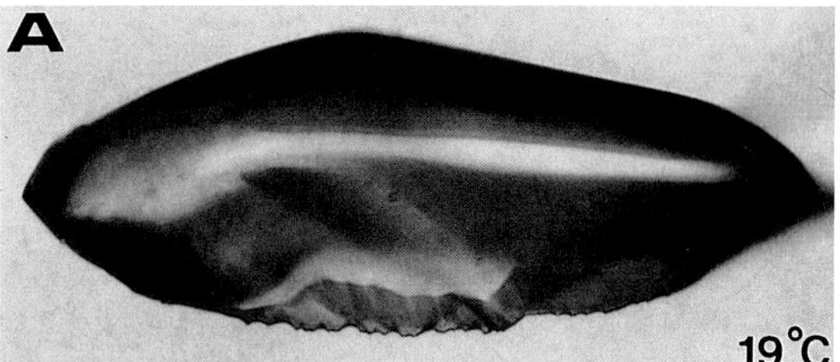

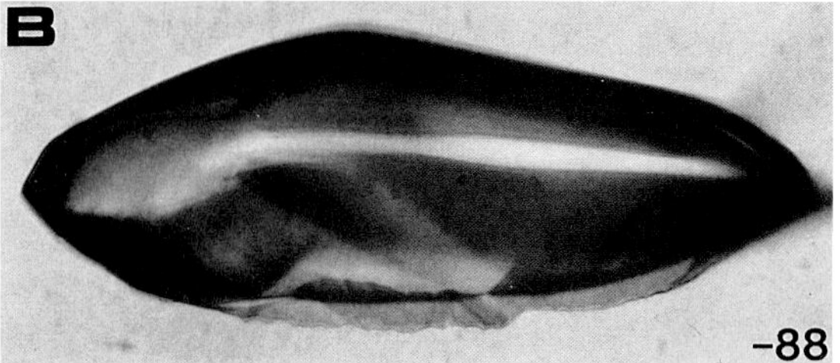

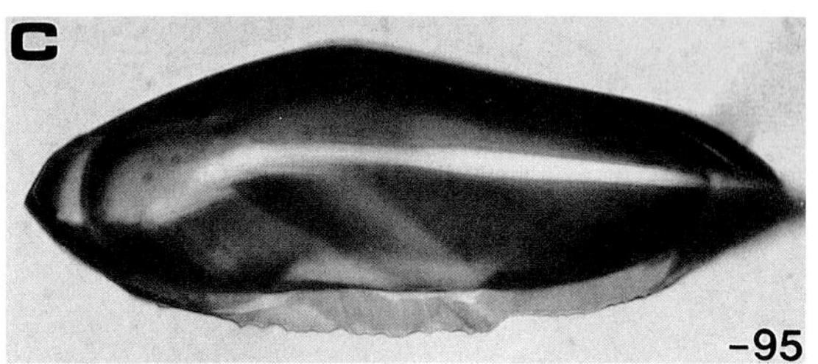

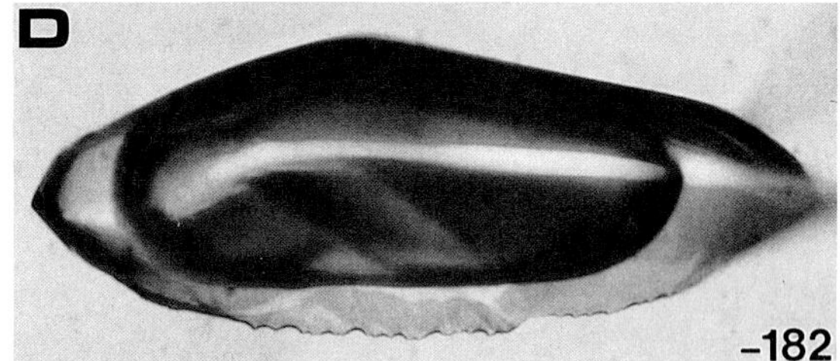

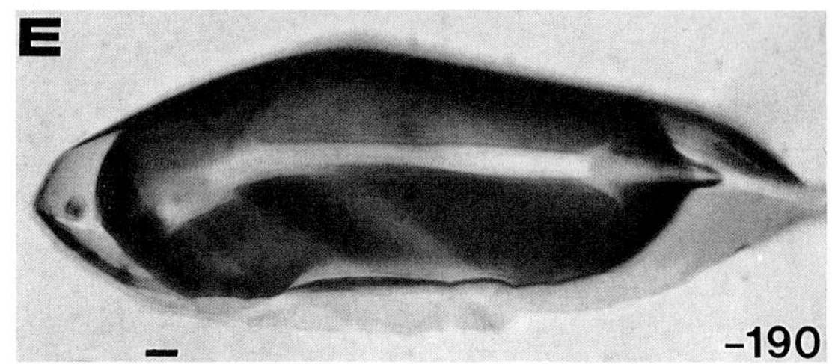

Fig. 7.20. *Photomicrographs of phase changes observable in a dark, low-density (below the critical density) CH_4 inclusion. At room conditions the fluid in the inclusion is supercritical, but once cooled below the critical temperature, liquid and vapor coexist. Below -190°C, CH_4 freezes as indicated by deformation of the bubble in E. The bubble "snaps" back to a smooth surface at -182.5°C (the triple point temperature of pure CH_4) and the inclusion homogenizes at -87.0°C. So, never ignore dark inclusions! Low-density CH_4 inclusions do not always appear as dark as the inclusion shown here: they may also be clear like the inclusion in Figure 7.21. This darkness is caused by refraction of the light at the surfaces of the inclusion. Bar scale is 7 μm.*

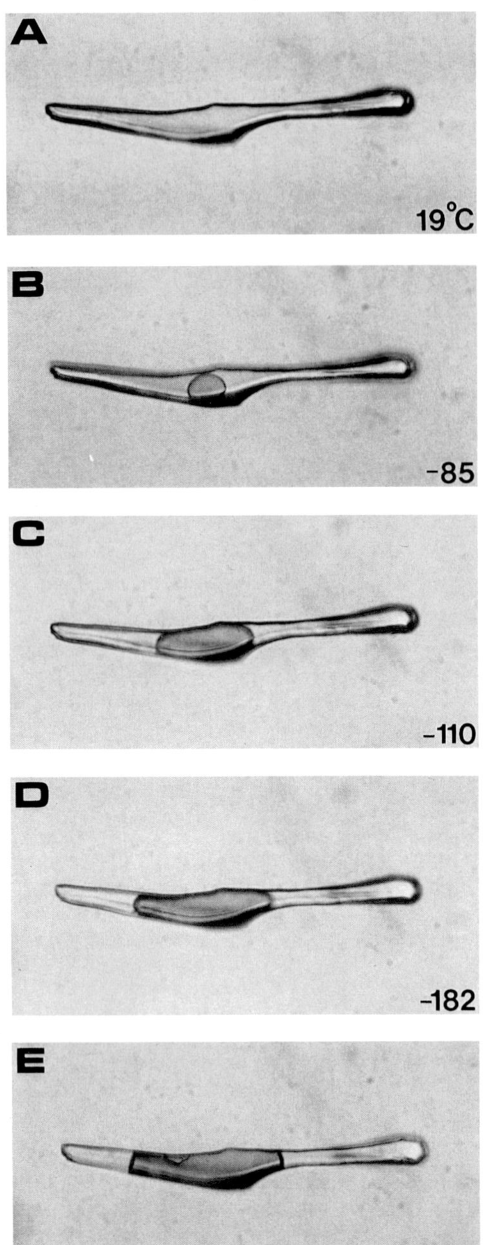

Fig. 7.21. *Photomicrographs of phase changes observable in a clear, high-density (above the critical density) CH_4 inclusion. At room conditions the fluid in the inclusion is supercritical, but once cooled below the critical temperature, liquid and vapor coexist. Below -190°C, CH_4 freezes as indicated by deformation of the bubble (E). The bubble "snaps" back to a smooth shape at -182.5°C (the triple point temperature of pure CH_4) and the inclusion homogenizes to liquid at -83.2°C. So, never ignore vacuoles without bubbles! CH_4 inclusions with densities above the critical density do not always appear clear at room conditions. They may also be very dark like the inclusion shown in Figure 7.20. The amount of light that passes through the inclusion to the observer depends on the direction that the light is refracted as it passes through the surfaces of the inclusion. Bar scale is 7 μm.*

of an inclusion, or look at areas where sharp tails protrude from the body of an inclusion to perhaps see a meniscus appear, which would demonstrate that the inclusion is dominated by gas, but contains a thin rim of liquid around its exterior below the critical temperature (Fig. 7.20B). At lower temperatures, the amount of liquid will increase (Fig. 7.20C, D). Inclusions with densities near the critical density will show a marked "turbulence" when the fluid is cooled to become liquid+vapor. In methane inclusions of densities above the critical density, a bubble will form within the inclusion (Fig. 7.21B), and with continued cooling, the bubble will grow in size as the liquid contracts (Fig. 7.21C, D).

As methane inclusions are cooled to liquid nitrogen temperature (-196°C), any liquid CH_4 in the inclusion may freeze, normally denoted by an instant deformation of the bubble (Fig. 7.20E, Fig. 7.21E). For pure CH_4 such an event will occur at temperatures below -190°C, but may not be visible if liquid nitrogen in the stage has obscured the image of the fluid inclusion. Also, for smaller inclusions and ones with lower densities, such an observation will be difficult to make. If the freezing event (bubble deformation) is not observed, do not give up yet: watch for a melting event upon warming above -196°C. As soon as the heating/cooling stage begins to warm, one should look for a "snap" of the bubble back into its normal shape upon melting of CH_4 solid. Pure CH_4 (a geologically unusual fluid) would melt invariantly at -182.5°C. Melting could occur at higher or lower temperatures with the addition of other volatile components.

With continued warming, the gas bubble either shrinks (if the bulk density of the inclusion is above the critical density; Fig. 7.21) or expands (if the bulk density of the inclusion is below the critical density; Fig. 7.20) until one of the phases disappears at Th. Just as with the aqueous inclusions discussed earlier, the Th defines the density of the inclusion fluid if it is known to be pure CH_4 (Fig. 3.2). Unlike the aqueous inclusions from the diagenetic realm, the methane inclusions from the diagenetic realm may homogenize by disappearance of the vapor phase or of the liquid phase.

The descriptions summarized above can often be observed in large, pure CH_4 inclusions. But inclusions from the diagenetic realm are not commonly >7 μm in size, and are even more rarely composed of pure CH_4. In fact, it is common for gases in sedimentary systems to contain significant proportions of many other gases (i.e., ethane, propane, CO_2, etc.). Presence of the other gases in methane-rich fluid inclusions would alter the low-temperature phase behavior summarized above. There are several clues that one might look for to identify the presence of other gaseous components in inclusions. During the freezing run of a gas-dominated inclusion, a freezing event at a higher temperature than the triple point of CH_4 indicates the presence of another component. Once the inclusion has been frozen, if initial melting occurs at a temperature other than the triple point of CH_4, the inclusion is not pure CH_4, as mentioned above. After the CH_4 has melted, if other solid phases are observed to melt at higher temperatures, the inclusion is not pure CH_4. Homogenization of the liquid and vapor phases at a temperature higher than the critical temperature of CH_4 (-82.1°C) would indicate the presence of other components. As clathrate melting is so pressure and salinity dependent, and as its melting occurs over such a wide range of temperatures, the temperature of clathrate melting typically would not be indicative of gas composition. In addition to microthermometry, there are other ways to identify the presence of CH_4 and other gases in fluid inclusions such as crushing, Raman microprobe, infrared spectroscopy, and extraction of inclusion gases followed by mass spectrometry or gas chromatography (see Chapter 12).

The H_2O-NaCl-CH_4 compositional model will not be used as a model to quantify the salinity determined from Tm ice. Nevertheless, this model can help to evaluate the error in interpreting salinity from Tm ice when CH_4 is present, but when another model is used. For instance, if a seawater model were to be employed to interpret Tm ice, but the inclusion was known to contain high amounts of CH_4, then the salinity determination could be in error by up to about 15 ppt, possibly an unacceptable amount of error for some studies. Of course, for inclusions containing CH_4, this model is crucial for the best possible interpretations of temperature and pressure conditions of formation as explained in Chapters 9 and 10.

SUMMARY

Philosophical approaches and practical procedures for conducting microthermometric analyses were first discussed in this chapter, followed by a thorough treatment of the more commonly observable microthermometric phase changes for some simple, but potentially applicable, fluid systems common in the diagenetic realm. Included are many helpful, practical aspects and tricks for collecting microthermometric measurements. However, it must be remembered that all of the fluid systems discussed are much more simple in composition than actual fluids from the diagenetic realm. Each system discussed should therefore be viewed as a "model system" that can be employed under the appropriate circumstances and with the realization that some error is inherent. Dealing with more complex fluid compositions may be possible, but this will necessitate analyzing the composition of the fluid in fluid inclusions through other analytical techniques — all of which seem to have some weaknesses (see Chapter 12) — and being able to identify all of the low-temperature phases that are present in fluid inclusions during phase changes by optical or other analytical methods such as Raman spectroscopy. If such data can be gathered, the phase equilibria of complex systems can be interpreted using a modeling approach. There are different styles of modeling that may prove useful in understanding the phase equilibria of more complex (actual) systems from the sedimentary realm. One model (Spencer and others, 1990) has proven successful in modeling the low-temperature phase equilibria of complex brines. Modeling of various gases in equilibrium with brines can be attempted using models such as Peng and Robinson (1976). A detailed discussion of these models, and of modeling in general, is beyond the scope of this text, but suffice it to say that the models are not without error: they are only as good as the experimental or theoretical data on which they are based, and should also be viewed as approximations. Understanding and studying fluid inclusion phase behavior is a useful endeavor that provides valuable information, but one should never forget that the interpretations are always limited.

Chapter 8

DATA PRESENTATION

INTRODUCTION

If readers have taken heed of our recommendations for an appropriate mental framework for selecting inclusions for microthermometry (see Chapter 7), then the method for proper presentation of data will be straightforward. Those who have not collected data with respect to petrographically related assemblages of inclusions (FIAs) have chosen an inappropriate methodology in which they must assume that any variability in their data is due to the vagaries of nature. Such an assumption leads to a philosophy that maintains that it is both appropriate and sufficient to portray variability to a reader by lumping all data into a few encompassing histograms, and then to interpret the data from statistically determined means and modes. It is very understandable that such logic seems reasonable, but the fallacy is the assumption that any variability is "natural" and to be expected. Consider the numerous natural cases where inclusions along a single microfracture or a single growth zone yield very consistent microthermometric results, or the cases in which data from synthetically produced inclusions vary less than a few tenths of a degree! For these groups of inclusions trapped at a certain P-T-X condition, there is, in fact, littlc "natural" variability. So, in contrast to what a geoscientist might initially think, the more appropriate philosophical viewpoint should be that data are *expected to be consistent,* but only if the following criteria are met:

1. The data come from a single petrographically related population (FIA) of inclusions *that samples a single event.*

2. Only a single homogeneous fluid was entrapped in each inclusion in the FIA.

3. All inclusions in the FIA remain chemically closed throughout their histories.

4. All inclusions in the FIA maintain a constant volume throughout their histories.

Any inconsistencies in the data would mean that one or more of the above criteria did not hold true. Note that the data must come from a single FIA. It then follows that the data must be presented for a single FIA in order to assess which of the other criteria are responsible for any variability shown within a single FIA. Proper data presentation involves assigning each FIA its own symbol, so that variability within each FIA may be assessed by a reader. This means that data from all measured inclusions from within a single healed fracture or growth zone should be plotted, and all should be plotted with a symbol dedicated to that single petrographically constrained group of inclusions.

Other compositional or petrographic factors may also deserve separate categorization when presenting data. For instance, the separation of petroleum inclusions, gas inclusions, and aqueous inclusions is essential. Data from primary and secondary inclusions are commonly shown with contrasting symbols. All-liquid inclusion salinity data might be separated from two-phase inclusion data, or inclusions containing certain daughter minerals could be separated from others that do not. Even groups of inclusions of different shapes may prove to be important in an interpretation, so a grouping of data with respect to shape is possible as wcll.

To summarize, fluid inclusion data should be collected, presented, and interpreted within a framework of individual fluid inclusions assemblages (FIAs). Complete information on collection of data is presented in Chapter 7, interpretation of data is covered in Chapters 9 and 10, and presentation formats are described below.

FREQUENCY HISTOGRAMS

The graphical approach which visually shows the range and variability of a particular thermometric parameter (e.g., Th or Tm ice) is the frequency histogram (Fig. 8.1). For diagenetic applications, the class interval for Th's should usually be 5°C. The class intervals selected for Tm ice will depend on the nature

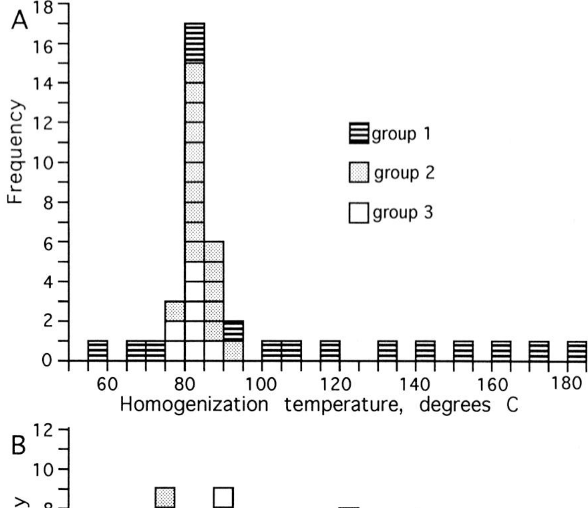

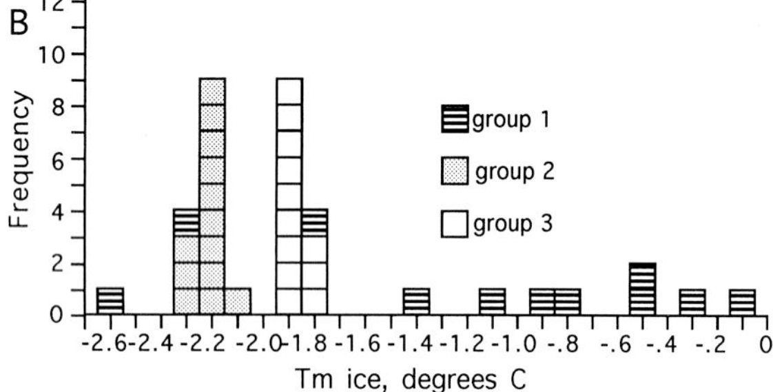

Fig. 8.1. *Frequency histograms.* ***A)*** *Homogenization temperature data.* ***B)*** *Tm ice data. Notice that data from each defined group of inclusions is given a different symbol. Typically, each group represents data from a separate fluid inclusion assemblage. Also, each group could indicate different modes of origin for inclusions, different localities, all-liquid versus two-phase inclusions, inclusions of different shape or size, or inclusions of different composition (e.g., petroleum versus aqueous).*

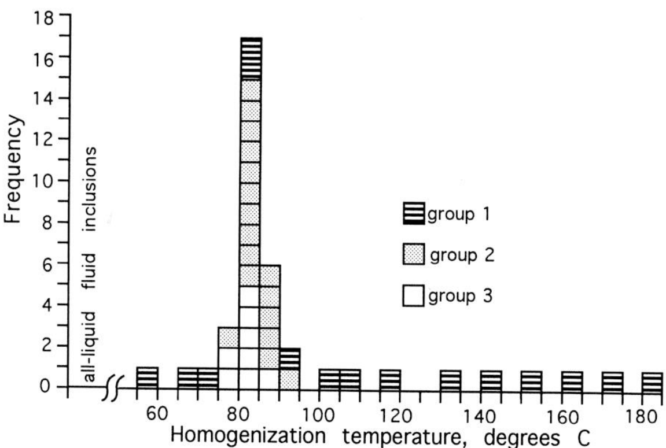

Fig. 8.2. *Frequency histogram of Th data that illustrates the presence of all-liquid fluid inclusions at room temperature. Notice that the temperature scale is broken at the lower end. The presence of all-liquid inclusions is not illustrated as individual data points because their position along the temperature scale and their abundance cannot be compared to Th data. Notice that data from each defined group of inclusions is given a different symbol. Typically, each group represents data from a separate fluid inclusion assemblage. Also, each group could indicate different modes of origin for inclusions, different localities, inclusions of different shape or size, or inclusions of different composition (e.g., petroleum versus aqueous).*

of the study. For example, if an inclusionist is attempting to assess variability of seawater salinity due to mixing or some other process, then intervals as small as 0.1°C will be appropriate. On the other hand, if the variability in the data in a single FIA ranges from 0.0°C to -20°C, and all that one wishes to display is this variability, then intervals of 2 to 5°C are perfectly fine. Use of any histogram involves generalization of data to some degree; therefore, the distribution of data in histogram format will depend on the width and boundaries of the class intervals chosen. (For this reason, be careful not to overinterpret subtle shifts in distributions.) Basically, when constructing frequency histograms there are two rules to follow: choose a small enough class interval to allow for analysis of variability of data, and use different symbols for inclusions of each different FIA, origin, composition (e.g., oil versus aqueous), original phase ratio (e.g., all-liquid versus two-phase), or any other parameter that would enhance interpretation of data.

Figure 8.2 shows one approach for displaying all-liquid inclusions on frequency histograms. Note that the lowest Th data should always be presented, and to display to a reader that single-phase inclusions exist (that do not result from metastability or from necking down; see Chapters 4 and 6), then a break in the temperature axis and the words "all-liquid inclusions" are important to show. One should not fabricate a low temperature (e.g., 50°C) and attempt to plot something to represent all-liquid inclusions because such is not truthful, and the quantity of observations may be misleading relative to the quantity of measurements on two-phase inclusions.

Frequency histograms may be vertically arranged to display other paragenetic (variability in time) or geologic (variability in space) contrasts as shown in Figure 8.3. Their vertical spacing can be arranged to reflect inclusion host differences, like depth or distance between samples.

The purpose of frequency histograms is to express fully all of the data, as opposed to expressing only its interpretation. However, the salinity converted from Tm ice (with statements in regard to the viability of the chemical model assumed — see Chapter 7), or the temperature of entrapment converted from Th (with statements justifying the pressure correction assumed — see Chapters 3 and 10) can be easily shown with additional axes.

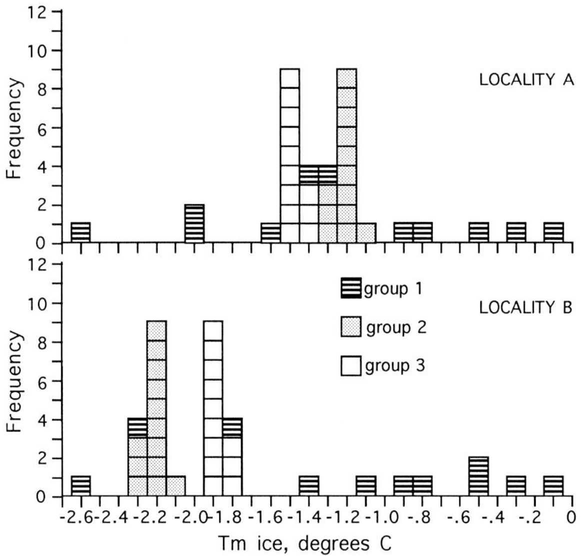

Fig. 8.3 Two frequency histograms of Tm ice data illustrating that separate histograms with the same scale can be stacked for comparison. Notice that data from each defined group of inclusions is given a different symbol. Typically, each group represents data from a separate fluid inclusion assemblage.

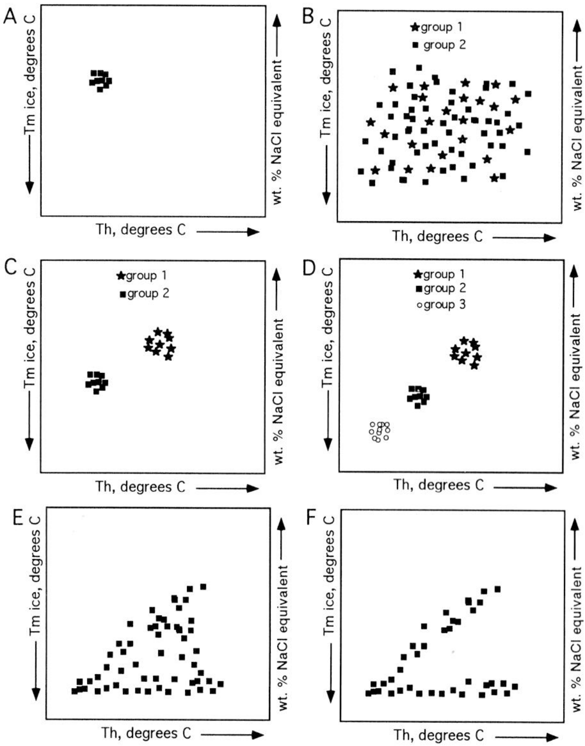

Fig. 8.4. *Bivariate plots illustrating different plausible distributions of Th and Tm ice data. Typically, each group represents data from a separate fluid inclusion assemblage.* ***A)*** *Tight distribution of Th and Tm ice data.* ***B)*** *Variably distributed data from two different groups.* ***C)*** *Distinct, tight clusters of data for inclusions of different groups.* ***D)*** *Distinct, tight clusters of data that indicate preservation of conditions for inclusions of different groups. If inclusions of different groups represent fluid inclusion assemblages that can be put in paragenetic sequence, then trends in evolution of temperature and salinity can be evaluated.* ***E)*** *Wedge-shaped distribution of Th and Tm ice data from a single fluid inclusion assemblage. The boundaries of the distribution converge on a point of low Th and low salinity. These data may result from reequilibration of some of the fluid inclusions by stretching, and/or leakage and refilling during conditions of increasing temperature and salinity.* ***F)*** *Two linear trends of Th and Tm ice data taken from a single fluid inclusion assemblage. The trends converge on a point of low Th and low salinity. These data may result from reequilibration of some of the fluid inclusions by stretching, or leakage and refilling during conditions of increasing temperature and salinity.*

BIVARIATE PLOTS

When Th and Tm ice can both be collected on an individual inclusion, then data from many such inclusions can be presented on a single plot to show exact positions of the data and possibly to display how the two parameters might covary (Fig. 8.4). Different symbols for inclusions of different FIAs should always be employed. As above, transformations of the data can be shown to display trapping temperatures and salinities if necessary.

Interpretation begins with well-presented (and, of course, appropriately collected!) data, as is readily illustrated in Figure 8.4. Interpretation of data presented in 8.4A, B, C, and D is rather obvious. Figure 8.4A shows a single FIA with very consistent data, whereas figures 8.4C and 8.4D show multiple FIAs with very consistent data. Something has happened to the two FIAs in Figure 8.4B as the data are not consistent. The trend of the data in figures 8.4E and 8.4F can readily be interpreted as FIAs showing different styles of reequilibration, which is explained in greater detail in Chapter 9.

SUMMARY

If fluid inclusion data are collected with respect to individual FIAs, then the presentation of the data is a relatively straightforward procedure. A proper presentation of data will readily portray results for each FIA studied, and will preserve all observations rather than generalize. Keep in mind, too, that interpretations can change with the paradigms of the times, whereas sound observations can be reinterpreted forever. The utility of recording sound observations is thus of utmost importance in the progress of science.

Chapter 9

PRACTICAL ASPECTS OF GEOTHERMOMETRY

INTRODUCTION

Ascertaining thermal information from fluid inclusions is known as fluid inclusion geothermometry. In Chapter 6, crude petrographic methods for determining thermal conditions of entrapment and post-entrapment history are introduced. The appropriate procedures for collecting microthermometric data from inclusions are presented in Chapter 7, in which the approach of collecting data from fluid inclusion assemblages (FIAs) is stressed. In Chapter 3 the basic principles for understanding the significance of microthermometric data are covered, and in Chapter 4 some processes that can modify inclusions and affect the resulting microthermometric data are treated. Thus, at this point, it should be relatively clear to readers that there is a sound theoretical basis for interpreting microthermometric data as meaningful indicators of some aspects of the thermal history that a rock has experienced. In addition, readers should have developed an appreciation for the fact that there are natural processes that occur which may render the interpretation of Th data less straightforward than they might have originally thought! The purpose of this chapter is to present a systematic framework for interpreting fluid inclusion microthermometric Th data in order to obtain valid information regarding temperature of mineral precipitation, minimum temperature of mineral precipitation, or merely temperatures that a rock has experienced. The limitations of various procedures will be stressed throughout this presentation.

Correct utilization of fluid inclusions as geothermometers begins with observations of fluid inclusion assemblages (FIAs) — groups of fluid inclusions that come from the most finely discernible event of inclusion entrapment. The reason for making observations with respect to individual FIAs is that variability within FIAs is the only means for assessing if an inclusion could have trapped more than one phase, or if the mass or volume of the fluid enclosed could have changed in an inclusion's history through some sort of reequilibration. FIAs of inclusions that trapped fluids of a single phase from a single geologic event should be expected to yield very consistent data. If data from an FIA is not consistent, then this is either an indication that inclusions within the FIA were entrapped over a period of time when a range of geologic conditions existed (inclusions not subsequently altered), or is an indication that one or more processes (natural or artificial) resulted in a violation of the fundamental requirements of trapping a homogeneous fluid and maintaining constant mass and volume throughout an inclusion's history. Thus, the systematics of interpreting Th's is based on the variability, or consistency, of data exhibited within individual FIAs, so this chapter is organized within a framework of the relative consistencies of microthermometric data from single FIAs.

FIAs WITH CONSISTENT DATA

What constitutes an FIA with consistent data? Based on considerable experience with various authigenic minerals, the authors choose to call an FIA consistent if two criteria are met:

1. The inclusions measured in the FIA must be of variable sizes and shapes (see Chapter 7).

2. 90% of the Th data from the FIA fall within a 10-15°C interval.

This rather arbitrary cut-off is based on our experience that FIAs yielding more variable data can often be correctly interpreted as having experienced some process that could have caused some degree of thermal reequilibration. The importance of satisfying the first criterion is that it is possible that inclusions of the same sizes and shapes could reequilibrate to the same degree, thus yielding consistent data that could not be distinguished from data of FIAs that had not reequilibrated. Inclusions of different sizes and shapes are expected to respond differently to processes of thermal reequilibration (see Chapter 4), and therefore are expected to yield a range of data if the processes had occurred. Thus, for FIAs showing consistency in Th data from inclusions of variable size and shape, the consistency in the data is strong evidence that the FIA had not undergone thermal reequilibration or significant necking after a phase change. In these cases, Th and salinity data should be a reliable record of entrapment conditions.

Minimum Entrapment Temperature

If an FIA yields consistent Th data among inclusions of various sizes and shapes (hereafter referred to as a consistent FIA), then the Th of the inclusions are a good measure of the minimum temperature at which they were entrapped (see Chapter 3). This interpretation of the Th data from a consistent FIA is very conservative, and has almost no chance of being incorrect! The next question appropriate to ask, however, is how much lower can the Th data be below the true entrapment temperature? An answer will depend on the composition of the fluid inclusion, its density, and the thermobaric gradient along which it was entrapped. For example, Th data of aqueous inclusions that are rich in CH_4 commonly closely approach the entrapment temperature (Fig. 3.7). For aqueous systems trapped at hydrostatic pressures, the pressure correction (amount of underestimation) is typically on the order of a few tens of degrees (see Fig. 4.4). For inclusions with very low Th's (<50°C; only very large inclusions will have Th's <50°C because smaller inclusions will not nucleate a bubble due to metastability) these amounts may increase because of some nonlinearity of high density isochores at these conditions (see Fig. 3.1B). Increasing salinity leads to decreased slopes of isochores (Fig. 3.3) so the underestimation increases slightly with increasing salinity. Alternatively, for aqueous inclusions trapped at pressures approaching lithostatic, the pressure correction may be on the order of 75°C (see Fig. 4.8). For some petroleum inclusions, isochoric slopes may be so low (see Fig. 3.8) that pressure corrections of great magnitudes may be required. Hence, the Th of petroleum fluid inclusions may be much lower than entrapment temperature if they were trapped well into the one-phase petroleum equilibrium field. Despite these limitations, the Th of aqueous inclusions provides a valid measure of the minimum entrapment temperature with an extremely low risk of being incorrect.

Entrapment Temperature

A less conservative approach for Th interpretation is to determine the temperature of entrapment by applying a pressure correction to the Th data from a consistent FIA. Correction of Th data to entrapment temperature (Tt) is a valid endeavor, but it is also a risky procedure that may be prone to some error if not applied carefully. To apply the correct pressure correction, first, enough must be known about the composition of the inclusion fluid so that the appropriate phase diagram can be selected. Then the pore fluid pressure that existed at the time of inclusion entrapment must be determined independently. Natural aqueous systems have variable salt contents, have various major ions, and some may be undersaturated or saturated with respect to various gases. Therefore, knowing composition is not always straightforward. Furthermore, phase diagrams with isochores of gas-bearing compositional systems may be based on equations of state constructed from scanty data, and may be difficult to evaluate. The hurdle of knowing the inclusions' compositions and their P-V-T relations may be insignificant in comparison to the requirement of an independent means of knowing the pressure. Procedures for determining pressure corrections are markedly different for those aqueous inclusions that contain gaseous components versus those that do not.

Gas-poor inclusions.—
Many of the techniques for determining the composition of the fluid in the inclusion have been summarized in Chapters 3, 6, and 7. There are well-known P-V-T relations for systems without gas, and there are some simple approaches that would identify these as applicable systems. Initially, one must rule out the presence of gases such as CH_4 and show that the only phase present in the bubble is water vapor. This is easily accomplished by crushing the inclusion while immersed in a liquid. Upon crushing a sample and opening an inclusion to one-atmosphere pressure, the inclusion's bubble will instantly collapse if it is just water vapor. Typically, the bubble will expand or at least not collapse completely if a gas component is present. The low-temperature phase behavior of a fluid inclusion can also be used to determine the composition of the fluid inclusion. For the gas-poor inclusions, no evidence of clathrate formation or melting, or evidence of freezing of the gas phase should be observed that would indicate the presence of gases other than water vapor (see Chapter 7). If gases are not detected, the low-temperature phase behavior can be used to constrain further the composition of the system. Observation of eutectic melting temperatures may be employed for determining which model water-salt system to assume (see Chapter 7). The final melting temperature (typically Tm ice) and, rarely, intermediate melting temperatures then can be used to interpret the salinity of the system. These combined data should constrain the composition of the fluid so that some P-V-T relations

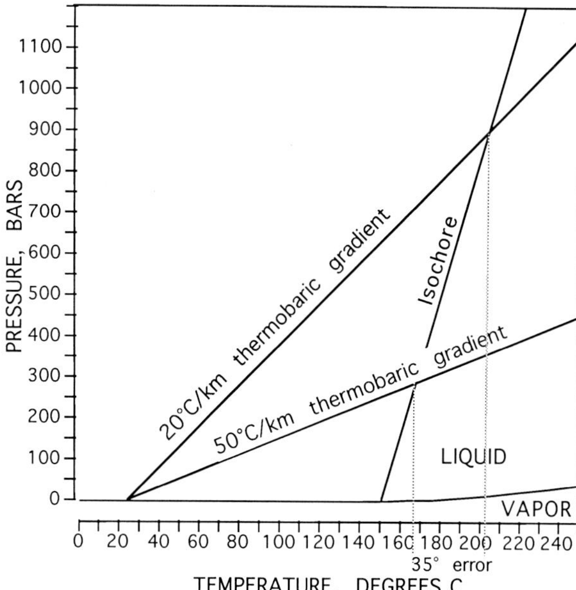

Fig. 9.1. Phase diagram for pure water with one of the isochores for pure water and the liquid-vapor field boundary. If a fluid inclusion homogenizes at 150°C, and is composed of pure water, it must have been entrapped along the isochore drawn. If one were to apply a pressure correction based on a present day thermobaric gradient of 20°C/100 bars, one would interpret entrapment temperature to be 203°C. However, if the true thermobaric gradient during inclusion entrapment were 50°C/100 bars, the assumption of present day thermobaric gradient would lead to a 35°C overestimation of entrapment temperature.

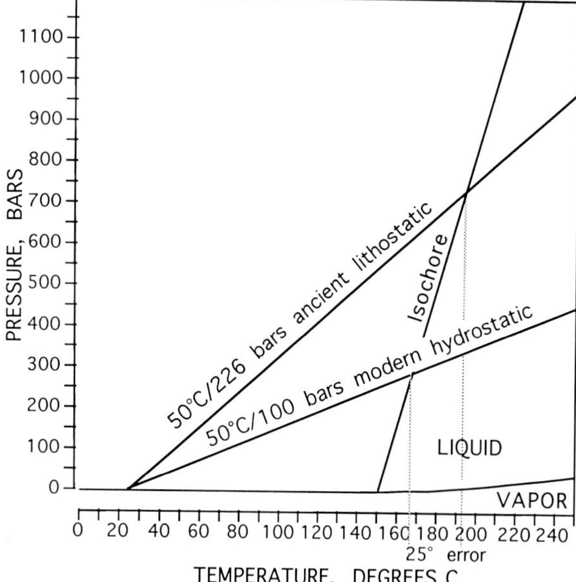

Fig. 9.2. Phase diagram for pure water with one of the isochores for pure water and the liquid-vapor field boundary. If a fluid inclusion homogenizes at 150°C, and is composed of pure water, it must have been entrapped along the isochore drawn. If one were to apply a pressure correction based on a present day hydrostatic thermobaric gradient of 50°C/100 bars, one would interpret entrapment temperature to be 167°C. However, if the true thermobaric gradient during inclusion entrapment were a lithostatic gradient of 50°C/226 bars, the assumption of present day thermobaric gradient would lead to a 25°C underestimation of entrapment temperature.

can be adopted to apply the pressure correction. Knowing which model system, its salinity, and Th defines the density of the fluid and thus the isochore along which the fluid inclusion was trapped. Isochores for various water-salt systems can be found in Fisher (1976), Potter (1977), and Potter and Brown (1975, 1977). Potter and Clynne (1978) illustrated that the $NaCl-H_2O$ system serves as a valid model for more complex brines. Their analyses showed that with atomic ratios less than 0.5 for Ca/Na, less than 0.3 for K/Na, and less than 0.2 for Mg/Na, the isochores determined for the $NaCl-H_2O$ system are very similar to those of more complex systems. Above these ratios, minor corrections might be needed.

The task of the geologist is now to determine independently the pressure at the time of inclusion entrapment. The P-T point at which this pressure intersects the isochore established above, yields the unique temperature of entrapment. Determining the pressure at the time of fluid inclusion entrapment is not a trivial matter. Pressure determination involves geologic interpretations or assumptions that are prone to error. To establish an ancient pressure at the time of inclusion entrapment, some employ the present-day to significant error. Different methods of pressure determination yield different sources and magnitudes of thermobaric gradient and assume that it is the gradient that existed at the time of inclusion entrapment. Others make some educated guess as to what the geothermal gradient might have been in the past. It is well known that heat flow and geothermal gradients in most areas change through time. Figure 9.1 illustrates the error inherent in misinterpreting the ancient thermobaric gradient. Similarly, if the pressure gradient has changed through time, error also can be introduced. For example, a hydrostatic thermobaric gradient might be assumed, based on a present-day system, but if the system actually had been under lithostatic pressure at the time of inclusion entrapment, the temperature of entrapment would be in error (Fig. 9.2). In summary, assumptions of ancient thermobaric gradients based on present day conditions may be valid, or may lead to error of tens of degrees when reconstructing the entrapment temperatures of FIAs. Other geothermometers such as vitrinite reflectance or illite crystallinity can be used to reconstruct an ancient geothermal gradient, and the burial history reconstructions can be used for reconstructing pressure. However, such an approach may also involve potentially significant errors due to limitations of the individual techniques and any assumptions required for the burial history reconstruction.

Another option is to simply make an educated guess as to what the pressure might have been at the time of fluid inclusion entrapment. A close guess may be possible in a simple basin in which core was taken from strata currently at maximum burial, and if there is evidence that the inclusions were entrapped at maximum burial. In such a case the current pressure may be appropriate, assuming there has not been a change in pressure (hydrostatic/lithostatic) since inclusion entrapment. From a geologic standpoint, one may never actually be able to know that the inclusions were trapped at time of maximum burial or that the pressure gradient has not changed. However, diagenetic events can be dated based on compaction features, cross-cutting relationships, and even may be dated radiometrically, so the timing of inclusion entrapment may be constrained within certain limits. In addition, there may be other cases in which the present-day pressure is not employed, but in which a relatively sound argument can be made for knowing the burial depth at the time of entrapment. For these instances, determining the ancient pressure from the burial history requires knowing the ancient pressure gradient (hydrostatic, lithostatic, or something in between the two), and making the wrong assumptions can cause potentially significant errors (see above). So, in the rare event that a close guess of pressure of entrapment can be hypothesized, the trapping temperature (Tt) is given by the intersection of the pressure of entrapment with the isochore determined from the Th and the composition of an FIA.

It should be clear from the above discussion that applying pressure corrections to FIAs requires that the researcher make geologic inferences, potentially causing significant errors. Many inclusionists choose not to apply such a correction because the Th data from a single FIA with consistent data is such a reliabie measurement of minimum entrapment temperature, and because this temperature in some cases closely approaches the true entrapment temperature.

Gas-rich inclusions.—

Aqueous fluid inclusions from the diagenetic realm commonly contain significant CH_4 as an additional component. In terms of applying pressure corrections, it is useful to separate inclusions with low methane content from those with high methane content. If a bubble does not expand beyond the volume of an inclusion upon crushing, the methane content can be shown to be less than about 1000 ppm; inclusions with bubbles that expand beyond the confines of the inclusion contain greater than about 1000 ppm CH_4. Each case is treated separately below.

Once crushing has identified that the gas bubble did not expand beyond the confines of a fluid inclusion, the same microthermometric procedures discussed in Chapter 3 and Chapter 7 for determining salinity and major ions should be applied to constrain the composition of the system. Entrapment temperature (Tt) for inclusions containing less than 1000 ppm CH_4 can be determined in a manner just as described above, ignoring the effects of CH_4 on the phase relations. Figure 9.3 demonstrates that using this method can cause an overestimation of the Tt by as much as 30°C for a 15 wt.% NaCl solution with 1000 ppm CH_4. However, this error does not include the additional errors that come with determining pressure, methane content, salt compositions, and construction of the phase diagrams! Furthermore, FIAs with consistent data and small amounts of CH_4 do yield Th's even closer to Tt than those of simple water-salt systems because homogenization takes place at higher pressures, closer to those of entrapment (less pressure correction is needed). Therefore, with all of the potential errors involved in applying a pressure correction to FIAs with small amounts of CH_4, some inclusionists are satisfied with not applying a pressure correction to this system.

For inclusions whose bubbles expand past the volume of the inclusion cavity upon crushing, the methane contents are generally higher, above about 1000 ppm. In our experience, it is difficult to determine reliably the volume of gas evolved upon crushing such an inclusion, which therefore precludes one from determining the amount of CH_4 from crushing. The phase equilibria of systems with such high gas contents differ significantly from those of simple H_2O-NaCl systems in that the liquid-vapor curves (bubble point curves) are at much greater pressures (Fig. 3.6). Therefore, applying a pressure correction by using the H_2O-NaCl system rather than using the correct phase relations of H_2O-NaCl-CH_4 can lead to more significant overestimation of entrapment temperature than the previous example with low methane content (Fig. 9.4). Applying a pressure correction using the H_20-NaCl-CH_4 system can prove to be a useful endeavor, but an inclusionist should have lots of time and money, or preferably a co-worker next door with a state-of-the-art Raman microprobe and a GC-MS to determine the gas content so that an appropriate phase diagram can be constructed or selected. These instruments are required to determine the

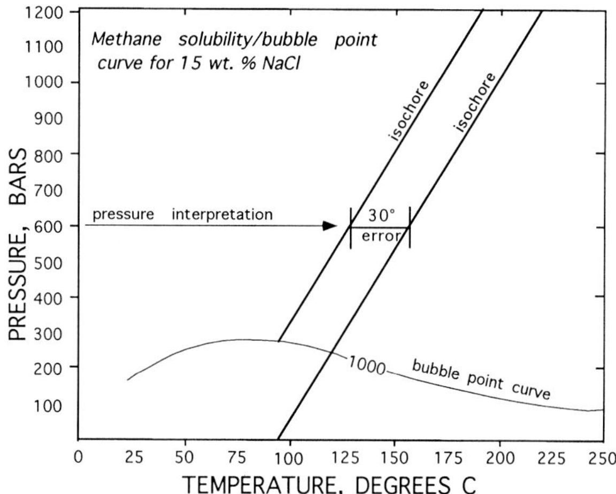

Fig. 9.3. Phase diagram illustrating bubble point curve for 1000 ppm methane in 15 wt.% NaCl solution. The two isochores drawn are for Th at about 90°C. The isochore that projects from the 1000 ppm bubble point curve is for an interpretation of 1000 ppm methane present in the inclusion. The other isochore extends down to a pressure near 0 and is for the purely aqueous system lacking methane. Note that an incorrect assumption of the purely aqueous system could lead to a 30°C overestimation of the trapping temperature when the pressure correction is applied.

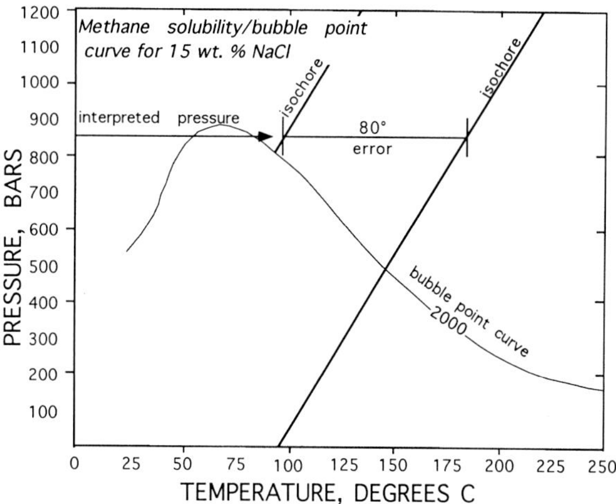

Fig. 9.4. Phase diagram illustrating bubble point curve for 2000 ppm methane in 15 wt.% NaCl solution. The two isochores drawn are for Th at about 90°C. The isochore that projects from the 2000 ppm bubble point curve is for an interpretation of 2000 ppm methane present in the inclusion. The other isochore extends down to a pressure near 0 and is for the purely aqueous system lacking methane. Note that an incorrect assumption of the purely aqueous system could lead to a 80°C overestimation of the trapping temperature when the pressure correction is applied.

compositions of inclusions so that the appropriate phase equilibria can be determined.

Generally, it is difficult to acquire or calculate phase equilibria for complex systems containing multiple gas components, water, and various salts of various concentration. Therefore, for most work, one must be lucky enough that the composition of the FIA under study is simple enough to be able to reconstruct the P-V-T relations. The salts present and their concentrations are determined by observations of eutectic melting phenomena, intermediate melting events and final melting temperature as described in Chapter 7. The salinity determination from final melting temperature is only slightly affected by high methane contents to produce a small source of error (Chapter 7). Several techniques are useful for determining if inclusions contain pure CH_4 or contain other components that could complicate the phase equilibria. Low-temperature phase equilibria may be used as described in Chapter 7, and deviations from ideal methane behavior indicate the presence of other gases. However, the most commonly available tool to use as a first approximation to test for the purity of CH_4 is the crushing stage. When crushing a sample in kerosene, organic gas bubbles that emanate from the inclusion will dissolve in the kerosene, whereas non-organic gases (N_2, CO_2, H_2S, etc.) will persist. A bubble remaining in kerosene after crushing would indicate that the system is not pure CH_4, so simple phase equilibria would not apply. If all the gas dissolves in kerosene, then the next step is to attempt to physically isolate a single FIA and subject it to bulk gas analysis (GC, GC-MS, and MS, see Chapter 12). Again, if considerable amounts of other gas components are present (probable), then the relatively simple phase relations of the H_2O-CH_4-NaCl system would not apply. If only CH_4 is observed through bulk analytical techniques, then the next step is to attempt to use the laser Raman microprobe to determine the methane pressure in the bubble. The laser Raman microprobe has inherent limits of detection which are a function of CH_4 internal pressure, bubble size, bubble shape, proximity to surface of sample, inclusion shape, optical characteristics of the mineral, and interference with fluorescence and other Raman bands caused by the host mineral. Although the Raman microprobe is capable of identifying other gas species in an inclusion bubble, these variable controls on detection limits require that bulk gas analyses be done to demonstrate their absence. Luckily, typical limits of detection for CH_4 are quite low and it is readily identifiable using the Raman microprobe (Wopenka and Pasteris, 1987). If one is able to obtain a Raman spectrum of CH_4 in the inclusion bubble (Raman spectra of other gas species should be missing), then a pressure can be approximated from the relative wave number position of the Raman band (Fig. 9.5; Fabre and Couty, 1986). Given an accuracy of ±0.5 wave numbers, the accuracy of CH_4 the pressure determination is about ±20 bars at lower pressures (50-100 bars), and about ±300 bars at higher pressures (above about 500 bars). Once the pressure has been determined using the Raman microprobe, one must petrographically measure the volume of the bubble and the volume of the aqueous phase in the inclusion at room temperature, to be able to calculate the amount of CH_4 dissolved in the aqueous phase at the time of entrapment. Because inclusions are of complex shapes, there will be some error in the volume determinations that could affect the resulting concentration of CH_4 by as much as 25%. This error can be minimized by choosing regularly shaped inclusions with spherical bubbles. Use of the spindle stage (Anderson and others, 1992) coupled with image analysis (Itard and others, 1989) could decrease the error, but the effort required in using these tools may not be justified, as the errors due to volume determination are surprisingly small when compared to the errors caused by making the geologic interpretation of pressure of entrapment!

With the gas compositional information of ppm CH_4 in the inclusion determined from the Raman microprobe, and with the salt contents determined from microthermometry, the composition of inclusions is constrained enough to construct or select an appropriate phase diagram in the system H_2O-NaCl-CH_4 (Haas, 1978; McGee and others, 1981; Blount and Price, 1982; Duan and others, 1992). Tabular data of CH_4 solubilities that can be used to construct bubble point curves are presented in Table 9.1 or can be calculated from the equation of state by Duan and others (1992). Drawing isochores for the system in the one-phase field can be simplified by using isochores for the H_2O-NaCl system discussed above, as the amount of CH_4 in fluids in the sedimentary realm <15,000 ppm is so small that the molar volumes of the aqueous fluids should not be affected significantly (R. Burruss, personal communication, 1993). To find the appropriate isochore: 1) use the determined bubble point curve to determine the pressure at homogenization; 2) using eutectic and final melting temperature data, look up the P-T data for the appropriate aqueous-salt system; 3) determine which isochore for that salt composition extends through the P-T condition at Th (from step 1) on the determined bubble point curve; and 4) extend the isochore above the bubble point curve to create the approximate isochore appropriate for the FIA.

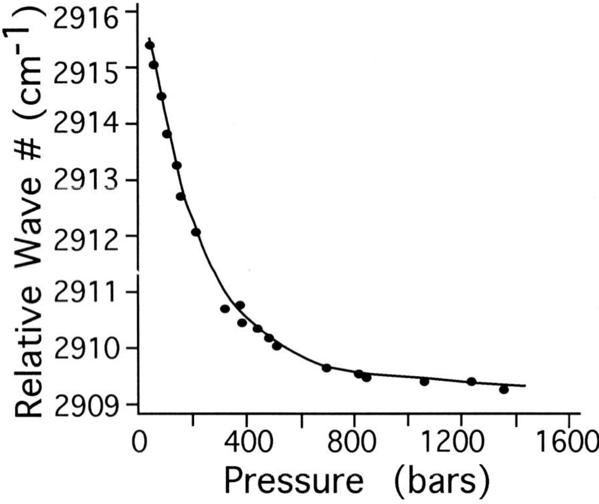

Fig. 9.5. Relationship between relative wave number and methane pressure determined from the laser Raman microprobe. Data from Fabre and Couty (1986).

The above exercise has only produced the appropriate phase diagram and isochore — one still needs the pressure! So, the same error-prone geologic interpretation of pressure must still be made in order to determine the pressure correction. Therefore, the authors highly recommend that a researcher first evaluate whether all of the above rigor is worth the effort. In the cases where the pressure is known, perhaps so. But in the many instances in which much interpretation is involved in assessing the pressure, then probably not. What course of action would be appropriate? Remember, the higher the CH_4 content, the higher the pressure will be at which homogenization takes place; thus, the Th's of inclusions with high CH_4 contents are even closer to the Tt's of inclusions with lower amounts of CH_4. In conclusion, an excellent approach would be to apply no pressure correction at all, and thus, to feel confident that the Th of a consistent FIA containing significant CH_4 is a reliable estimate of the minimum temperature of entrapment, and that Th may, in fact, be as close of an approximation to Tt as one could ever hope to achieve.

FIAs WITH VARIABLE DATA

Some FIAs yield homogenization temperatures that are geologically reasonable, but that are more variable than the "consistent" cutoff of 90% within 10-15°C. Such FIAs may represent either entrapment over a range of geologic conditions or thermal reequilibration of an originally consistent FIA. This section describes ways in which one might be able to extract useful geothermometry from some of these ambiguous FIAs. Other FIAs have inclusions with highly variable liquid-to-vapor ratios ranging from all-liquid to all-vapor one-

Table 9.1. *Methane solubilities (m/kg water) in NaCl aqueous solutions**

P (bar)	NaCl=0 T (°C)									
	0	30	60	90	120	150	180	210	240	270
1	0.0023	0.0012	0.0008	0.0000	0.0000	0.0000	0.0000	0.0000	0.0000	0.0000
50	0.0974	0.0547	0.0412	0.0380	0.0401	0.0455	0.0523	0.0556	0.0431	0.0000
100	0.1623	0.0949	0.0736	0.0697	0.0755	0.0892	0.1103	0.1368	0.1614	0.1625
150	0.2060	0.1249	0.0993	0.0957	0.1055	0.1271	0.1614	0.2092	0.2678	0.3245
200	0.2382	0.1481	0.1202	0.1177	0.1314	0.1603	0.2068	0.2742	0.3640	0.4715
300	0.2876	0.1840	0.1531	0.1530	0.1740	0.2160	0.2840	0.3859	0.5309	0.7280
400	0.3285	0.2129	0.1793	0.1813	0.2084	0.2615	0.3477	0.4788	0.6705	0.9435
500	0.3657	0.2381	0.2017	0.2053	0.2375	0.2999	0.4017	0.5576	0.7892	1.1267
600	0.4010	0.2613	0.2218	0.2264	0.2628	0.3332	0.4482	0.6255	.8913	1.2837
700	0.4351	0.2830	0.2402	0.2454	0.2854	0.3626	0.4891	0.6848	0.9799	1.4193
800	0.4684	0.3037	0.2574	0.2629	0.3057	0.3888	0.5252	0.7369	1.0573	1.5368
900	0.5012	0.3236	0.2735	0.2789	0.3242	0.4124	0.5574	0.7829	1.1252	1.6390
1000	0.5337	0.3429	0.2888	0.2939	0.3412	0.4338	0.5862	0.8237	1.1848	1.7278
1100	0.5659	0.3616	0.3034	0.3079	0.3568	0.4531	0.6120	0.8600	1.2372	1.8050
1200	0.5982	0.3799	0.3174	0.3211	0.3712	0.4708	0.6353	0.8922	1.2833	1.8718
1300	0.6303	0.3978	0.3308	0.3335	0.3846	0.4869	0.6562	0.9208	1.3236	1.9294
1400	0.6626	0.4154	0.3437	0.3452	0.3970	0.5016	0.6751	0.9462	1.3588	1.9788
1500	0.6949	0.4326	0.3561	0.3563	0.4086	0.5150	0.6920	0.9686	1.3894	2.0207
1600	0.7274	0.4497	0.3681	0.3668	0.4193	0.5273	0.7072	0.9884	1.4158	2.0560
1700	0.7601	0.4665	0.3798	0.3767	0.4293	0.5385	0.7207	1.0057	1.4384	2.0852
1800	0.7930	0.4831	0.3910	0.3862	0.4386	0.5487	0.7329	1.0208	1.4575	2.1090

P (bar)	NaCl=1 m T (°C)									
	0	30	60	90	120	150	180	210	240	270
1	0.0019	0.0010	0.0006	0.0000	0.0000	0.0000	0.0000	0.0000	0.0000	0.0000
50	0.0788	0.0442	0.0332	0.0306	0.0323	0.0366	0.0420	0.0446	0.0345	0.0000
100	0.1304	0.0762	0.0591	0.0559	0.0605	0.0714	0.0882	0.1093	0.1288	0.1296
150	0.1645	0.0997	0.0793	0.0764	0.0842	0.1014	0.1286	0.1666	0.2131	0.2579
200	0.1891	0.1177	0.0955	0.0935	0.1044	0.1274	0.1643	0.2176	0.2887	0.3737
300	0.2260	0.1449	0.1206	0.1207	0.1373	0.1705	0.2242	0.3045	0.4187	0.5739
400	0.2557	0.1662	0.1402	0.1420	0.1634	0.2052	0.2729	0.3758	0.5263	0.7404
500	0.2824	0.1845	0.1567	0.1598	0.1851	0.2341	0.3137	0.4357	0.6168	0.8805
600	0.3075	0.2012	0.1714	0.1754	0.2039	0.2589	0.3486	0.4868	0.6940	0.9997
700	0.3317	0.2168	0.1847	0.1893	0.2205	0.2807	0.3791	0.5312	0.7606	1.1021
800	0.3554	0.2317	0.1972	0.2020	0.2355	0.3001	0.4059	0.5702	0.8187	1.1906
900	0.3790	0.2461	0.2090	0.2138	0.2492	0.3176	0.4299	0.6046	0.8697	1.2676
1000	0.4027	0.2603	0.2203	0.2249	0.2618	0.3336	0.4516	0.6353	0.9147	1.3348
1100	0.4267	0.2742	0.2312	0.2355	0.2736	0.3483	0.4712	0.6629	0.9547	1.3937
1200	0.4510	0.2882	0.2419	0.2456	0.2847	0.3619	0.4891	0.6878	0.9903	1.4454
1300	0.4759	0.3021	0.2523	0.2553	0.2953	0.3746	0.5057	0.7104	1.0221	1.4909
1400	0.5015	0.3162	0.2627	0.2648	0.3053	0.3865	0.5210	0.7311	1.0508	1.5310
1500	0.5279	0.3304	0.2731	0.2740	0.3150	0.3978	0.5352	0.7499	1.0765	1.5664
1600	0.5553	0.3449	0.2834	0.2831	0.3244	0.4086	0.5486	0.7674	1.0999	1.5977
1700	0.5838	0.3598	0.2938	0.2922	0.3335	0.4189	0.5612	0.7835	1.1210	1.6254
1800	0.6136	0.3751	0.3043	0.3012	0.3425	0.4289	0.5731	0.7985	1.1403	1.6499

P (bar)	NaCl=2 m T (°C)									
	0	30	60	90	120	150	180	210	240	270
1	0.0016	0.0008	0.0005	0.0000	0.0000	0.0000	0.0000	0.0000	0.0000	0.0000
50	0.0645	0.0361	0.0271	0.0250	0.0263	0.0298	0.0341	0.0362	0.0280	0.0000
100	0.1061	0.0620	0.0480	0.0454	0.0491	0.0579	0.0714	0.0884	0.1041	0.1046
150	0.1330	0.0807	0.0641	0.0618	0.0680	0.0819	0.1038	0.1343	0.1717	0.2075
200	0.1520	0.0947	0.0769	0.0753	0.0840	0.1025	0.1321	0.1749	0.2319	0.2999
300	0.1797	0.1155	0.0963	0.0964	0.1097	0.1363	0.1792	0.2433	0.3344	0.4581
400	0.2015	0.1313	0.1110	0.1126	0.1297	0.1630	0.2169	0.2987	0.4182	0.5882
500	0.2207	0.1448	0.1233	0.1260	0.1462	0.1850	0.2481	0.3447	0.4880	0.6967
600	0.2387	0.1569	0.1341	0.1375	0.1602	0.2037	0.2745	0.3837	0.5471	0.7883
700	0.2560	0.1682	0.1438	0.1478	0.1726	0.2200	0.2975	0.4173	0.5979	0.8665
800	0.2731	0.1790	0.1530	0.1572	0.1837	0.2345	0.3177	0.4467	0.6419	0.9340
900	0.2903	0.1896	0.1616	0.1660	0.1939	0.2477	0.3358	0.4728	0.6807	0.9927
1000	0.3077	0.2000	0.1701	0.1743	0.2034	0.2598	0.3522	0.4962	0.7151	1.0442
1100	0.3257	0.2106	0.1784	0.1823	0.2125	0.2710	0.3673	0.5174	0.7459	1.0897
1200	0.3443	0.2213	0.1866	0.1901	0.2211	0.2816	0.3813	0.5369	0.7738	1.1302
1300	0.3638	0.2323	0.1949	0.1979	0.2295	0.2918	0.3945	0.5550	0.7993	1.1666
1400	0.3844	0.2437	0.2034	0.2056	0.2378	0.3016	0.4071	0.5719	0.8227	1.1995
1500	0.4061	0.2555	0.2120	0.2134	0.2459	0.3111	0.4192	0.5879	0.8446	1.2295
1600	0.4293	0.2679	0.2209	0.2213	0.2541	0.3206	0.4309	0.6033	0.8651	1.2571
1700	0.4541	0.2810	0.2302	0.2294	0.2624	0.3299	0.4424	0.6181	0.8846	1.2828
1800	0.4807	0.2948	0.2399	0.2378	0.2708	0.3394	0.4538	0.6325	0.9033	1.3069

after Duan and others (1992)

Table 9.1. *Continued from previous page*
*Methane solubilities (m/kg water) in NaCl aqueous solutions**

P (bar)	NaCl=4 m T (°C)									
	0	30	60	90	120	150	180	210	240	270
1	0.0011	0.0006	0.0004	0.0000	0.0000	0.0000	0.0000	0.0000	0.0000	0.0000
50	0.0448	0.0251	0.0188	0.0173	0.0181	0.0205	0.0234	0.0248	0.0191	0.0000
100	0.0728	0.0425	0.0329	0.0311	0.0335	0.0395	0.0486	0.0601	0.0706	0.0707
150	0.0902	0.0548	0.0436	0.0419	0.0461	0.0554	0.0702	0.0907	0.1156	0.1395
200	0.1020	0.0636	0.0517	0.0507	0.0565	0.0689	0.0887	0.1173	0.1553	0.2005
300	0.1181	0.0761	0.0637	0.0638	0.0727	0.0904	0.1188	0.1612	0.2215	0.3031
400	0.1299	0.0852	0.0723	0.0735	0.0849	0.1068	0.1422	0.1959	0.2742	0.3855
500	0.1401	0.0925	0.0792	0.0813	0.0946	0.1199	0.1611	0.2240	0.3172	0.4529
600	0.1494	0.0990	0.0852	0.0878	0.1027	0.1309	0.1768	0.2474	0.3530	0.5089
700	0.1584	0.1050	0.0905	0.0935	0.1097	0.1403	0.1902	0.2673	0.3834	0.5561
800	0.1674	0.1109	0.0955	0.0988	0.1160	0.1487	0.2020	0.2847	0.4097	0.5967
900	0.1767	0.1167	0.1004	0.1038	0.1219	0.1564	0.2127	0.3001	0.4329	0.6320
1000	0.1865	0.1227	0.1053	0.1086	0.1275	0.1635	0.2225	0.3142	0.4537	0.6633
1100	0.1970	0.1289	0.1102	0.1135	0.1330	0.1704	0.2317	0.3273	0.4727	0.6915
1200	0.2083	0.1355	0.1153	0.1184	0.1384	0.1771	0.2406	0.3397	0.4905	0.7174
1300	0.2207	0.1426	0.1207	0.1235	0.1439	0.1838	0.2493	0.3516	0.5073	0.7414
1400	0.2344	0.1503	0.1265	0.1288	0.1496	0.1906	0.2581	0.3634	0.5236	0.7643
1500	0.2495	0.1586	0.1327	0.1344	0.1556	0.1976	0.2669	0.3751	0.5397	0.7864
1600	0.2663	0.1678	0.1394	0.1404	0.1619	0.2049	0.2760	0.3871	0.5557	0.8080
1700	0.2851	0.1779	0.1467	0.1469	0.1686	0.2125	0.2855	0.3993	0.5719	0.8295
1800	0.3063	0.1891	0.1546	0.1539	0.1757	0.2206	0.2954	0.4120	0.5885	0.8513

P (bar)	NaCl=6 m T (°C)									
	0	30	60	90	120	150	180	210	240	270
1	0.0008	0.0004	0.0003	0.0001	0.0015	0.0018	0.0022	0.0030	0.0041	0.0058
50	0.0328	0.0183	0.0137	0.0126	0.0132	0.0148	0.0169	0.0178	0.0137	0.0058
100	0.0526	0.0307	0.0237	0.0223	0.0241	0.0283	0.0348	0.0429	0.0503	0.0503
150	0.0644	0.0391	0.0311	0.0299	0.0329	0.0395	0.0499	0.0643	0.0819	0.0986
200	0.0719	0.0450	0.0366	0.0358	0.0400	0.0487	0.0626	0.0827	0.1093	0.1409
300	0.0815	0.0528	0.0442	0.0444	0.0507	0.0630	0.0828	0.1123	0.1541	0.2107
400	0.0881	0.0581	0.0495	0.0505	0.0584	0.0735	0.0980	0.1350	0.1890	0.2656
500	0.0934	0.0622	0.0535	0.0551	0.0643	0.0818	0.1099	0.1530	0.2168	0.3095
600	0.0982	0.0657	0.0569	0.0589	0.0692	0.0884	0.1196	0.1676	0.2395	0.3453
700	0.1030	0.0690	0.0599	0.0622	0.0733	0.0941	0.1278	0.1800	0.2585	0.3752
800	0.1079	0.0722	0.0627	0.0653	0.0771	0.0991	0.1350	0.1907	0.2749	0.4007
900	0.1131	0.0756	0.0656	0.0682	0.0806	0.1038	0.1416	0.2003	0.2894	0.4230
1000	0.1189	0.0791	0.0685	0.0712	0.0840	0.1082	0.1477	0.2092	0.3025	0.4430
1100	0.1253	0.0830	0.0716	0.0742	0.0875	0.1126	0.1537	0.2176	0.3149	0.4613
1200	0.1325	0.0872	0.0749	0.0775	0.0911	0.1171	0.1596	0.2259	0.3268	0.4786
1300	0.1408	0.0920	0.0786	0.0809	0.0949	0.1217	0.1657	0.2342	0.3385	0.4954
1400	0.1502	0.0974	0.0827	0.0847	0.0990	0.1266	0.1720	0.2427	0.3503	0.5119
1500	0.1611	0.1035	0.0873	0.0890	0.1035	0.1319	0.1787	0.2516	0.3625	0.5287
1600	0.1737	0.1105	0.0924	0.0936	0.1084	0.1376	0.1858	0.2610	0.3752	0.5459
1700	0.1882	0.1184	0.0982	0.0988	0.1138	0.1439	0.1936	0.2711	0.3886	0.5639
1800	0.2051	0.1276	0.1048	0.1047	0.199	0.1508	0.2021	0.2821	0.4030	0.5829

**after Duan and others (1992)*

phase inclusions with all possible phase ratios in the associated two-phase inclusions. Such FIAs would typically yield some homogenization temperatures that are higher than temperatures geologically reasonable, and may contain inclusions with low liquid-to-vapor ratios at temperatures exceeding 300°C. These types of FIAs are also treated in this section.

Highly Variable Liquid-to-Vapor Ratios

As was discussed in Chapter 3 and Chapter 7, there are three potential explanations for FIAs that have such variable liquid-to-vapor ratios that they produce geologically unreasonable Th data. They could result from necking down after a phase change, heterogeneous entrapment in a low-temperature two-phase system such as the vadose zone, or heterogeneous entrapment of elevated-temperature (>50°C) gas-aqueous fluid inclusions that existed as a two-phase immiscible system during inclusion entrapment. The Th data from FIAs that have necked down after a phase change will not yield useful information on temperature of entrapment. Similarly, homogenization temperatures of fluid inclusions trapped from low-temperature, two-phase systems such as the vadose zone will be meaningless. However, if an inclusionist can determine that an FIA shows variable liquid-to-vapor ratios due to entrapment at a time when a elevated-temperature gas-aqueous fluid system contained two phases, then some useful Th data may be attainable from the FIA: such useful data may be obtained from nearby (petrographically related) FIAs that fortuitously trapped end-member (all one-phase) compositions of the immiscible fluid.

Several characteristics of highly variable liquid-to-vapor ratio FIAs have proven useful in distinguishing between the three modes of origin. The presence of some one-phase, all-liquid aqueous inclusions (which do not result from significant metastability) are common in low-temperature, two-phase systems such as the vadose zone and also may be common in populations of necked-down fluid inclusions. These all-liquid inclusions should be absent in elevated-temperature, gas-aqueous immiscible FIAs. Crushing analysis also can be used to distinguish between the three modes of origin. Only bubbles of inclusions in FIAs that are trapped at depth from a fluid containing gas will expand on crushing in the lab. All other bubbles will contract and disappear except those from the vadose zone which will remain the same. It may be difficult to distinguish an elevated-temperature, gas-aqueous immiscible FIA from an elevated-temperature gas-rich FIA that trapped a liquid in the one-phase field and then necked down after cooling and nucleating another gas-rich phase, but that has no all-liquid inclusions that could have been a clue to necking down. Other petrographic characteristics that may be helpful in distinguishing between these two possibilities are:

a. If within a single FIA, the overall volume of the gas phase in the inclusions is higher (typically >20%) than would be expected from necking down (after a phase change) of elevated-temperature inclusions trapped in the one-phase field, then heterogeneous entrapment of an immiscible fluid is a possibility, and in this case, more gas was entrapped in the FIA.

b. When an elevated-temperature fluid inclusion trapped in the one-phase field necks down after it has nucleated a bubble, upon further cooling, a bubble may nucleate in the separated inclusion. One fluid inclusion will have a high liquid-to-vapor ratio relative to the other and occur in petrographic association with the other. If one observes such petrographic association (petrographic pairing) then necking down is a possible explanation for the variability in liquid-to-vapor ratio.

Once the above petrographic characteristics have been investigated, and if the inclusionist has decided that an FIA with variable liquid-to-vapor ratios formed by elevated-temperature immiscibility from a gas-aqueous system, then one should search for other nearby, petrographically related FIAs with consistent liquid-to-vapor ratios. These FIAs may have inclusions with all gas (commonly dark in color and no visible meniscus), or have inclusions dominated by liquid H_2O with gas bubbles and exhibiting consistent liquid-to-vapor ratios and yielding consistent Th. Rarely, one may be fortunate enough to find both gas- and liquid-rich end-members in a variable FIA, with the liquid-rich aqueous inclusions yielding consistent Th data. The homogenization temperatures of these liquid-rich end-members (yielding consistent Th within a single FIA) are equal to the temperature of entrapment. No pressure correction is required (see Chapter 3). Similarly, petroleum-gas immiscibility is possible in nature, and if demonstrated petrographically in fluid inclusions, the Th of oil-rich inclusions also is equal to Tt.

Even though the following is not specifically related to geothermometry, it is the appropriate point to discuss the validity of compositional data collected from FIAs of highly variable liquid-to-vapor ratios. For all-liquid inclusions that have been demonstrated to be useful, and not the result of necking down or metastability, the all-liquid inclusions must contain the fluids present at the time of entrapment and have not been refilled during thermal reequilibration, so the salinity data are a faithful record of the properties of the fluid entrapped at temperatures below about 50°C. For two-phase inclusions that record one-atmosphere pressure (vadose zone), salinity determinations also reveal composition of the fluids at the time of entrapment, because these fluid inclusions could not have leaked and refilled during thermal reequilibration; if they had, their internal pressures would not remain one atmosphere. This should be true for both liquid-rich and vapor-rich inclusions. Finally, the FIAs yielding consistent Th's, but that formed from elevated-temperature immiscible fluids, must yield salinity data that are representative of the fluids present at the time of initial vacuole formation, because if they were reequilibrated in any way, the Th's for various sized and shaped inclusions would not be consistent. The other fluid inclusions in the FIAs discussed above may or may not record salinity reliably at time of initial entrapment, because leakage and refilling cannot be ruled out for them. For example, inclusions yielding variable Th data only because of necking will preserve a reliable record of salinity only if they have not reequilibrated. But since the Th data are inconsistent, reequilibration could have occurred to some degree before or after necking was completed. Inclusions that have undergone simple stretching during thermal reequilibration also should record salinity reliably, as well as those FIAs yielding variable Th only because they were trapped from an immiscible high temperature gas-aqueous system. However, for highly variable FIAs, there is no general test (similar to consistency of Th's among inclusions of various sizes and shapes) to determine whether or not inclusion vacuoles leaked and refilled during thermal reequilibration to alter the fluid composition filling the vacuoles. Nevertheless, it is important to attempt to identify such altered inclusions because they may preserve a valuable record of fluids in which a rock was bathed during thermal reequilibration. Salinity determinations (mostly Tm ice) on fluid inclusions may provide some useful clues. If salinities of some inclusions in an FIA can be shown to be unaltered, they can be compared to other fluid inclusions in the FIA: if all-liquid inclusions (not from necking or metastability), consistent end-member FIAs of immiscible systems, or inclusions with internal pressure of one atmosphere are preserved, and their salinities differ from other inclusions in the same FIA, then leakage and refilling of part of that FIA may have taken place during thermal reequilibration. However, if the entire FIA contains the same salinities as the other unaltered fluid inclusions in the FIA, then the entire FIA likely preserves original conditions; that is, leakage and refilling has not taken place.

Moderately Variable Th Data

Moderately variable Th data within an FIA is characterized easily as data which fall between the two end-members discussed above. Variability is outside the (90% within 10-15°C) cut-off, but all data are geologically reasonable. Moderately variable Th data for individual FIAs may be due to thermal reequilibration, undetectable necking down, or may represent unaltered fluid inclusions collected from FIAs with real variability; that is, FIAs that formed over a range of P-T-X conditions through time. Our experience with data from fluid inclusions in sedimentary systems is that a significant proportion of FIAs yield such moderately variable data. Whereas many of these data will yield potentially useful information about P-T-X conditions that a rock may have experienced, in many such FIAs, the original formation conditions may never be determined from the microthermometric data. As already discussed in Chapter 4, this limitation is due to the fact that one may not be able to distinguish thermal reequilibration of the fluid inclusions from an originally consistent FIA, from those inclusions that faithfully record variable initial conditions of entrapment. There are several approaches that may help discriminate between these two possible scenarios.

All-liquid inclusions present.—
The first approach is to look for all-liquid inclusions in the FIA. Try to determine if these all-liquid inclusions provide a useful indication of low-temperature entrapment, or if they result from necking down or significant metastability. Remember that the petrographic pairing of all-liquid inclusions with vapor-rich inclusions, and the overall ratio of liquid to vapor in the FIA can be used as evidence for necking down after a phase change. Significantly metastable all-liquid inclusions are those that have a density that should yield homogenization above about 50°C. Commonly, such significantly metastable inclusions may yield vapor bubbles during cooling, or are only the smaller inclusions in the FIA. If all-liquid inclusions do not generate bubbles during cooling, if they are in the same size range as the two-phase inclusions of the FIA (all-liquid inclusions are not only the small ones), and if they do not result from necking down, then all-liquid inclusions are not significantly metastable and were entrapped below *about* 50°C. If such inclusions are found in an FIA (perhaps along with two-phase inclusions), then conditions of initial entrapment (for the all-liquid inclusions) were below *about* 50°C (Chapters 3, 4, 7; Goldstein, 1990, 1993). In addition, these all-liquid inclusions should provide reliable records of the salinity. Unless there is some petrographic evidence to somehow further subdivide the FIA (already defined as the finest subdivision), it may be difficult to interpret the origin of any two-phase inclusions that co-occur with the all-liquid fluid inclusions. They may have formed from thermal reequilibration of low-temperature fluid inclusions, or represent conditions of original entrapment at conditions higher than about 50°C. Nevertheless, even if the salinities of these two-phase fluid inclusions may not necessarily record conditions at initial entrapment, they do record fluid compositions experienced by the rock during its history. The same may be said for the Th values unless high Th values result from undetected necking down or stretching from high internal gas pressures. Of course, relative consistency of data within a closely-spaced, petrographically associated cluster of inclusions of variable size and shape within one FIA, would argue against high Th values due to undetected necking down.

Another approach to understanding moderately variable Th data is to compare the data from different FIAs in the same sample, in which one assemblage contains all-liquid fluid inclusions. The FIAs to be compared should be in the same mineral (in the same sample) so that each could be assumed to respond similarly to overheating. This would be an appropriate assumption if both FIAs contain inclusions of similar ranges of sizes and shapes. For instance, if an early FIA contains very consistent Th data of lower temperature than a later FIA (Fig 9.6A) , or contains all single-phase liquid inclusions (Fig. 9.6B), the variable data from the later FIA record real variation in the temperature of entrapment and have not been altered by thermal reequilibration; otherwise, the earlier FIA should show variation also. Similarly, if an FIA that is trapped late in the petrographic sequence contains only two-phase inclusions that yield moderately variable data, one can use the presence of all-liquid inclusions in an earlier FIA to show that at least the lower Th data in the later assemblage record conditions of initial entrapment by the following logical reasoning. If the paragenetically earlier FIA contains some all-liquid inclusions that are not a result of metastability or necking down, together with some two-phase inclusions exhibiting variable Th, then it is apparent that at least some of the lowest temperature inclusions (the all-liquid inclusions) have survived overheating without reequilibration (Fig 9.6C). If the lowest temperature inclusions have survived overheating without reequilibration in the earlier FIA, then the lowest temperature inclusions in the later FIA also would have survived overheating without reequilibration. Therefore, the lowest Th data in the later FIA must be recording conditions of initial entrapment. The remainder of the Th data from the later FIA may record original conditions of entrapment, but also may result from thermal reequilibration.

Convergent trends.—

Given moderately variable Th data from a single FIA, there are other possibilities for distinguishing reequilibrated inclusion data from unaltered fluid inclusions that preserve reliable temperatures and salinities. If an FIA were formed at some burial temperature (above 50°C) and then were to be buried deeper, during the increase in temperature, some of the inclusions may stretch their walls, which would increase the volume of cavity and hence result in an increase of Th when measured in the lab. Because variables such as size, shape and position within the crystal affect the amount and timing of stretching, the altered population may yield variable Th but still would retain the same salinity. Additionally, some of the inclusions would tend to leak and refill during burial heating. In many basins, pore fluid salinity increases monotonically or increases and levels off with increasing depth (Hanor, 1984). Fluid inclusions that have leaked and refilled could contain fluids of any of the salinities or temperatures experienced during progressive burial. Therefore, if both mechanisms of thermal reequilibration were acting, and if once a fluid inclusion stretches, it continues to do so sporadically as burial depth increases, and once a fluid inclusion leaks, it continues to do so sporadically, then the Th and salinity data of aqueous inclusions would define two trends on a bivariate plot, a stretching trend and a leakage and refilling trend (Fig. 9.7A). Both trends

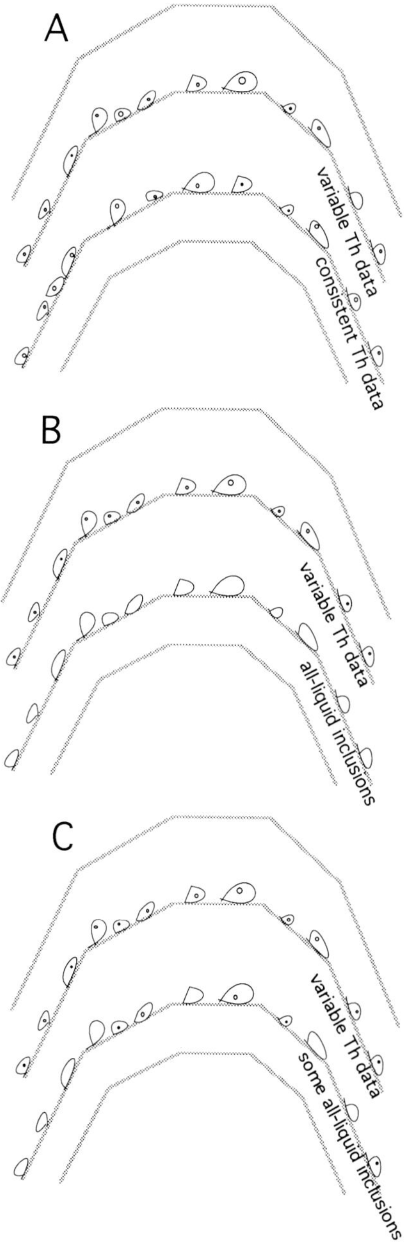

Fig. 9.6. *Schematic representations of FIAs in consecutive growth zones.. Each case illustrates different reasoning employed in the evaluation of the degree of preservation of inclusions.* ***A)*** *If an early FIA contains very consistent Th data of lower temperature than a later FIA then the variable data from the later FIA records real variation in the temperature of entrapment and has not been altered by thermal reequilibration.* ***B)*** *If an early FIA contains all single-phase liquid inclusions, then the variable data from the later FIA records real variation in the temperature of entrapment and has not been altered by thermal reequilibration.* ***C)*** *If some all-liquid inclusions in an early FIA have survived overheating without reequilibration, then the lowest Th inclusions in the later FIA also must have survived, and thus record conditions of initial entrapment.*

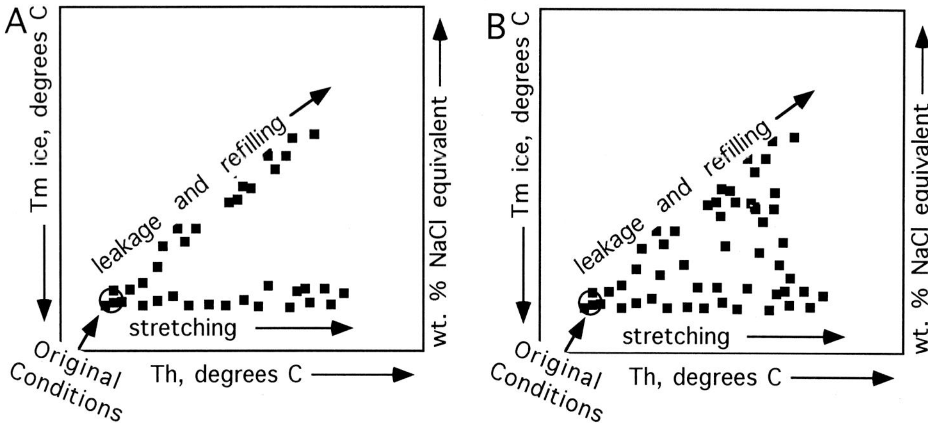

Fig. 9.7. Bivariate plots of Th and Tm ice from moderately variable FIAs showing trends that converge on the conditions of initial FIA entrapment. These leakage and refilling trends reflect the conditions experienced during thermal reequilibration of inclusions during progressive burial heating. ***A)*** *Mechanism of reequilibration for each inclusion does not change.* ***B)*** *Mechanism of reequilibration for each inclusion varies.*

converge on a single point: this point of convergence indicates the original Th and original salinity of the unaltered FIA. Alternatively, if fluid inclusions stretch or leak at random during burial heating, the same convergent trends are produced, but the data scatter between the two trends to define a wedge of fluid inclusion data on a bivariate plot. The point of convergence still is valid for interpreting the original conditions of inclusion entrapment (Fig. 9.7B). This approach only applies to FIAs that have moderate variability and that lack all-liquid fluid inclusions. Other methods (discussed above) are useful for interpreting populations that contain all-liquid fluid inclusions.

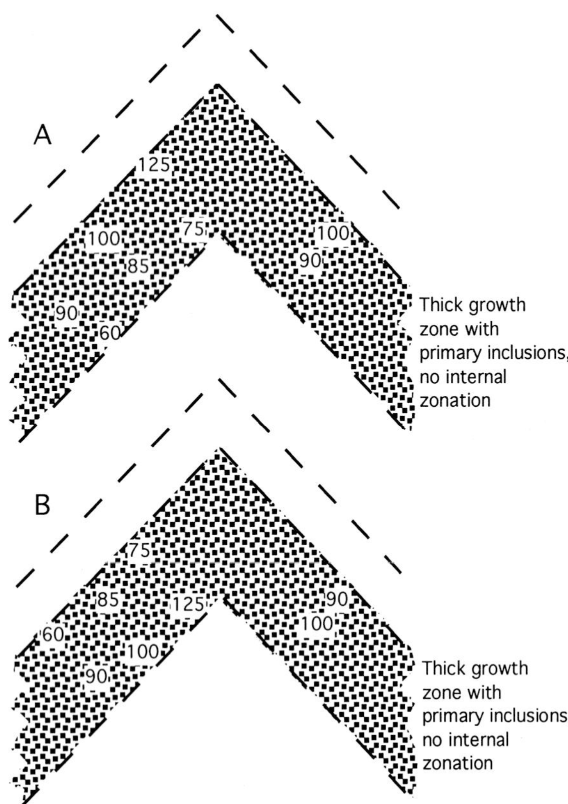

Fig. 9.8. Schematic illustration of fluid inclusion Th data distributed throughout a single thick growth zone. ***A)*** *Increase in Th outward indicates increase in temperature through time.* ***B)*** *Decrease in Th outward indicates trend of decreasing temperature through time.*

Petrographic trends.—

Other geothermometric information can be gleaned from FIAs with moderately variable Th data, in which some petrographic distribution of Th data within the assemblage suggests a monotonic temperature trend during entrapment of fluid inclusions in the assemblage. For instance, for a single thick growth zone, it may be possible to determine the relative paragenetic (temporal) position of groups of Th data within the zone. Given an FIA with a range of sizes and shapes, consistent increase in Th outward in a growth zone signifies that fluids were increasing in temperature as the growth zone precipitated (Fig. 9.8A). For these cases, it should be stressed that the actual Th values are not necessarily indicative of the fluid densities at the time of initial vacuole formation. Inclusions may have reequilibrated during or after precipitation of the growth zone, but the evidence of *increasing but perhaps unknown temperature through time* is valuable information nonetheless. Given an assemblage with a range of sizes and shapes, consistent decrease in Th outward in a growth zone signifies that fluids were decreasing in temperature as the growth zone precipitated (Fig. 9.8B). For this example also, it must be stressed that these inclusions may have been altered by later thermal events, so that the decrease in Th indicates a trend of decreasing but perhaps unknown temperature through time.

Th DATA THAT APPROACH MAXIMUM TEMPERATURE

Up to this point we have been presenting a rigorous approach for gathering and interpreting fluid inclusion data to maximize confidence in the resulting geologic interpretation. The procedure can be time-consuming and the interpretation may require significant expertise. An alternative rapid approach for utilizing inclusion Th data as a *minimum estimate* of how hot a rock has been in its history is to measure Th's in aqueous inclusions in calcite and find the highest temperature mode in randomly gathered data of various origins (not by FIAs), or find the highest temperature Th from a consistent FIA in any mineral in a sample.

The basis for the first approach is really one of statistical probability (not our normal or favored approach). If the single mode or uppermost mode of randomly gathered aqueous Th data from calcite, taken from subsurface wells that are interpreted currently to be at peak temperature are plotted against that present peak temperature, there is a (surprisingly) good correlation between the two (Fig. 9.9; Barker and Goldstein, 1990). Typically the upper mode of the Th data lies just below the present peak temperature in the wells. The reason this Th mode appears to closely approach peak temperature must be that some fluid inclusions are entrapped at peak temperature or that low-temperature inclusions have reequilibrated to approach peak temperature. The single or uppermost mode seems to most closely approach peak temperature in high-temperature (mostly geothermal) systems and Th mode has more scatter below peak temperature in the non-geothermal systems. Other support for the validity of this approach is indicated by comparing the vitrinite reflectance geothermometer to the upper or single mode of aqueous Th data in calcite. The correlation of Th to vitrinite reflectance (Fig. 9.10) is very similar to published correlations of peak temperature to vitrinite reflectance (Barker and Pawlewicz, 1986). Even though this correlation supports the use of this Th technique for obtaining a value that closely approaches peak temperature in many studies, it is not professed to be a calibration for the vitrinite reflectance geothermometer.

As with all empirical methods, this method of determining a minimum measure of peak temperature is subject to ambiguities. The validity of the empirical Th measure of minimum peak temperature is based on an assumption that aqueous fluid inclusions in calcite are trapped, initially, during or before the maximum thermal event. As inclusions are naturally overheated, many inclusions tend to reequilibrate to higher temperature conditions. It is important to remember that a significant amount of overheating may be required for reequilibration of a large portion of the population, and that some of the original fluid inclusions may survive overheating altogether. Obviously, if no inclusions are trapped before or during peak temperature,

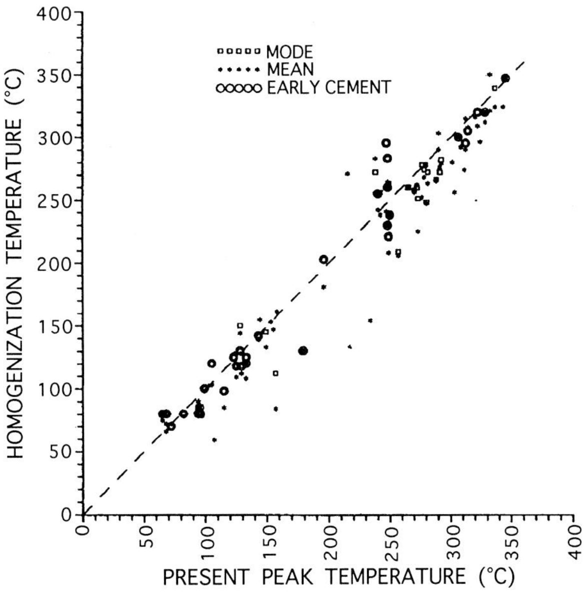

Fig. 9.9. Plot of aqueous Th data from calcite (means or upper mode) versus the present peak temperature. Dashed line of 1:1 correlation is for reference. The early cements are noted for comparison with later cements and cements of unclear origin. Notice how Th data closely approach present peak temperature. After Barker and Goldstein (1990).

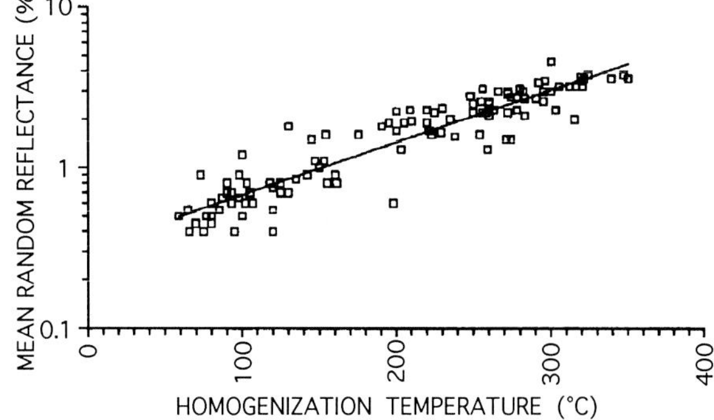

Fig. 9.10. Plot of mean or uppermost mode of Th data from calcite compared to vitrinite reflectance. For the most part, paired vitrinite and fluid inclusion data are from closely associated samples. Line of correlation included. After Barker and Goldstein (1990).

this approach will provide data that may significantly underestimate minimum peak temperature. Other conditions may cause scatter or a spurious determination of minimum peak temperature from the empirical technique of interpreting Th data: 1.) variable amounts of methane trapped as a component dissolved in the aqueous fluid; 2.) variable correction for pressure which may depend on methane content, inclusion composition, or pressure of entrapment; 3.) variable strength of the crystal in which the inclusion is contained; 4.) fluid inclusion size, shape, and orientation; 5.) amount of natural overheating, as a small degree of overheating will not initiate significant reequilibration; 6.) amount of confining pressure, as high pressure inhibits reequilibration; 7.) distribution of the original Th population; 8.) composition of the fluid and its control on pressure increase of the inclusion; 9.) entrapment of fluid inclusions during cooling below maximum temperature; 10.) significant amount of unrecognized necking down after a phase change; 11.) heterogeneous entrapment of liquid and gas from a two-phase system; 12.) measurement of a data set that is too small to be representative of the fluid inclusion population. With all of these caveats, one might wonder why such a technique should be applied. The technique is valid, but only in a statistical sense: the correlation shown (Fig. 9.9) indicates a high probability that in a significant number of studies, the result will provide a close approach to maximum temperature. With all of the weaknesses of using this statistical approach, it has the advantage of being rapid by sidestepping many of the procedures we have outlined (and recommend), and of requiring little expertise.

An alternative but related approach can be applied to fluid inclusions in any mineral to achieve a minimum measure of maximum temperature experienced by the rock. This method involves searching for the FIA that preserves the highest consistent Th. Again, this method circumvents many of the important procedural steps presented in this volume; however, for those who only desire a minimum measure of the maximum temperature that a rock has experienced, it is a valid technique. Furthermore, one can be confident that the highest Th obtained was in fact experienced by the rock because the data are from a consistent FIA with inclusions of different sizes and shapes. For this approach, as the sample is heated on the stage and fluid inclusions are homogenized, the sample is searched for FIAs that still have bubbles and that appear to have consistent ratios of liquid to vapor. The FIA that still contains two-phase inclusions after other inclusions have homogenized becomes the source of data. All inclusions in the assemblage are measured for Th, and if these data are consistent, they provide a "quick and dirty" minimum measure of maximum temperature.

The methodologies discussed above for determining a minimum measure of maximum temperature are methods that tend to *approach* maximum temperature. Just how closely they approach the maximum temperature may be unknown for a particular sample because of unknown pressure correction and unknown degree of entrapment of fluid inclusions during the maximum temperature event. These procedures are shortcuts that circumvent many of the careful petrographic procedures discussed earlier, but that can provide useful, albeit sometimes equivocal data. As is normally the case, more effort will be required for higher confidence levels.

CAN Th's BE HIGHER THAN Tt's?

We are often asked if Th data can provide anomalously high values that are actually higher than Tt or even higher than the maximum temperature reached by the rock. These questions originate because fluid inclusions sometimes yield Th values that are higher than values suspected in many sedimentary systems from vitrinite reflectance and burial history considerations. In nearly every such case, we have found that the Th data are valid and that the previous assumptions on which the thermal history was based were incorrect. Nevertheless, it is possible for *some* fluid inclusion Th data to yield values higher than trapping temperature, but such cases are of minor significance when taken within the context of the many other fluid inclusion observations within a single study. Some mechanisms which may cause Th to exceed Tt are: internal generation of methane from dissolved organics or cracking of hydrocarbons in oils, possible unstable crack growth during fluid inclusion decrepitation, reequilibration from internal overpressure of methane-rich fluid inclusions during uplift, stretching or expansion of methane-rich fluid inclusions during measurement of Th in the lab, undetected necking down after a phase change, and undetected heterogeneous entrapment. To date, the comprehensive data gathered indicate that such problems appear to be of minor importance for the inclusionist who carefully follows the procedures outlined in this text.

SUMMARY

Fluid inclusions provide a useful and commonly unequivocal geothermometer for the diagenetic realm. Temperature information can be gleaned from simple petrographic examination of liquid-to-vapor ratios (see Chapter 6), and fluid inclusion Th's can be used as measurements of minimum entrapment temperatures. Determining entrapment temperature from Th may involve a pressure correction which requires that a pressure be independently known. The assumptions necessary for determining a pressure potentially can cause large errors. Even with burial and reequilibration of inclusions, in many cases, temperatures of mineral precipitation as well as subsequent thermal histories can still be deciphered.

Chapter 10

PRACTICAL ASPECTS OF GEOBAROMETRY

INTRODUCTION

Fluid inclusions provide a good tool to determine pressure at the time of fluid inclusion entrapment, but the ability to use this tool is extremely dependent on the fluid inclusion assemblage (FIA) preserved and the state of knowledge of the composition and P-V-T properties of the fluid system trapped in the inclusions. The best method for using inclusions to determine true pressure of entrapment requires that FIAs preserve a record of gas-water immiscibility. Without this petrographic evidence of immiscibility, guesses must be made about the temperature of fluid inclusion entrapment to estimate the pressure of entrapment. For some fluid inclusions, the best that one can accomplish reliably is to determine a minimum pressure of fluid inclusion entrapment, and there are several possible approaches. Many of the methods developed in the previous chapter to determine entrapment temperature have parallels for determining entrapment pressure. Another approach using the intersection of isochores from petroleum and water systems (Narr and Burruss, 1984) is not covered in this text because recent work suggests a need for careful reevaluation of its theoretical and practical basis (Burruss, 1992).

MINIMUM PRESSURE OF ENTRAPMENT

From Crushing

Given a suite of fluid inclusions trapped at greater than earth surface temperatures, one simple method for obtaining a measure of the minimum pressure of entrapment is to determine the pressure of the gas phase in a fluid inclusion at room temperature by either performing a crushing run (Chapter 6) or by Raman spectroscopy (Chapter 9, 12). The pressure of the gas phase determined at room temperature must be lower than the pressure that existed at the time of formation of the inclusion (Fig. 3.7), and is potentially considerably lower. Thus, this method will significantly underestimate the pressure at the time of entrapment, but at least provides a minimum value.

From Determination of the Bubble Point Curve

Another method for determining the minimum pressure of fluid inclusion entrapment is to determine the composition of an inclusion and either find an existing phase diagram or construct a P-T phase diagram with bubble point curve and isochores for the appropriate water ± salt ± gas system. If one measures Th's from FIAs that yield consistent values among inclusions of various sizes and shapes, the pressure corresponding to the Th on the bubble point curve is a measure of the minimum entrapment pressure. Minimum pressures obtained in this way for gas-poor systems will far underestimate entrapment pressures, but those obtained for gas-rich inclusions will be more reasonable minimum pressures that more closely approach entrapment pressures. The procedure for constructing bubble point curves for gas-bearing systems was presented in Chapter 9. Applying this technique involves first working with an FIA that yields consistent Th data. The composition and concentration of the salts in solution must be determined by using low-temperature microthermometry or some other analytical technique. The internal pressure must be determined by crushing or Raman microprobe, and its volatile composition must be evaluated by crushing, Raman microprobe or GC (gas chromatography), MS (mass spectrometry), or GC-MS. The bulk composition of the inclusion fluid can then be determined by measuring the volumetric ratios of the two phases present in the fluid inclusion. Once all of this information is obtained, and if the composition of the system is simple enough, a bubble point curve can be constructed for the fluid inclusion chemical system. The pressure corresponding to the Th on the bubble point curve represents a minimum pressure of entrapment, and because many subsurface fluids are close to saturation with methane, these minimum pressures may be close to the true pressure of entrapment.

ENTRAPMENT PRESSURE

Using a Consistent Fluid Inclusion Assemblage

One approach for determining entrapment pressure is to make some assumptions about the temperature of entrapment of fluid inclusions from an FIA yielding consistent Th data. This methodology is valid, but it is also a risky procedure that may be prone to some error. The procedure first is to determine the isochore on which the FIA was trapped and then to assume a trapping temperature to determine the corresponding pressure on the isochore. To accomplish this, one must know enough about the composition of the fluid to know its P-V-T relations and must be able to make a valid assumption about temperature conditions of entrapment.

Water-salt system.—
Many of the techniques for determining the composition of the fluid in the inclusion have been summarized in Chapters 3, 6, and 7. The methods for finding a unique point on an isochore in the simple water-salt system have been presented in Chapter 9. The main points that relate to determining pressure of entrapment are summarized briefly here. Initially, one must rule out the presence of gases such as methane and show that the only phase present in the bubble is water vapor, by crushing the inclusion sample and showing that the bubble collapses upon exposure to one atmosphere pressure. Low-temperature phase equilibria during microthermometry should show no evidence of clathrate formation or melting, or evidence of freezing of the gas phase. Observations of eutectic melting temperatures are useful for determining which model water-salt system to assume. The final melting temperatures (typically Tm ice) can then be used to interpret the salinity of the system. These combined data should constrain the composition of the fluid so that some P-V-T relations can be adopted to determine the appropriate isochore at the consistent Th measured for the fluid inclusion assemblage.

The geologist must now determine independently the temperature at the time of inclusion entrapment. The point in P-T space at which this temperature intersects the isochore established above, yields the unique pressure of entrapment. Determining the trapping temperature involves geologic interpretations or

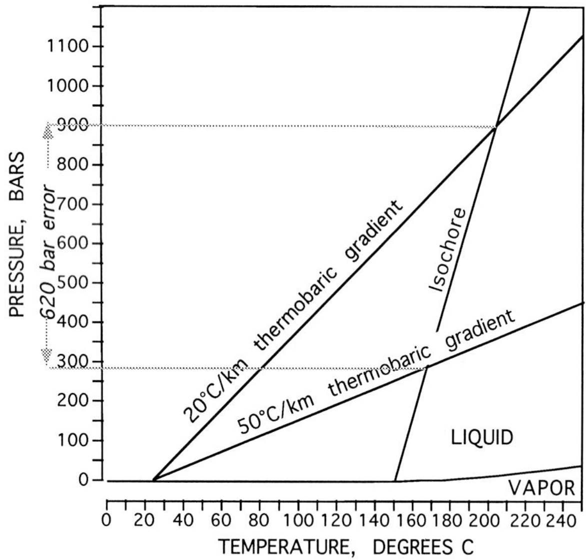

Fig. 10.1. Phase diagram for pure water with one of the isochores for pure water and the liquid-vapor field boundary. If a fluid inclusion homogenizes at 150°C, and is composed of pure water, it must have been entrapped along the isochore drawn. If one were to interpret pressure of entrapment based on a present day thermobaric gradient of 20°C/100 bars (20°C/km, 100 bars/km), one would interpret entrapment pressure to be 900 bars. However, if the true thermobaric gradient during inclusion entrapment had been 50°C/100 bars (50°C/km, 100 bars/km), the assumption of present day thermobaric gradient would have led to a 620 bar overestimation of entrapment pressure!

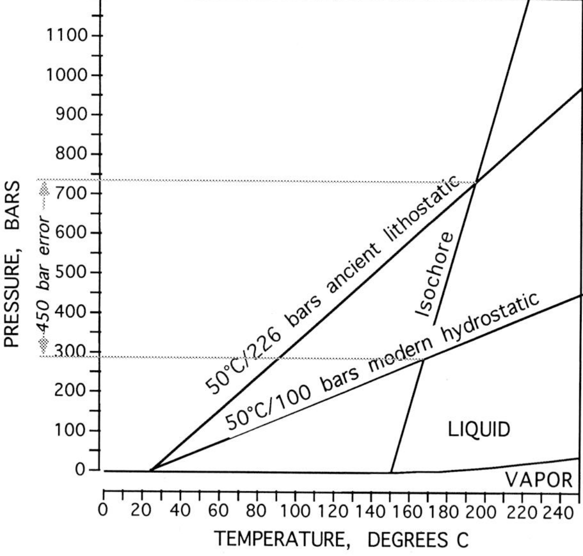

Fig. 10.2. Phase diagram for pure water with one of the isochores for pure water and the liquid-vapor field boundary. If a fluid inclusion homogenizes at 150°C, and is composed of pure water, it must have been entrapped along the isochore drawn. If one were to interpret pressure of entrapment based on a present day hydrostatic thermobaric gradient of 50°C/100 bars, one would interpret entrapment pressure to be 280 bars. However, if the true thermobaric gradient during inclusion entrapment had been a lithostatic gradient of 50°C/226 bars, the assumption of present day thermobaric gradient would have led to a 450 bar underestimation of entrapment pressure!

assumptions that are prone to significant error. To establish an ancient temperature at the time of inclusion entrapment, some inclusionists employ the present-day thermobaric gradient and assume that it is the gradient that existed at the time of inclusion entrapment. Others make an educated guess as to what the thermobaric gradient might have been in the past. Assumption of the incorrect thermobaric gradient may lead to error of hundreds of bars when reconstructing the entrapment pressure of an FIA. It is well known that heat flow and geothermal gradients in most areas change through time. Figure 10.1 illustrates the error inherent in misinterpreting the ancient thermobaric gradient because of a great change in geothermal gradient. Also, if the pressure gradient has changed through time, error also can be introduced. One might assume a thermobaric gradient based on the present-day pressure gradient. For example, if one were to assume that the ancient system active during inclusion entrapment had been under hydrostatic pressure, (like the present-day system of this particular example), but if the system actually had been under lithostatic pressure at the time of inclusion entrapment, the pressure of entrapment would be in error (Fig. 10.2).

Another option is to make an educated guess as to what the temperature might have been at the time of fluid inclusion entrapment. Unfortunately, it is extremely difficult to know the exact trapping temperature at the time of inclusion entrapment without knowing the pressure. However, if one knows the temperature of entrapment (perhaps from another geothermometer), one may determine the pressure by the intersection of that temperature with the isochore (determined from the Th and the inclusion composition).

Methane-water-salt system.—

Given consistent Th data from an FIA of aqueous inclusions containing CH_4, pressure at the time of entrapment can be estimated after the isochore and bubble point curve have been determined. Initially, the composition must be determined by the techniques discussed above and in Chapter 9 so that the bubble point curve can be constructed. Then, the isochore for the Th determined is constructed (see Chapter 9 for instructions). After construction of the isochore, one must assume some thermobaric gradient that existed at the time of fluid inclusion entrapment. Where the thermobaric gradient crosses the isochore determines the pressure of entrapment (Fig. 10.3).

When There is Petrographic Evidence of Immiscibility

Low-temperature systems.—

For systems with all-liquid aqueous fluid inclusions, indicating trapping at temperatures close to laboratory temperatures, one can get a good estimate of pressure of entrapment if immiscible bubbles of gas were trapped in the inclusions along with the all-liquid inclusions. An example already discussed is identification of fluid inclusions trapped in the vadose zone in which upon crushing the sample, the bubbles of immiscible gas remain the same size, indicating one-atmosphere pressure at the time of entrapment. Another application of this approach could be the determination of water depth from fluid inclusions trapped in marine or non-marine basins, or shallow groundwater, at temperatures near that of laboratory temperatures. For example, if a marine basin were 100 m deep and submarine cements were actively growing on the sea floor, it is possible that bubbles of immiscible gas could be exsolving at depth and sticking to the surface of the growing crystal. If these bubbles were trapped as inclusions in the growing cement, when brought to the surface they would still exist at the pressure of entrapment (because the temperature of entrapment and the temperature of the laboratory are nearly the same). When the sample is crushed, the bubbles would expand to indicate an internal pressure of about 11 bars, the hydrostatic pressure at about 100 m water depth.

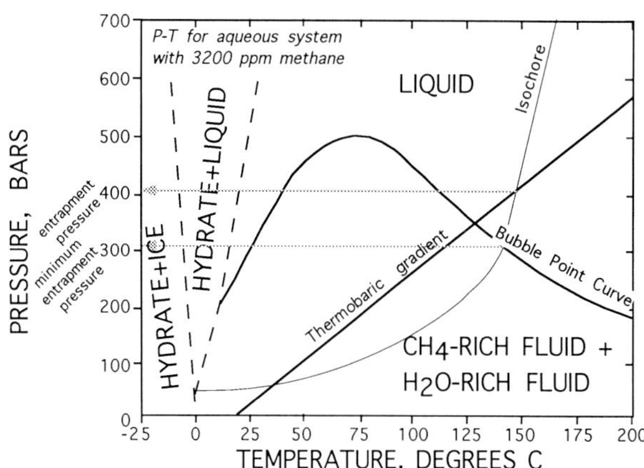

Fig. 10.3. P-T phase diagram for H_2O-CH_4 system with 3200 ppm CH_4. An entrapment pressure can be interpreted by assuming a thermobaric gradient (shown as 28°C/100 bars here), and knowing the inclusion composition (3200 ppm CH_4 here) and the Th (here, 140°C): the intersection of the thermobaric gradient with the isochore of the inclusion yields an entrapment pressure of 410 bars. The intersection of the homogenization temperature and the bubble point curve yields a minimum entrapment pressure of 310 bars. Modified from Hanor (1980).

Elevated-temperature systems.—

When there is petrographic evidence for immiscibility in an elevated-temperature water-salt-gas system, there are two methods for determining the pressure of fluid inclusion entrapment, both of which require that the composition of the fluid phase be determined. Evidence of such immiscibility should be preserved as inclusions of the water-rich phase, inclusions of the gas-rich phase and mixtures of the two. One method requires that the inclusionist determine Th of the aqueous-rich end-member fluid, determine the composition of the gas phase, and determine the Th of the gas-rich end-member. The other requires determination of several parameters on the water-rich end-member FIA, including the Th of a consistent water-rich FIA, the identification of the salts present in solution and their concentration, the composition of the gas phase, and the molar ratio of gas to aqueous phase. The discussion below will focus on immiscibility in a simple system in which the gas phase is pure CH_4; if the gas phase can be shown not to be pure CH_4, then more complex phase equilibria than those used below would apply.

The first method requires that one be able to observe homogenization of the gas-rich (CH_4 in this example) end-member and the water-rich end-member. One must also document that the gas-rich end-member is pure CH_4 (see Chapter 9) in order to use the following simple phase equilibria. If a more complex gas composition is determined, it may still be possible to determine pressure of entrapment, but different phase equilibria for the more complex system must be employed. Measuring homogenization of the CH_4 (ignoring the invisible H_2O in the inclusion) of the CH_4-rich end-member inclusions will be fairly straightforward for methane inclusions with bulk density greater than the CH_4 critical density (i.e., homogenize to a liquid phase at T <-82.1°C), but for CH_4-rich inclusion fluids lower than the critical density (i.e., homogenize to the gas phase at T <-82.1°C), the disappearance of the thin rim of liquid CH_4 upon homogenization may be difficult to observe in small inclusions. If the gas-rich inclusion is shown to be essentially pure CH_4 (in actuality it should still contain some H_2O), then the homogenization temperature of the methane inclusion can be used to find the methane isochore on which the inclusion was entrapped. As discussed in Chapter 3, when a fluid inclusion entraps a single phase of an immiscible fluid system, the homogenization temperature (Th) yields the true temperature of trapping (Tt); that is, there is no pressure correction for Th. The Th of the CH_4-rich end-member is not Tt because of the invisible H_2O that an inclusionist cannot see, that would not homogenize until Tt. But, the readily observable Th of the aqueous-rich end-member must be the Tt. Thus, if the methane isochore is extended to cross the trapping temperature (determined from Th of the aqueous-rich end-member FIA with consistent data among inclusions of different sizes and shapes), then a unique pressure of entrapment is defined (Fig. 10.4; see Mullis, 1979).

Even though there is a little water present in the CH_4-rich end-member inclusion (because it was trapped in equilibrium with an aqueous phase), the water in the gas-rich inclusion should not affect the Th significantly. Moreover, the presence of water should not significantly change the isochore position in P-T space compared to the isochore for pure CH_4, as shown by R. Burruss (personal communication, 1993). To evaluate a worst case scenario, taking a CH_4-rich fluid inclusion in equilibrium with an aqueous phase, with *actual* entrapment conditions of 1,150 bars, 200°C, density of 0.311 g/cm^3, and mole fraction H_2O of 0.059, Burruss used a Peng-Robinson equation of state to approximate the maximum effect of ignoring the presence of H_2O in the gas-rich inclusion. If one takes the homogenization temperature that would be generated by this fluid, and finds the pure CH_4 isochore that intersects it, the pure CH_4 isochore would cross the trapping temperature just 100 bars below the actual trapping pressure of 1,150 bars. Thus, it appears that ignoring the small amount of H_2O that exists within the CH_4-rich inclusion has a relatively minor effect on the pressure determination.

The second method to determine pressure of entrapment involves determination of position on the bubble point curve for an elevated-temperature (>50°C) gas-water immiscible system. This was explained in detail in Chapter 9 and should only be applied to systems in which there is solid petrographic evidence of gas-water immiscibility. Initially, the composition of the system must be constrained well enough to construct a bubble point curve. For most phase equilibria currently available, this requires that the gas composition be relatively simple (e.g., nearly pure CH_4). The data needed to determine the bubble point curve are the salts present and their concentration, and the molar ratio of gas to aqueous phase in a consistent aqueous-rich end-member FIA. Once this bubble point curve has been determined, the pressure on the bubble point curve corresponding to the Th of the consistent aqueous-rich end-member FIA yields the pressure of entrapment (Fig. 10.5).

SUMMARY

Fluid inclusions potentially provide one of the most powerful tools for determining the pressures of diagenetic fluids. Minimum pressures can often be

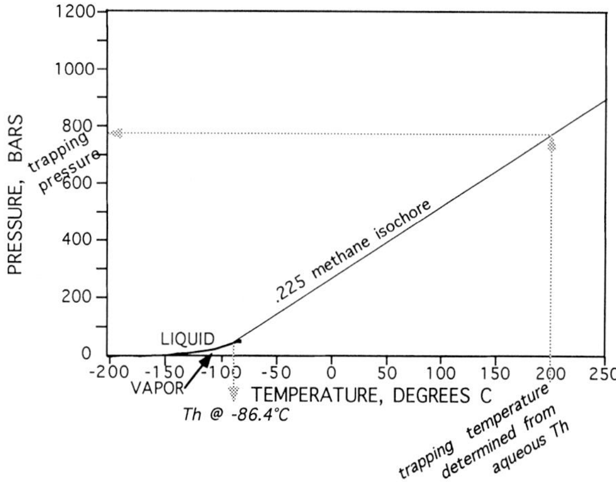

Fig. 10.4. P-T phase diagram for pure CH_4 showing how, with petrographic evidence of methane-water immiscibility, the unique P and T of fluid inclusion entrapment can be determined. If a consistent aqueous-rich end-member FIA yields consistent Th's at 200°C, then the Tt is 200°C. If the methane phase of the methane-rich end-member FIA homogenizes to liquid at -86.4°C (ignoring the invisible water), then the inclusions must have been entrapped along the .225 methane isochore. The pressure at which this isochore intersects the entrapment temperature is the trapping pressure for the FIA.

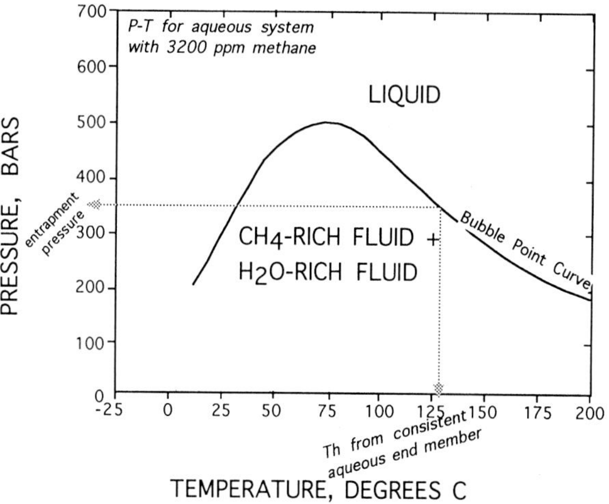

Fig. 10.5. P-T phase diagram for H_2O-CH_4 system with 3200 ppm CH_4. If there is petrographic evidence of immiscibility, and if a consistent aqueous-rich end-member FIA with 3200 ppm CH_4 yields consistent Th's of about 130°C, the pressure of fluid inclusion entrapment must have been 350 bars. Modified from Hanor (1980).

determined, but will usually be far lower than true entrapment pressures unless significant amounts of gases are present in the inclusions. Without fluid inclusion evidence of immiscibility of a rather simple system, determining pressure of entrapment requires assumptions that may lead to significant error if incorrect. But, when there is definitive petrographic evidence of immiscibility, and if the gas-dominant end-member inclusion is pure CH_4, then a very accurate pressure can be determined from the intersection of the methane isochore with the temperature of trapping determined from the aqueous-rich end-member homogenization temperature. Otherwise, if the composition of the inclusions can be accurately described so that the P-V-T relationships can be modeled, then pressure can be determined from the intersection of the homogenization temperature with the appropriate bubble point curve. So, if nature provides the necessary inclusions, pressures of entrapment will be possible to obtain.

Chapter 11

CASE HISTORIES

INTRODUCTION

In this chapter a series of case histories are employed to illustrate the methodology that should be applied in properly conducting a study of fluid inclusions in diagenetic minerals. Rather than attempting to make an exhaustive listing of fluid inclusion studies that have appeared in the literature, particular studies were selected to illustrate important points about the preservation of fluid inclusions in diagenetic minerals, proper approaches for organizing fluid inclusion studies, and methods of analyzing fluid inclusion data from the diagenetic realm.

EVALUATING A FLUID INCLUSION STUDY

When faced with the task of evaluating a study of fluid inclusions, the techniques and methods covered in this book will prove useful in determining the validity of the conclusions reached. In such endeavors, a good critic will take an approach that experienced inclusionists would take: to be an habitual skeptic who will never take fluid inclusion data at face value without first considering all of the possibilities that could cause the interpretations to be invalid.

A properly conducted study of fluid inclusions from the diagenetic realm normally involves a structured, methodical approach. First, a geologic problem is defined that the researchers hope fluid inclusions will solve. Once the question has been posed, it should be obvious what fluid inclusions must be preserved to answer the question. For example, if the question deals with the origin of a diagenetic mineral, primary fluid inclusions that have not totally reequilibrated are necessary. If the question simply deals with a record of temperature and fluid composition the rock has experienced, secondary, primary, and reequilibrated fluid inclusions may be sufficient. If information is desired about pressure, fluid inclusions recording gas-water immiscibility are useful. This text should be an aid to inclusionists deciding which fluid inclusions must be preserved to answer the geologic problem. As always, the study should have field and stratigraphic control. Typically, the inclusionist will conduct a broad reconnaissance petrographic survey of many samples to determine the paragenetic framework in which representative samples will be placed. All samples must be handled properly to avoid reequilibration of samples in the laboratory. The inclusionist determines petrographically if fluid inclusions are present that will answer the question posed. If they are not present, the study may be justifiably abandoned.

When conducting a study of fluid inclusions or when evaluating another study, it is important to remain a skeptic, to look for common mistakes, and to follow the logical procedures outlined in this text. In evaluating various fluid inclusion studies over the years the authors have compiled a listing of recurrent questions that must be asked during evaluation:

1. Were samples prepared in such a way to avoid alteration of the fluid inclusion population? Could samples have been overheated during preparation? Were the samples studied with cathodoluminescence before data collection?

2. Are the data placed in a geologic, stratigraphic, and petrographic context?

3. Is there convincing evidence for fluid inclusion origin? For primary origin, are there photomicrographs, sketches, or detailed descriptions that show fluid inclusions are related to mineral growth?

4. Were microthermometric measurements conducted by a methodology that would avoid alteration of the fluid inclusions? Were low-temperature fluid inclusions overheated during heating runs or were inclusions altered by freezing them before heating runs were made? A good inclusion study must document proper methodology to show that the data are reliable.

5. Were the data collected and presented with respect to individual fluid inclusion assemblages, or were they petrographically overgeneralized? If they were presented as fluid inclusion assemblages, are data from fluid

inclusion assemblages internally consistent or variable? If variable, have heterogeneous entrapment, necking down, and thermal reequilibration been evaluated as an explanation for the variability?

6. Are there enough data to answer the questions posed? Are there enough data to evaluate spatial and temporal variability?

7. Has there been an attempt to explain every item of fluid inclusion data, or have authors improperly interpreted the means of fluid inclusion populations?

8. If a pressure correction was applied, was it applied to a consistent fluid inclusion assemblage or improperly applied to variable data? If a pressure correction was applied to a consistent fluid inclusion assemblage, was the composition of the fluid inclusion assemblage constrained well enough for this to be valid? Was the pressure constrained well enough to make the pressure correction meaningful?

9. What assumptions did the researcher make? Are they stated? Are they valid? If it is possible that the assumptions are invalid, what potential effect does this have on the interpretation?

These questions are justifiable for any fluid inclusion study. Furthermore, if a researcher makes a pointed effort to ask these questions throughout the course of a study, then the chances of success will be enhanced.

CASE HISTORIES

Case histories of fluid inclusion studies from the diagenetic realm are summarized in the following pages. The presentation is designed to illustrate the methodology the authors have recommended for conducting fluid inclusion studies in a wide range of diagenetic settings and to show the degree of preservation of fluid inclusions in the diagenetic realm. Rather than attempting to be all-inclusive, the focus is more on what an inclusionist needs to know to apply fluid inclusions as a tool in the diagenetic realm. In the interest of brevity, many of the studies have been pared to the minimum required to instruct the reader in the fluid inclusion approach, and we have not included the many important implications of these studies with respect to petroleum generation, migration, and basin evolution. Sample preparation and measurement techniques have not been included, but before Tm ice was determined from originally all-liquid inclusions, overheating in the laboratory was used to produce vapor bubbles. For the following studies, one should assume that the samples had been treated properly unless stated otherwise. As an aid, logic flow charts (Figs. 11.1-11.5) are presented to help the reader interpret the most common fluid inclusion phase assemblages. Use these flow charts as companions in understanding the logic employed in each case history, and as guides to interpreting fluid inclusion assemblages (FIAs) in your own fluid inclusion research.

The case histories from each diagenetic environment are organized to teach the reader the logic for dealing with fluid inclusions from the diagenetic realm. They should not be considered to be the norm; many "problem" data sets have been included along with the well-preserved fluid inclusion data to illustrate the proper approaches for dealing with various types of fluid inclusion data sets. Thus, each study of a young low-temperature diagenetic mineral that has never been heated during burial is followed by ancient examples that have experienced a longer history of changing pore fluid chemistry and heating during burial. This will

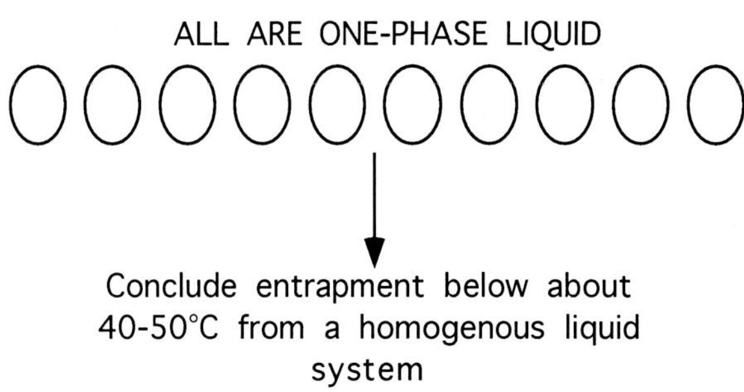

Fig. 11.1. Flow chart illustrating procedures for interpreting a fluid inclusion assemblage consisting of all-liquid fluid inclusions. An inclusionist should always check for "significant" metastability by attempting to nucleate bubbles in all-liquid inclusions (see Chapters 4 and 6).

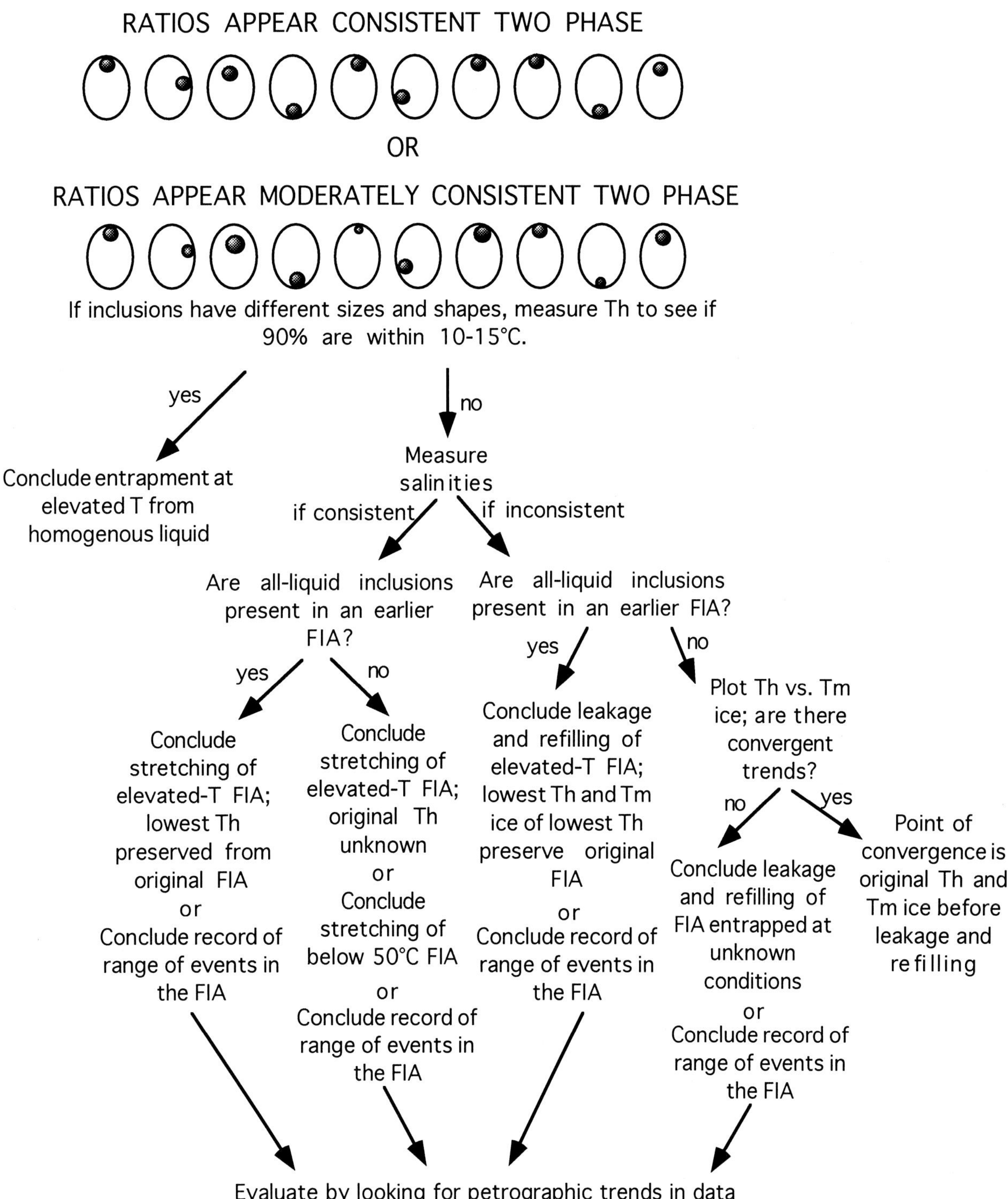

Fig. 11.2. Flow chart illustrating the procedure for interpreting a fluid inclusion assemblage with apparently consistent or moderately consistent phase ratios. Inclusions are of various shapes and sizes.

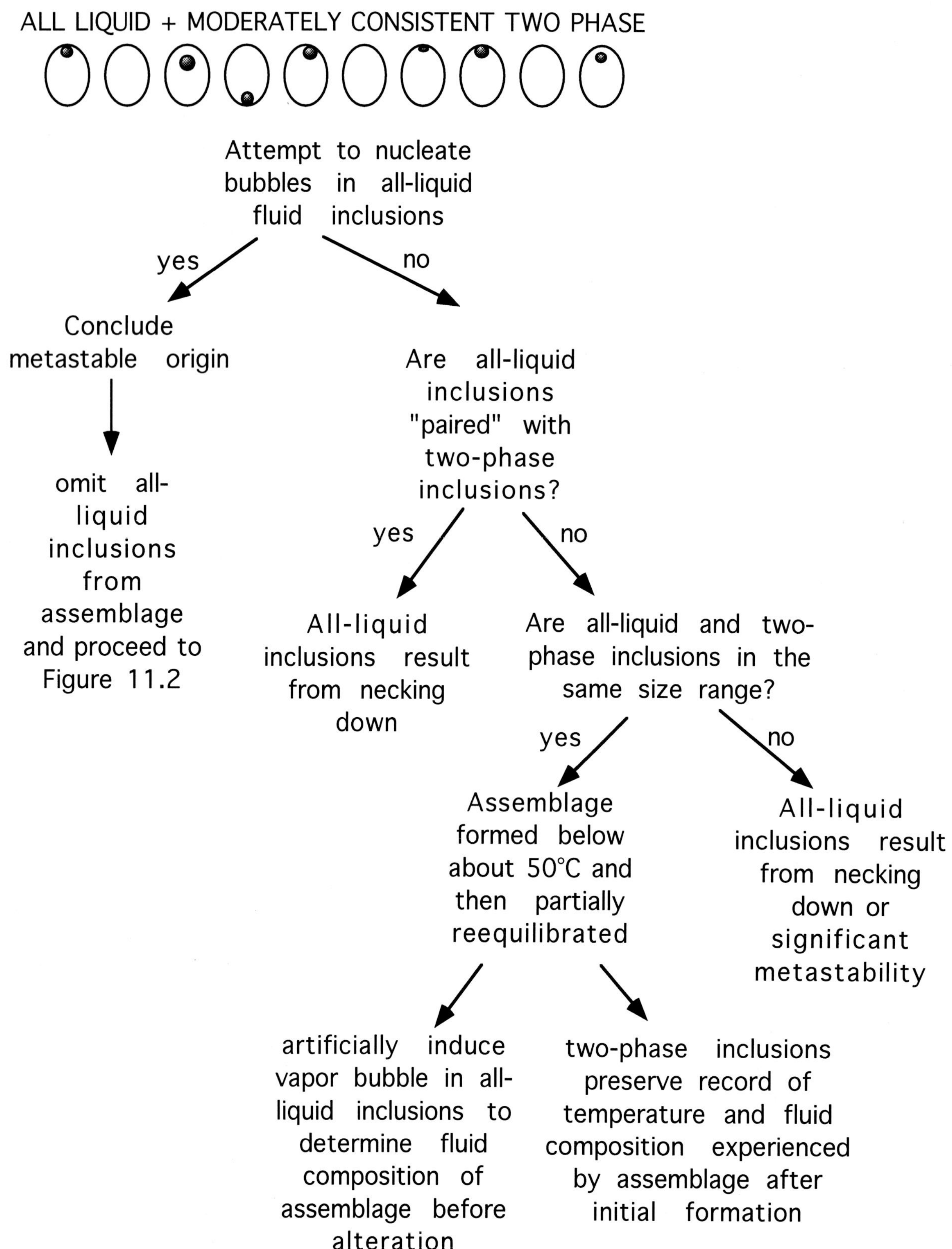

Fig. 11.3. *Flow chart illustrating procedure for interpreting a fluid inclusion assemblage with moderately consistent phase ratios together with all-liquid fluid inclusions. Inclusions are of various shapes and sizes.*

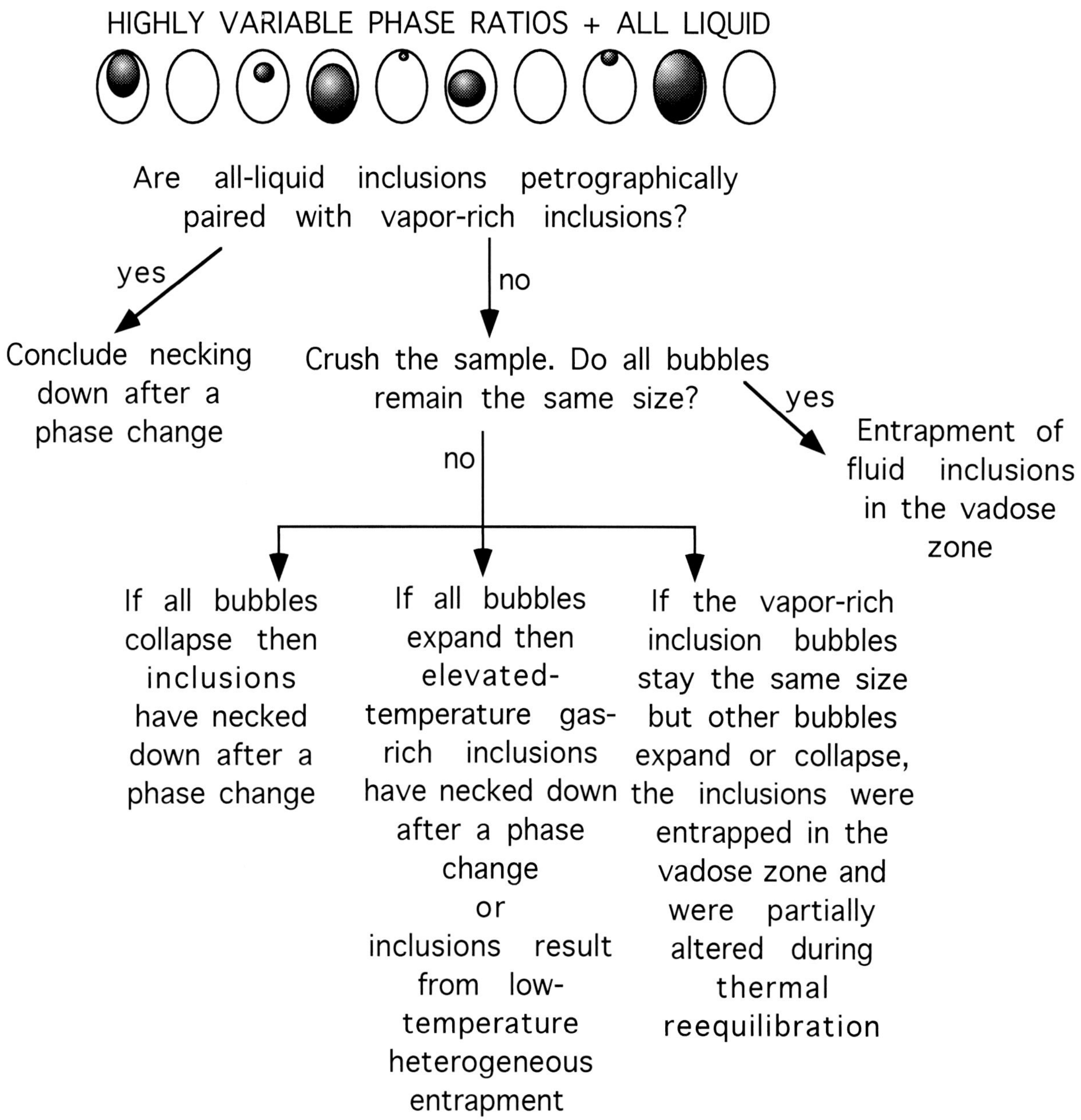

Fig. 11.4. *Flow chart illustrating procedure for interpreting a fluid inclusion assemblage consisting of all-liquid fluid inclusions together with inclusions of highly variable phase ratio. Inclusions are of various shapes and sizes.*

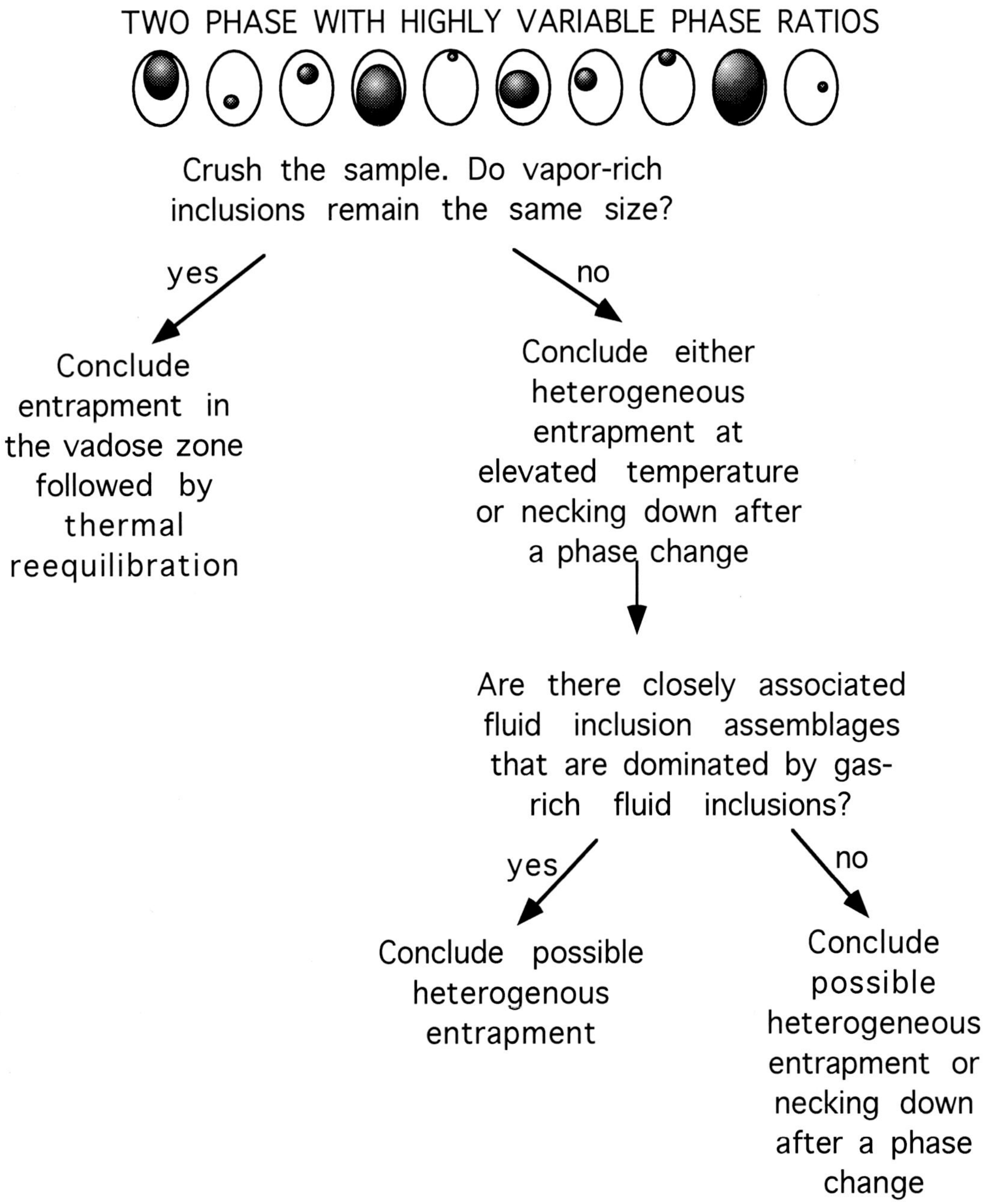

TWO PHASE WITH HIGHLY VARIABLE PHASE RATIOS

Crush the sample. Do vapor-rich inclusions remain the same size?

yes

Conclude entrapment in the vadose zone followed by thermal reequilibration

no

Conclude either heterogeneous entrapment at elevated temperature or necking down after a phase change

Are there closely associated fluid inclusion assemblages that are dominated by gas-rich fluid inclusions?

yes

Conclude possible heterogenous entrapment

no

Conclude possible heterogeneous entrapment or necking down after a phase change

Fig. 11.5. *Flow chart illustrating procedure for interpreting a fluid inclusion assemblage with highly variable phase ratios and without all-liquid fluid inclusions. Inclusions are of various shapes and sizes.*

allow the reader to deal with the methodology for evaluation of the problem of thermal reequilibration of low-temperature fluid inclusions. These groupings fit within an organization by diagenetic environment that begins with presentation of studies from the vadose zone, moves down into the low-temperature freshwater phreatic zone, continues into the freshwater-marine mixing zone, then the marine phreatic zone, low-temperature hypersaline environments, and finally records of elevated-temperature diagenesis, thermal history, and fluid history of basins.

Origin of Late Fracture-Filling Banded Flowstone, Nevada

Near Winnemucca, Nevada, open fractures in Triassic strata are partially filled by banded calcite cement. Structurally, fractures appear to be related to the Basin and Range uplift. Fracture fills are the most recent diagenetic mineral and post-date all elevated-temperature diagenetic events. The geologic setting indicates that the samples have not been heated naturally after formation. This study was discussed in more detail in Goldstein (1986a, b).

Objective.—
The objective of this study is to determine the temperature, diagenetic environment, and salinity of formation for the banded fracture filling calcite. The case history will illustrate the characteristics of fluid inclusions that were entrapped in the freshwater vadose zone and that never have been subject to burial heating.

Petrography.—
The calcite in question is growth banded. Bands are defined by solid inclusions and fluid inclusions. Many fluid inclusions are spike-shaped and oriented parallel to growth direction. The concentration of fluid inclusions define growth zone boundaries and indicate a primary origin for fluid inclusions (Fig. 11.6). Some fluid inclusion assemblages (FIAs) consist of only all-liquid inclusions. Other closely related FIAs contain inclusions with highly variable ratios of gas to liquid (Fig. 11.7).

Microthermometry.—
Only some of the primary fluid inclusions in this sample homogenize when heated to a maximum temperature of 250°C. These homogenize to the liquid phase. The other inclusions do not homogenize below 250°C. The homogenization temperatures reflect the variability in phase ratios detected with the petrographic study (Fig. 11.8) and range from all-liquid at room temperature to above 250°C.

Cold stage work illustrates consistent Tm ice of all fluid inclusions at 0.0°C.

Crushing.—
When samples are crushed, all bubbles remain approximately the same size.

Interpretation.—
The calcite formed below about 50°C. The evidence for this is the presence of FIAs that contain all-liquid fluid inclusions. Necking down of elevated-temperature,

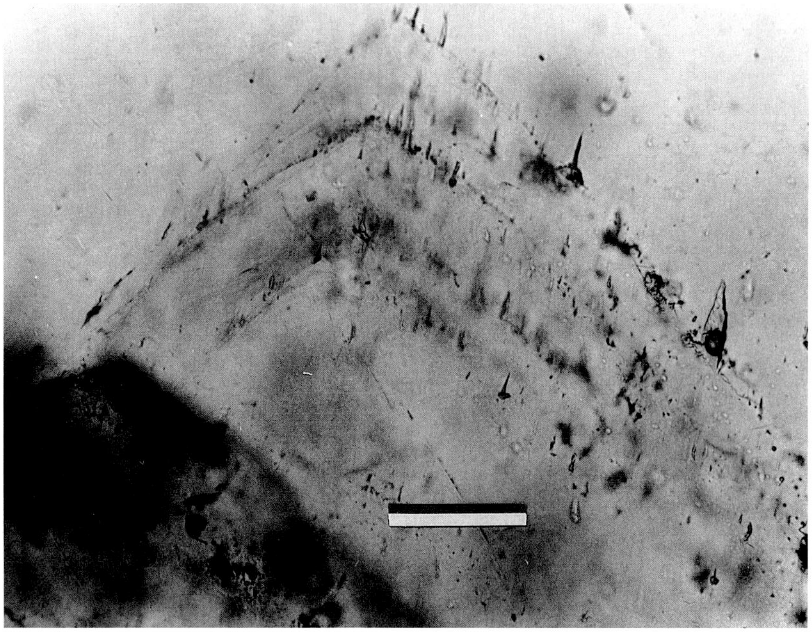

Fig. 11.6. *Transmitted light photomicrograph of primary fluid inclusions in calcite flowstone from Nevada. Notice that fluid inclusions define growth zones and are oriented in the direction of crystal growth. Large inclusion on the right contains a large vapor bubble. Most of the other inclusions contain only liquid. Bar scale is 100 μm.*

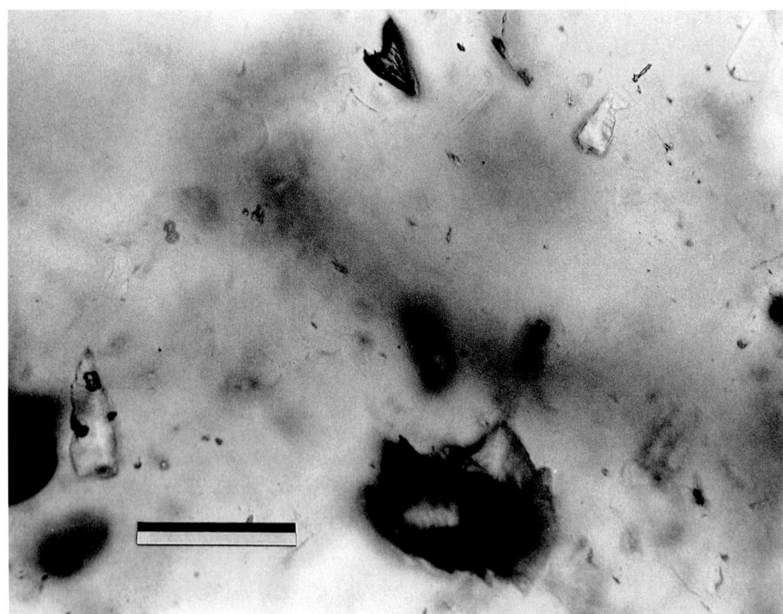

Fig. 11.7. Transmitted light photomicrograph of large fluid inclusions in calcite flowstone from Nevada. Notice that ratios of liquid to vapor vary from all liquid (upper right) to mostly vapor (right of scale). Bar scale is 50 μm.

two-phase inclusions is ruled out because of field evidence that the sample had never been heated. The variability in phase ratios comes from heterogeneous entrapment in a low-temperature, two-phase system. The variable homogenization temperatures have no significance for interpreting temperature of entrapment. They merely reflect the relative amounts of gas and water trapped in the fluid inclusions. The all-liquid inclusions remain the key in interpreting a low-temperature origin. The fluid inclusions are fresh water because there is no depression of Tm ice. The pressure in the fluid inclusions is one atmosphere because there was no change in the bubble size upon crushing and exposure to one-atmosphere pressure. Thus, the inclusions were entrapped in the low-temperature, fresh-water vadose zone.

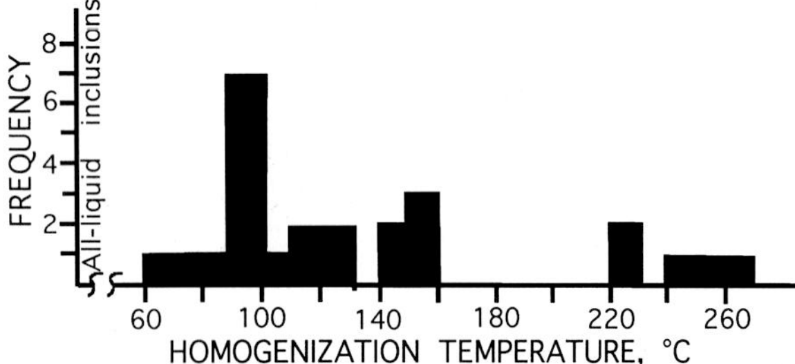

Fig. 11.8. Homogenization temperatures of fluid inclusions of calcite from Nevada. These Th data merely reflect heterogeneous entrapment in the vadose zone and carry no meaning with respect to entrapment temperature.

Origin of Calcite Cement in Miocene Strata, SE Spain

This study uses field relationships, transmitted light petrography, cathodoluminescence petrography, and fluid inclusion analysis to identify an event of Late Miocene subaerial exposure of carbonate strata at Mesa Roldan in southeast Spain. Miocene carbonate strata of southeastern Spain, about 100-150 m thick, covered topographic highs of a Middle to Late Miocene archipelago. The Miocene carbonate exposures indicate minimum shelf-to-basin relief of 50-200 m, generally over a lateral distance of 1-2 km. The Miocene carbonates of southeast Spain display little structural deformation, and no apparent thermal alteration. Armstrong and others (1980) suggested offshore reef complexes in the subsurface would be covered by 350 m or more of Miocene (?) and Pliocene fine-grained sedimentary rocks. Onshore, exposed reef complexes (such as Mesa Roldan) likely would have been covered by less than 350 m. The Miocene strata of interest here were collected just above present day Mediterranean sea level and are overlain by a "curious" erosion surface that is, in turn, overlain by marine strata that have been mapped as Pliocene in age. More details of this study are available in Goldstein and others (1990).

Objective.—
The objective of this study is to determine the origin of calcite associated with the "curious" erosion surface between Miocene and Pliocene strata. Also, this case history will prove useful in documenting the utility of

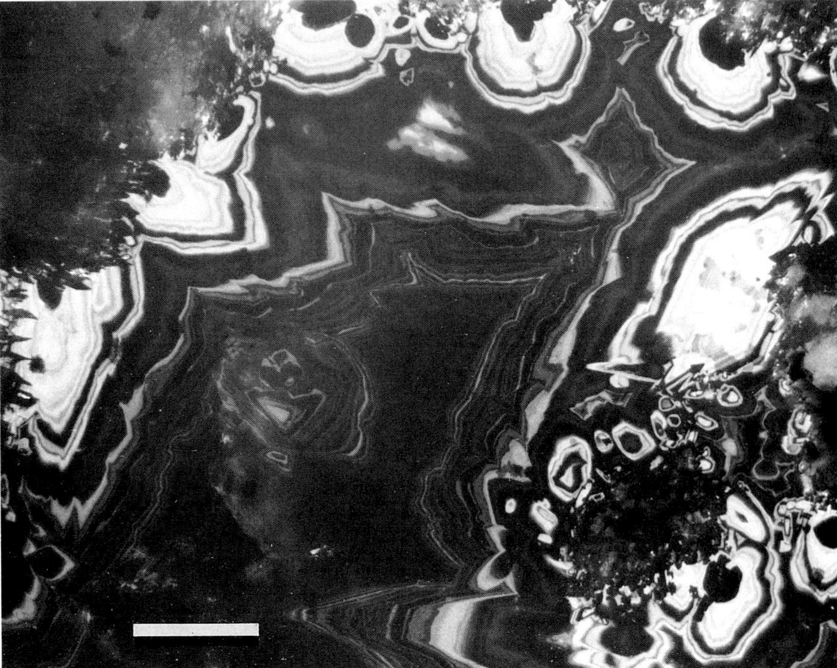

Fig. 11.9. Cathodoluminescence photomicrograph of luminescently banded calcite from Miocene strata of SE Spain. Fluid inclusion-rich bands are confined to zones defined by luminescent growth zonation. Bar scale is 100 μm.

highly accurate Tm ice data in low-temperature diagenetic systems, the identification of fluid inclusions entrapped in the vadose zone, and the record of fluid inclusions that can be preserved in low-temperature systems despite a later history of bathing fluid inclusion-rich cements in other fluids.

Petrography.—

The Miocene strata directly below the erosion surface consist of sedimentary breccia, which is cemented with calcite cement. The cement postdates deposition of the clasts because they encrust reoriented geopetal fabrics in the clasts. The cement predates deposition of the overlying Pliocene strata because it is covered by a later geopetal sediment that is associated with the erosion surface and maintains present-day "up" directions. The early phase of cement that predates deposition of the Pliocene consists of bladed or equant calcite that appears cloudy because of abundant fluid inclusions and some solid inclusions. This cloudy cement is overgrown by a later phase of clear, equant or bladed calcite cement. The cloudy zones become abruptly clear at former growth zone boundaries as defined by fluid inclusion distribution and cathodoluminescence zonation (Fig. 11.9). The inclusions are interpreted as primary on this basis.

The primary fluid inclusions have highly variable ratios of liquid to vapor. In any field of view, one may find one-phase all-liquid, one-phase all-gas, and two-phase fluid inclusions (Fig. 11.10). Two-phase inclusions consist of a liquid and a gas phase; the volume ratio of liquid to gas is highly variable from very liquid rich to very liquid poor. No fields of view contained two-phase inclusions with consistent phase ratios. The petrographic distribution of one-phase and two-phase inclusions indicates that most of the variable

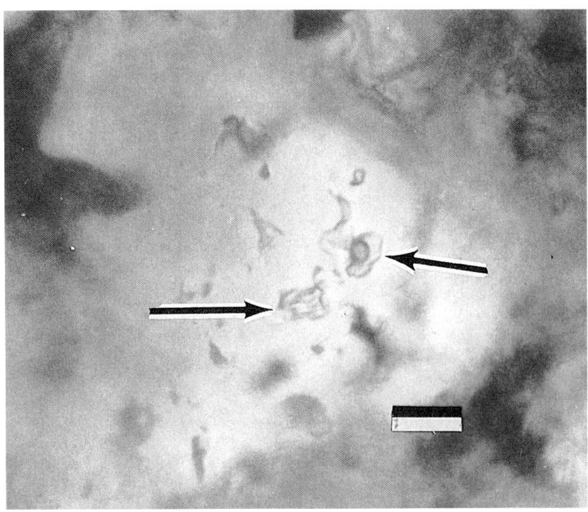

Fig. 11.10. Transmitted light photomicrograph of fluid inclusions in calcite from Miocene strata of SE Spain (arrows). Photo illustrates variability in ratio of liquid to vapor with all-liquid inclusions alongside inclusions with large bubbles. Bar scale is 10 μm.

phase ratios are not the result of necking down of fluid inclusions with originally consistent phase ratios.

Microthermometry.—
The primary fluid inclusions were heated to a maximum temperature of 200°C to determine homogenization temperatures of two-phase fluid inclusions. Many fluid inclusions decrepitated before the Th was reached. Most of the two-phase fluid inclusions had not yet homogenized upon reaching 200°C. Some inclusions homogenized to liquid, with the Th values ranging from 52°C to greater than 200°C. No mode was noted; the Th data reflect the variable ratios of water and gas enclosed during growth of the calcite cement.

The data from the cloudy calcite indicate most fluid inclusions contain fresh water (Fig. 11.11) because Tm ice is 0.0°C; for a small percentage of the measurements, minor depression of Tm ice is as low as -0.7°C.

Crushing.—
Crushing of calcite to release inclusion fluids results in little or no change in the size of the vapor bubble. Therefore, fluid inclusions are under one atmosphere pressure.

Interpretation.—
Primary fluid inclusions were trapped in calcite before deposition of Pliocene strata. The all-liquid inclusions indicate formation of the cement below about 50°C. The homogenization temperatures are meaningless in terms of temperature of cement formation; they merely reflect the trapping of fluids from a heterogeneous system. The variable phase ratios, one-atmosphere internal pressure, and all-liquid inclusions indicate entrapment in the vadose zone. If cements precipitated from a freshwater vadose zone, one would expect Tm ice at 0°C. If cements precipitated from other vadose zone environments, different Tm ice values would be expected (Tm ice about -1.9°C for marine vadose; lower than -1.9°C for evaporitic; variable Tm ice for mixed). Most of the Tm ice measurements are 0.0°C. A few values are as low a -0.7°C indicating slightly brackish salinities. These lower values reflect salinities as high as 1.2 wt.% NaCl equivalent. Although these slightly brackish fluids suggest there is a minor source of salt or saltwater and could indicate proximity to the shoreline, the bulk of the data support cementation from the freshwater vadose zone. Each data point is meaningful though, and the slightly brackish salinities were present in the vadose zone before deposition of the Pliocene.

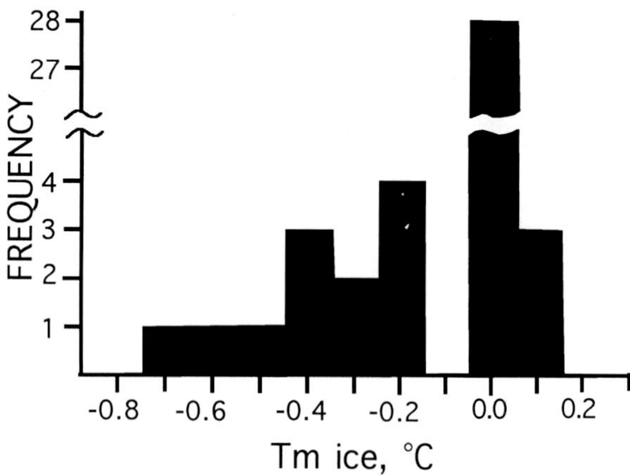

Fig. 11.11. *Tm ice data from aqueous fluid inclusions in calcite cement from Miocene strata of Spain. Most inclusions indicate freshwater compositions but some indicate brackish salinities.*

Because freshwater vadose conditions occurred near present day sea level before deposition of marine Pliocene strata, the "curious" erosion surface likely is a surface of subaerial exposure, and the record of the vadose zone indicates that relative sea level fell at least 200 m to expose the Miocene strata. Furthermore, when the marine Pliocene strata were deposited directly on top of the exposure surface, the Miocene strata must have been bathed in Pliocene marine waters. It is important that despite being bathed in later marine fluids, the pre-Pliocene fluid inclusions did not leak and refill with marine water that would have yielded Tm ice of -1.9°C. This indicates that low-temperature fluid inclusions in calcite maintain their integrity while bathed in later low-temperature fluids. In other words, fluid inclusions are effective containers of fluid and did not leak and refill.

Origin of Masses of Calcite Spar, Lake Valley Formation, New Mexico

In the Sacramento Mountains of southern New Mexico, the top of the Mississippian Lake Valley Formation is rubbly in appearance and is unconformably, overlain by Pennsylvanian siliciclastic strata. Channel, vug, and breccia pores or cavities occur within the top of the Lake Valley Formation. These cavities either parallel or cut across stratification of the limestone, and are filled with gray, green, or maroon, crudely fissile shale in close association and apparently interstratified with masses of columnar sparry calcite which appears similar to speleothem material (Fig. 11.12). The cavities and their fillings appear to predate deposition of the overlying Pennsylvanian strata. Details of this study can be found in Goldstein (1990).

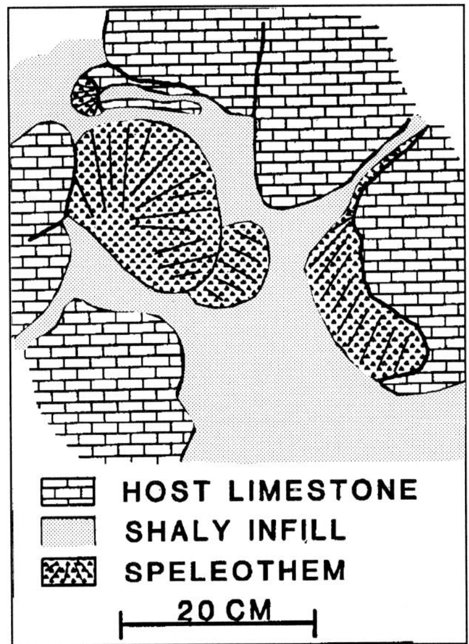

Fig. 11.12. *Sketch of field occurrence of masses of calcite spar (speleothem) that occur within large cavities in Mississippian limestones of the Lake Valley Formation.*

Objective.—

The objective of this study is to determine the origin of the masses of sparry calcite that partially fill the cavities in the Mississippian limestone. The goal is to determine the temperature of calcite precipitation, the salinity of the fluid from which it precipitated, the diagenetic environment, and the timing of its formation. This study also serves to illustrate the preservation of fluid inclusions entrapped at low temperatures in the vadose zone, and the methodology for handling such populations.

Petrography.—

All hand specimens of the calcite aggregates are growth banded with layers of orangish-brown, iron oxide-rich clay, crinoid fragments, argillaceous infill, or silty sediment (Fig. 11.13). Most of the crystals are columnar, composite crystals composed of parallel, bladed-to-fibrous subcrystals in optical continuity. Most growth zonation appears as alternating brownish (rich in fine, unidentified inclusions) and clear bands. The banding may be closely spaced, with as many as 10 alternations per 50 µm, or may be widely spaced or absent. Pendant and other gravity asymmetric spar are common with both upward and downward thickening represented, suggesting growth in the vadose zone. Some growth bands define crystal terminations of well-formed euhedra with as much as 5 mm of relief on the termination, suggesting growth within a water-saturated pore, such as that which would be present below a perched or permanent water table (Kendall and Broughton, 1978). Twinning of many of the crystals indicates significant deformation. Some of the sparry masses are cross-cut by stylolites. The stylolite occurrence and twinning are consistent with some burial after calcite precipitation. Staining with potassium

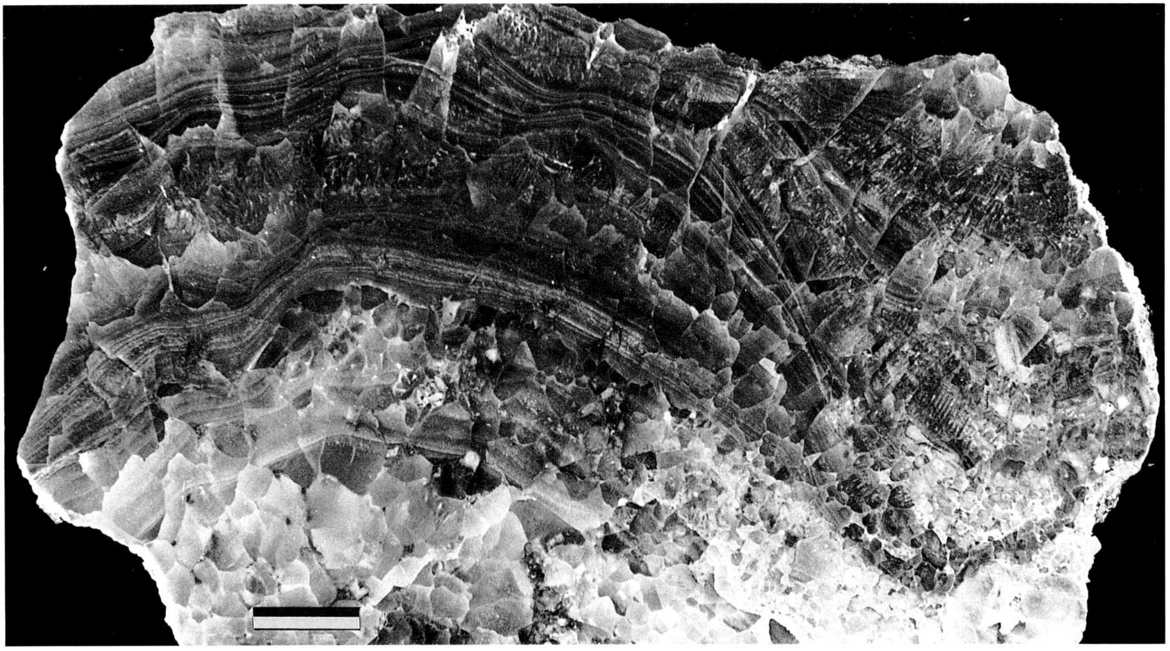

Fig. 11.13. *Photograph of polished slab of mass of sparry calcite from Mississippian Lake Valley Formation, New Mexico. Bar scale is 1 cm.*

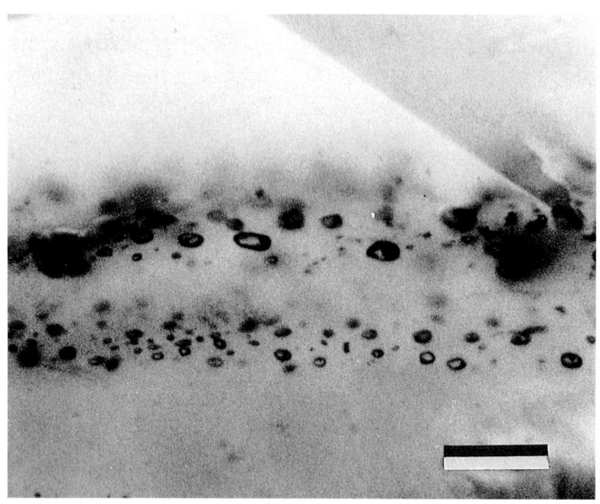

Fig. 11.14. Transmitted light photomicrograph of trains of secondary fluid inclusions in calcite from the Mississippian Lake Valley Formation. All inclusions contain vapor bubbles and the ratio of liquid to vapor appears consistent. Bar scale is 20 μm.

ferricyanide and alizarin red S indicates a nonferroan calcite composition for most of the spar, but reveals small patches of ferroan calcite that fill intercrystalline pores, later fractures, and breccia pores. The ferroan calcite is petrographically similar to the post-Mississippian, pre-Permian age cement zone 5 of the Lake Valley Formation (Meyers and Lohmann, 1985).

Many healed fractures contain secondary fluid inclusions that are two-phase with consistent liquid-to-vapor ratios (Fig. 11.14). Primary fluid inclusions are identified by their distribution along growth zones and between elongate subcrystals of the composite crystal (Fig. 11.15, 11.16). Primary one-phase all-liquid fluid inclusions are abundant (Fig. 11.17). In some FIAs, they may dominate, suggesting they are not the result of necking down of two-phase inclusions. These all-liquid inclusions do not appear to contain *significantly* metastable "stretched" liquid because they are in the same size range as many of the two-phase inclusions, and bubbles could not be nucleated during cooling. Two-phase, liquid-vapor fluid inclusions, which have small bubbles and somewhat consistent liquid-to-vapor ratios, are common in some FIAs, but typically occur along with the all-liquid inclusions. Two-phase, liquid-vapor inclusions with highly variable liquid-to-vapor ratios, ranging from mostly vapor to mostly liquid, are present. Some inclusions appear to contain all vapor. Some of these have leaked and can be shown to be open to the atmosphere because they are refilled easily with fluid by immersion in a liquid. Others cannot be refilled in this manner.

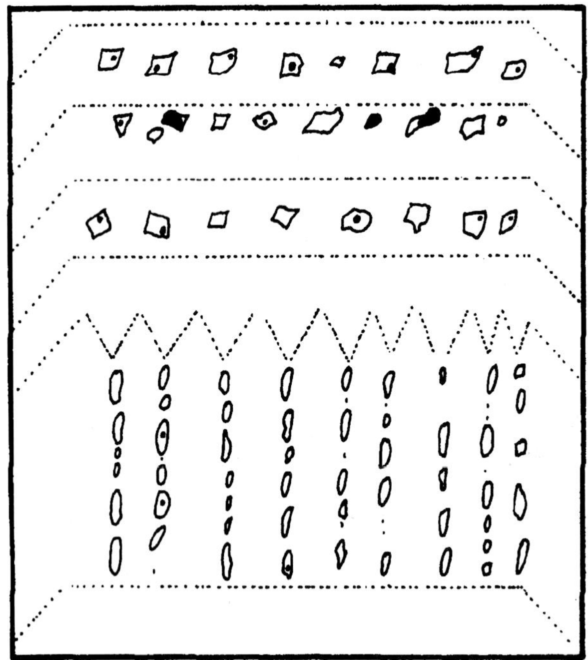

Fig. 11.15. Sketch illustrating occurrence of primary fluid inclusions within composite columnar calcite crystals. Trains of primary inclusions occur that outline subcrystal boundaries. Other primary fluid inclusions parallel growth zones. Width of field of view is approximately 1 mm.

Crushing.—

When samples containing two-phase inclusions with small vapor bubbles are crushed and exposed to kerosene at one-atmosphere pressure, many of the vapor bubbles collapse, indicating they are shrinkage bubbles caused by thermal contraction of the liquid during cooling to surface temperature. Normally, this behavior would suggest elevated-temperature entrapment of fluid inclusions, but as these inclusions occur in association with one-phase all-liquid inclusions (low-temperature), the two-phase inclusions must result from high-temperature reequilibration of one-phase fluid inclusions during later burial heating. For other two-phase inclusions with small bubbles, crushing and exposure to kerosene at one-atmosphere pressure results in little size change of the fluid inclusion's bubble, indicating that the bubble probably is composed of ancient air (vadose atmosphere) at one-atmosphere pressure. This suggests original entrapment of a liquid phase and a gas phase in the vadose zone. For a few of the two-phase inclusions with small bubbles, the bubbles expand and then partially dissolve when crushing exposes them to kerosene at one-atmosphere pressure. Such behavior indicates bubble pressures greater than one atmosphere and the presence of organic gases. This provides evidence for leakage of some low-temperature fluid

Fig. 11.16. *Transmitted light photomicrograph of fluid inclusion distribution relative to growth of calcite from Mississippian Lake Valley Formation. Calcite growth direction was from the base to the top.* I *denotes positions of vertical trains of primary fluid inclusions that were trapped at the boundaries between subcrystals. At the level of* T, *faint terminations are observed. At the level of* R, *ragged terminations and fluid inclusion rich growth zones are observed. Bar scale is 200 μm.*

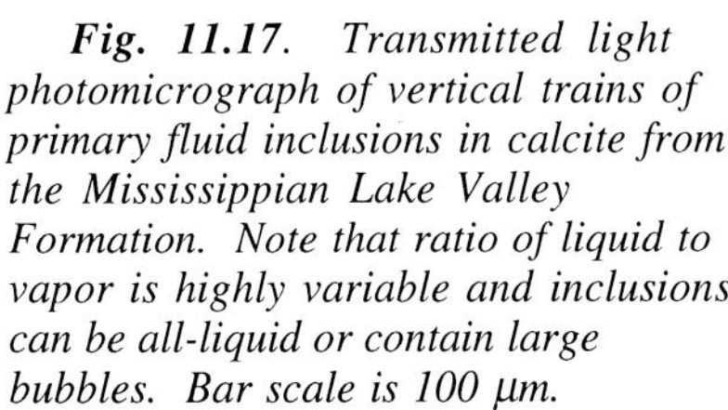

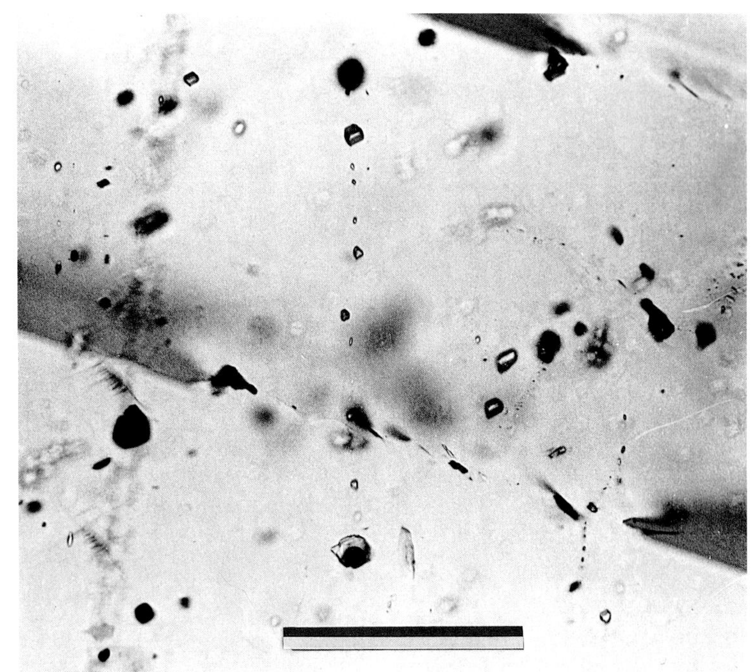

Fig. 11.17. Transmitted light photomicrograph of vertical trains of primary fluid inclusions in calcite from the Mississippian Lake Valley Formation. Note that ratio of liquid to vapor is highly variable and inclusions can be all-liquid or contain large bubbles. Bar scale is 100 μm.

inclusions and refilling with organic gas-rich fluids during burial heating.

When samples containing inclusions with large bubbles or inclusions from FIAs with highly variable phase ratios are crushed to expose inclusions to one-atmosphere pressure, most of these inclusions show little size change of the vapor bubble, indicating one-atmosphere pressure within the inclusions. These inclusions are characteristic of entrapment of vadose air and water within the vadose zone (Goldstein, 1986a; Barker and Halley, 1988).

When samples with inclusions that appear to be all vapor are crushed under one-atmosphere pressure, most of the gas phases exhibit little size change; they apparently trapped ancient air in the vadose zone.

Microthermometry.—

The Th measurements show a wide range of values. The population includes one-phase inclusions at room temperature (entrapment less about 50°C) and yields Th data from 43°C to above 150°C (data reported in Goldstein, 1986b). Of the 60 primary fluid inclusions measured for Tm ice, 65% had two phases at room temperature. For these two-phase inclusions, 26% yield Tm ice around 0.0°C indicating fresh water, and the rest contain concentrated brines (Tm ice down to -22°C) with salinities ranging from about 5 to 19 wt.% NaCl equivalent (Fig. 11.18). In contrast, every primary one-phase, all-liquid fluid inclusion yielded Tm ice at 0.0 ±0.1°C (Fig. 11.18).

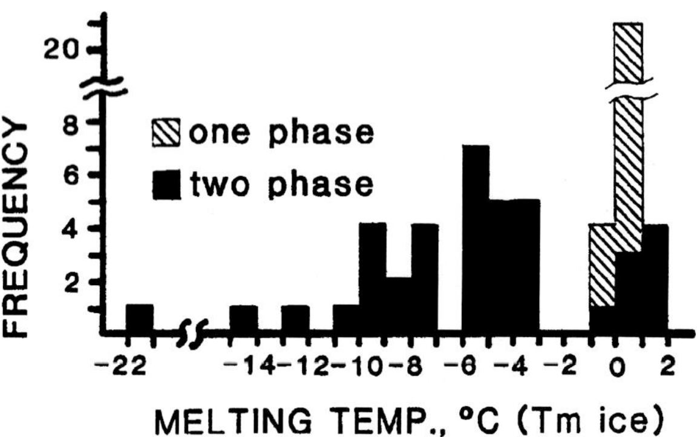

Fig. 11.18. *Tm ice data for primary fluid inclusions in masses of calcite spar from the Mississippian Lake Valley Formation. Every all-liquid inclusion yields Tm ice data indicative of fresh water whereas the two-phase inclusions preserve a record of later fluids. The measurements above 0°C represent "overshooting" of Tm ice from using an inappropriate warming rate. For pure water, all of the heat of melting must be transferred into an inclusion at 0°C, so overstepping of Tm ice by using too rapid a warming rate is a possible outcome.*

Interpretation.—

For the large aggregates of sparry calcite, fine but discontinuous growth banding, coarse and fine solid inclusions, smooth or flattened terminations, euhedral terminations, growth asymmetry, and mode of occurrence in solution pockets strongly support a vadose to possibly phreatic, calcite speleothem origin; most of these features are well known from speleothems that form in modern vadose zones (Kendall and Broughton, 1978). Therefore, the observations suggest filling of karst cavities took place in a low-temperature environment that was close to a subaerial surface. Because sediments of the overlying Pennsylvanian Gobbler Formation do not fill the solution pockets, the pockets and fills probably developed either before deposition of the Gobbler, and are pre-Pennsylvanian deposition in age, or well after deposition of the Gobbler. Superficially, the speleothems look similar to more recent speleothem material of the Sacramento Mountains. To demonstrate an ancient age for them unequivocally, evidence must indicate that initial precipitation near the surface was followed by deeper burial. The record of deeper burial suggests probable deposition of several km of post-Mississippian strata indicated by stratigraphic reconstructions, and heating to at least 100°C indicated by fluid inclusion homogenization temperature modes determined from nearby outcrops of Pennsylvanian strata at similar stratigraphic horizons (Pray, 1961; Goldstein, 1988). The stylolites, fractures, and twinning are consistent with some burial after calcite precipitation, and the ferroan calcite which fills the fractures is similar to the post-Mississippian, pre-Permian cement zone 5 of the Lake Valley Formation (Meyers and Lohmann, 1985). Thus, the observations indicate formation of karst and speleothems in a low-temperature, near-surface environment, followed by a period of deeper burial which began before the Permian.

The simple petrographic observations of fluid inclusions indicate that the calcite precipitated in the low-temperature (less than about 50°C) vadose zone, and then was heated. All liquid inclusions provide the record of initial formation at low temperature. The variable phase ratios and inclusions which retain one-atmosphere pressure indicate entrapment in the vadose zone. The pressure increase in fluid inclusions, caused by heating after calcite precipitation, resulted in reequilibration of some low-temperature fluid inclusions by leakage and refilling with elevated-temperature burial fluids, or stretching of the fluid inclusion cavities. Elevated-temperature secondary fluid inclusions also were formed during the elevated-temperature event. Some of the one-phase and two-phase inclusions survived burial heating without reequilibration and still provide a record of low-temperature, vadose-zone cementation. The all-liquid inclusions faithfully preserve the record of freshwater conditions of precipitation, whereas the two-phase inclusions contain fresh water and other fluids that filled vacuoles during thermal reequilibration. Therefore, in a general sense, these data indicate that although some of the original fluid inclusions reequilibrate when low-temperature calcites are subjected to burial heating, some of the original inclusions may be preserved. One-phase, all-liquid inclusions may survive burial heating to provide a record of the temperature and original salinity of low-temperature calcite precipitation. In similar fluid inclusion populations with mixtures of one-phase, all-liquid and two-phase fluid inclusions, the geologist should study the one-phase inclusions to provide the most reliable measure of the salinity of mineral precipitation and to indicate a low temperature of formation.

Origin of Low-Temperature Plio-Pleistocene Calcite Cement

Many samples of Plio-Pleistocene limestone from the Key Largo Limestone, Miami Formation, and Caloosahatchee Formation of south Florida contain bladed or equant calcite cement. These units generally have experienced complex histories, but never have been deeply buried or heated. They have been subjected to events of subaerial exposure that have allowed fresh water to circulate through the units to leach and recrystallize aragonitic grains and cement the units with low-Mg calcite. Some of this cementation took place in the freshwater phreatic zone and some took place in the vadose zone.

Objective.—

The objective of this study is to use fluid inclusions to verify the origin of calcite cement from relatively recent units that have not been deeply buried. The ultimate goal of presenting this case history is for comparison with later case histories in this chapter, so it will not be presented in detail. It is designed to show briefly (typically overgeneralizing) that calcite cements precipitated in the freshwater phreatic zone do entrap all-liquid fluid inclusions of freshwater.

Petrography.—

Calcite cements typically are equant or bladed. Commonly, they are clear and some may have a slight yellowish tint in hand specimens. The cements fill primary pores as well as moldic pores. Fabrics are suggestive of precipitation in the vadose zone. Fluid inclusions may occur randomly distributed in some crystals, but the best evidence for a primary origin are fluid inclusion-rich zones that outline former growth zone boundaries. Most inclusions in equant calcite are irregular and equant. Inclusions in bladed calcite are concentrated in inclusion-rich growth zones, but are spike shaped and oriented parallel to the growth direction. Typically, FIAs consist of all-liquid fluid inclusions that are less than 5 μm in size. Larger inclusions are rare and tend to be more closely associated with more coarsely bladed calcites (Fig. 11.19). Thus, the typical FIA associated with precipitation in the low-temperature freshwater, phreatic

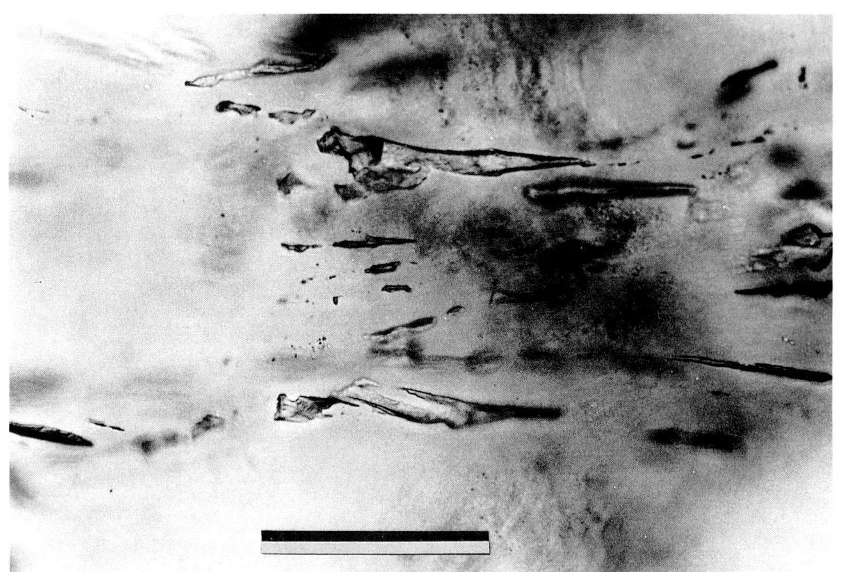

Fig. 11.19. *Transmitted light photomicrograph of primary fluid inclusions in calcite from the Pleistocene Miami Limestone, S. Florida. Spike-shaped inclusions are all-liquid which is a characteristic of the low-temperature phreatic zone. In reality, this cement originates in the vadose zone and other fluid inclusions (not pictured) are of highly variable liquid-to vapor-ratio. Sample provided by C. Barker. Bar scale is 20 μm.*

zone consists of all-liquid fluid inclusions.

Microthermometry.—
The lack of vapor bubbles precludes measuring homogenization temperatures. Once inclusions have been stretched artificially, Tm ice data can be gathered for the fluid inclusions. The data cluster at 0.0°C, and attest to the freshwater origin for the fluid inclusions.

Interpretation.—
The equant calcite of Plio-Pleistocene age from south Florida, precipitated from the low-temperature freshwater phreatic zone. Calcite precipitated from this environment traps freshwater inclusions that are all-liquid.

Origin of Pre-Pennsylvanian Calcite, SE Kansas

In SE Kansas, drill core has been recovered that spans the Mississippian-Pennsylvanian boundary. The strata at this depth may have experienced burial as deep as 1.8 km and have been heated by injection of hot fluids to a temperature as high as about 150°C (Barker and others, 1992; Wojcik and others, 1992, 1994). The Mississippian crinoidal limestone underlying the Pennsylvanian typically shows evidence of pre-Pennsylvanian karst formation. Further details of this study can be found in Wojcik (1991) and Wojcik and others (1994).

Objective.—
The objective of this study is to determine the origin of any calcite cement that precipitated before deposition of the overlying Pennsylvanian strata. To determine the origin of the calcite, the salinity, temperature and diagenetic environment must be determined. The goal of presenting this study is to illustrate the preservation of fluid inclusions from the freshwater phreatic zone despite a history of later heating. It will illustrate that an FIA from the low-temperature phreatic zone can be preserved despite a later history of partial thermal reequilibration.

Petrography.—
The calcite cement in the Mississippian limestones exists as syntaxial overgrowths on echinoderm fragments. No growth asymmetry is apparent. The cements have been reworked and truncated by reddish infills of insoluble residue associated with karst development before deposition of the Pennsylvanian strata. Thus, this simple cross-cutting relationship indicates that the syntaxial calcite precipitated at low temperature, near the surface, and in close association with karst development. The temporal and spatial relationship to karst and the lack of growth asymmetry is evidence that the calcite is associated with low-temperature phreatic zone cementation.

The calcites are rich in primary fluid inclusions. The primary origin is identified by cloudy, fluid inclusion-rich areas that terminate abruptly against growth boundaries which are overgrown by clear calcite (Fig. 11.20). A typical FIA is dominated by two-phase inclusions that appear to have relatively consistent ratios of liquid to vapor. However, these coexist with other inclusions that are all-liquid. Some of the all-liquid inclusions are smaller than the two-phase

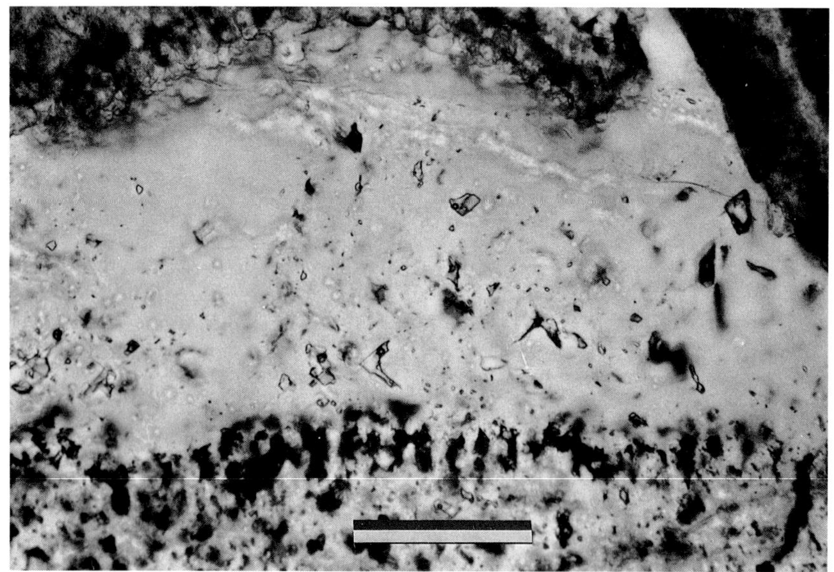

Fig. 11.20. *Transmitted light photomicrograph of fluid inclusions in calcite cement from Mississippian limestone from Kansas. Calcite cement is an overgrowth (center) on a crinoid fragment (below). Cloudy area in center is a growth-zone bounded area of fluid inclusions. Most of these primary fluid inclusions are two-phase with relatively consistent ratios of liquid to vapor and some are all liquid. Bar scale is 25 μm.*

inclusions, but many all-liquid inclusions and two-phase inclusions are in the same size range. The petrographic distribution of all-liquid inclusions is such that they do not appear to be paired with any vapor-rich fluid inclusions. During cooling, all-liquid inclusions did not generate bubbles unless they were stretched during freezing. Thus, the all-liquid inclusions do not appear to be significantly metastable and do not appear to have necked down from two-phase fluid inclusions; they preserve a record of low-temperature entrapment of the FIA. The coexisting two-phase inclusions are those that have reequilibrated during natural heating.

Microthermometry.—
We present the microthermometric data from a single FIA gathered by Wojcik (1991). Although these include measurements from only six fluid inclusions, the microthermometric data yield a remarkable amount of information. All inclusions, both two-phase and all-liquid provide Tm ice at about 0.0°C, indicating formation of the cement from fresh water (Fig. 11.21). The presence of all-liquid inclusions indicate precipitation below about 50°C. The two-phase fluid inclusions are those that have reequilibrated and range from 91°C-117°C (Fig. 11.22).

Interpretation.—
The cross-cutting relationships indicate precipitation of calcite before deposition of overlying Pennsylvanian strata. The presence of all-liquid fluid inclusions indicate that the calcite precipitated below about 50°C. Highly variable phase ratios associated with heterogeneous entrapment in the vadose zone are lacking. The associated two-phase inclusions are those that have reequilibrated during later heating. The all-liquid inclusions are filled with fresh water and indicate that the calcite precipitated from a low-temperature freshwater phreatic diagenetic environment. The two-phase inclusions still contain fresh water so they must have stretched during thermal reequilibration. Overall, the conclusion is that this single FIA has recorded precipitation of cement from the low-temperature phreatic zone followed by later heating.

Origin of Carbonate Cement, Plio-Pleistocene Hope Gate Formation, Jamaica

Plio-Pleistocene carbonates of the Hope Gate Formation, North Jamaica, have been well-studied with respect to their diagenesis (Land, 1973a) and have formed the basis on which many models of freshwater-marine mixing zone dolomitization have been based. These strata never have been deeply buried or heated. The samples analyzed from this study come from one outcrop locality of the Hope Gate Formation that contains dolomite and contains some high-Mg calcite that has not recrystallized. Samples were provided by Lynton S. Land. Fluid inclusion analyses were completed by Daniel Lehrmann.

Objective.—
The objective of this study is to determine the origin of carbonate cements that have precipitated in the Hope Gate Formation. This locality has provided an important example of diagenesis in the marine-meteoric mixing zone (Land, 1973). This fluid inclusion study should be useful in determining the origin of diagenetic minerals precipitating during migration of a proposed mixing zone. The goal of this study is to demonstrate that fluid inclusions provide a useful and unequivocal record of fluid inclusions trapped in a low-temperature phreatic marine-meteoric mixing zone. This relatively recent example can be used for comparison to diagenetic minerals that have been heated during burial in

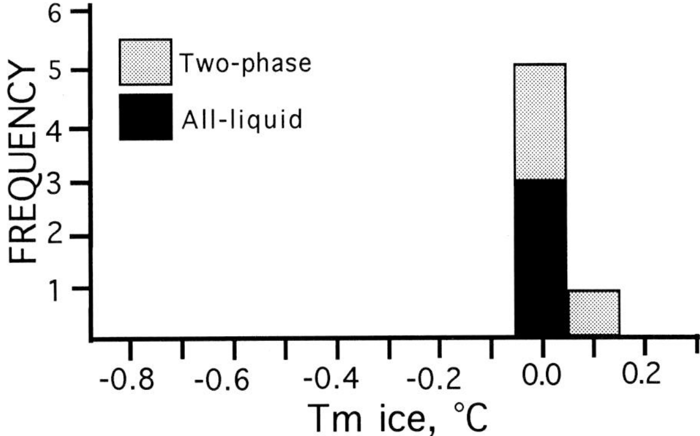

Fig. 11.21. Tm ice data from a single fluid inclusion assemblage from pre-Pennsylvanian calcite, SE Kansas. Data from Wojcik (1991).

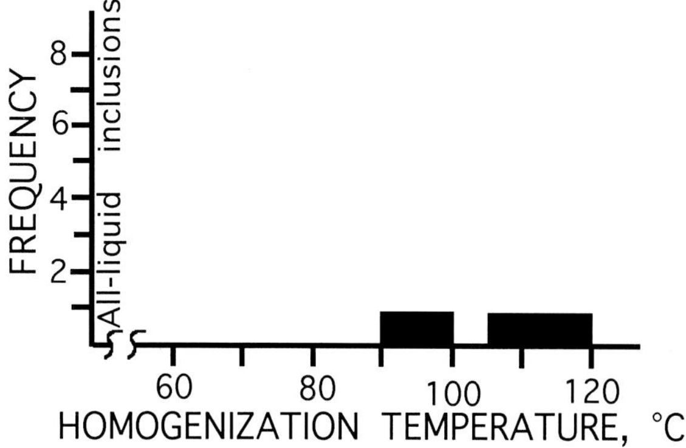

Fig. 11.22. Th data from a single fluid inclusion assemblage in Mississippian calcite cement, Kansas. Data from Wojcik (1991).

subsequent case histories.

Petrography.—

The pores from this locality are lined isopachously with cloudy, fibrous-to-bladed high-Mg calcite cement. This cement is overgrown by clearer blades (Fig. 11.23) that exhibit fine subzonation when etched. Elongate, all-liquid fluid inclusions oriented perpendicular to growth direction occur confined to this subzoned part of the cement. The fact that they are confined to the growth subzones and oriented perpendicular to growth direction is strong evidence for a primary origin. This fluid inclusion-rich cement is overgrown by bladed and equant low-Mg calcite cement. This later cement contains only rare, truly isolated fluid inclusions; all of these are liquid-filled. Typically, it would be dangerous to interpret these "isolated" fluid inclusions as primary, but it is clear that they must have been enclosed after entrapment of the earlier primary fluid inclusions because they occur in a later diagenetic phase.

Microthermometry.—

The elongate, primary fluid inclusions trapped in the growth-zoned phase are all-liquid and yield Tm ice values that range from -2.2 to 0.0°C (Fig. 11.24). The all-liquid inclusions isolated in the more equant later calcite are all 0.0°C.

Interpretation.—

The presence of all-liquid fluid inclusions in the zoned calcite indicates precipitation in the phreatic zone at a temperature below about 50°C. Fluid inclusions trapped in the later calcite are also all-liquid and indicate the existence of a low-temperature phreatic zone. The Tm ice values can be converted to salinity by assuming a model composition for the system. The most likely source of salty water in this system is seawater. The outcrops remained in a coastal setting, the initial cements are marine high-Mg calcite cements, and sea level apparently rose and fell several times since the deposition of the Hope Gate Formation. Thus, a seawater model composition will be assumed to interpret the Tm ice values. The highest salinity (Tm ice of -2.2°C) is 41 ppt seawater equivalent, a value well in line with slightly evaporated seawater. All of the other values vary between that and fresh water, indicating cementation of the zoned cement in a mixing zone that varied between its marine and freshwater end members. The later calcite cement contains only all-liquid fluid inclusions with Tm ice of 0.0°C, indicating a freshwater phreatic environment followed the mixing zone. The fluid inclusions of the Hope Gate Formation provide an unambiguous record of mixing zone cementation followed by a freshwater phreatic environment.

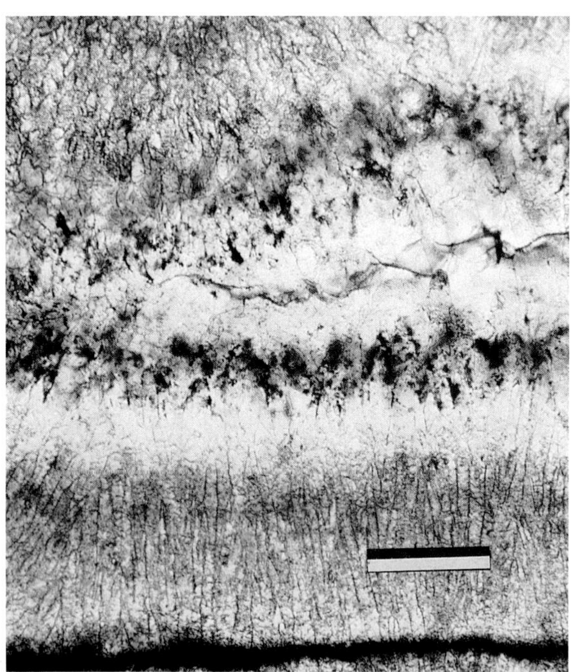

Fig. 11.23. Transmitted light photomicrograph of carbonate cements in the Plio-Pleistocene Hope Gate Formation, Jamaica. Fibrous high-Mg calcite gives way to later bladed cements that contain primary fluid inclusions. These are post-dated by equant, clear low-Mg calcite in the center of the pore. Bar scale is 100 μm.

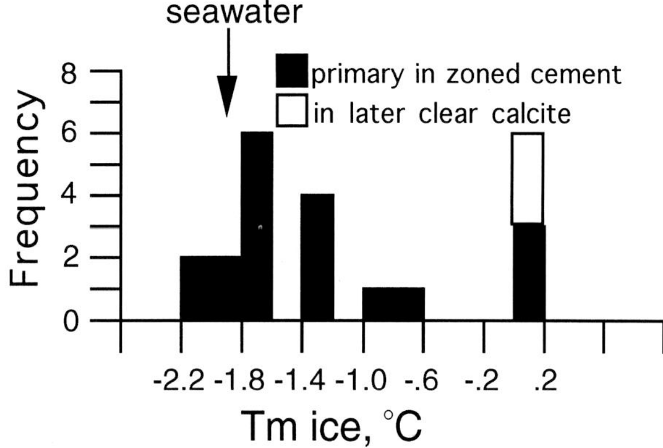

Fig. 11.24. Tm ice data from primary fluid inclusions in zoned carbonate phase and later calcite cement from the Hope Gate Formation, Jamaica. The data from the zoned cement indicates precipitation from a fluid of seawater and mixing zone salinity, whereas the later calcite precipitated from fresh water.

Origin of Early Calcite Cements, Pennsylvanian of Kansas

Pennsylvanian strata of Kansas consist of limestone-siliciclastic cyclothems that represent a history of relative rise and fall of sea level. The ideal Kansas cyclothem is capped by a shoal-water marine limestone that may be capped by a surface of subaerial exposure (Heckel, 1980). Thus, the early history of diagenetic fluids should record initial marine conditions, followed by migration of a marine-freshwater mixing zone, and followed by freshwater conditions. After these initial events, the Pennsylvanian strata were subjected to a complex history of later fluids and were heated to temperatures up to about 150°C (Wojcik and others, 1994). This study includes analyses from diagenetic calcite cement from the Pennsylvanian (Missourian) limestones of the Pen Oil Field, Graham County, Kansas (Phares, 1991), and analyses of Pennsylvanian (Desmoinesian and Missourian) limestone from the Cherokee Basin, Kansas (Wojcik, 1991; Wojcik and others, 1994).

Objective.—
The objective of this study is to determine the origin of the earliest (post-submarine) phases of calcite cement in Pennsylvanian cyclothems of Kansas. The goal of presenting this case history is to illustrate the preservation of fluid inclusion evidence of cementation in low-temperature marine-freshwater mixing zones of Pennsylvanian age, despite a later history of heating during burial and migration of multiple later pore fluids.

Petrography.—
These calcite cements are typically equant or bladed. They reduce primary pores in limestones or fill skeletal molds. In all cases, they predate compaction features of the limestone. In cathodoluminescence, they are either dully luminescent or contain internal brightly luminescent subzones. The luminescent zonation patterns are confined to only one cyclothem and cannot be correlated to cements in the overlying or underlying strata. Primary fluid inclusions occur in thin, fluid inclusion-rich growth zones that outline former crystal terminations and parallel cathodoluminescent zonation (Fig. 11. 25), or occur in cloudy, fluid inclusion-rich crystals that become clear at former growth boundaries. The fluid inclusions along these growth zones consist of two-phase fluid inclusions (with small bubbles) which appear to have relatively consistent ratios of liquid to vapor and occur along with all-liquid fluid inclusions. No vapor-rich inclusions are present so there is no pairing of all-liquid inclusions with vapor-rich inclusions to indicate that all-liquid inclusions originated from necking down. All-liquid inclusions are in the same size range as two-phase inclusions and cooling does not generate bubbles in all-liquid inclusions, so the all-liquid inclusions do not result from significant metastability.

Microthermometry.—
There is no need to measure homogenization temperatures of two-phase fluid inclusions to accomplish the objectives of this study. Because all-liquid inclusions are not *significantly* metastable and do

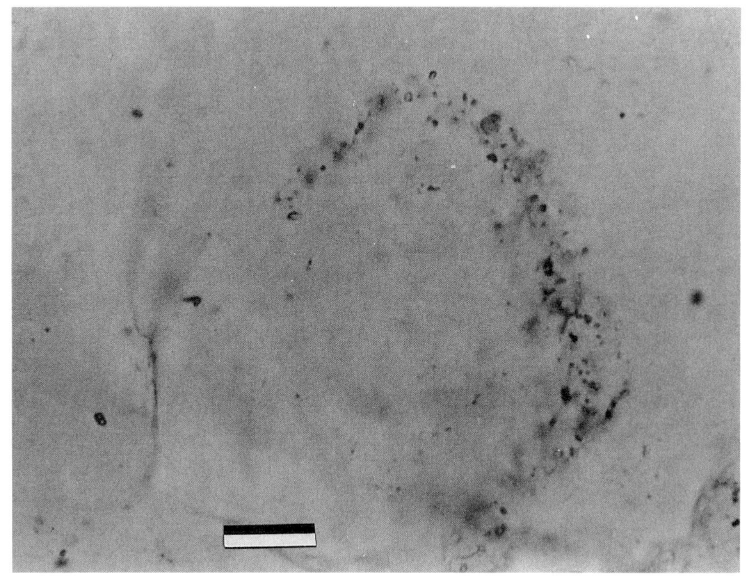

Fig. 11.25. *Transmitted light photomicrograph of calcite cement from the Pen Oil Field, Kansas. Notice cloudy growth zone that is rich in small fluid inclusions. This distribution provides strong evidence of primary origin for the fluid inclusions. Bar scale is 75 µm.*

not result from necking down, they provide a record of low-temperature of entrapment of fluid inclusions. The two-phase inclusions are those that have reequilibrated during burial heating, and the all-liquid inclusions should preserve the pristine record of cementation. Thus, the only Tm ice measurements are from all-liquid inclusions that were artificially stretched in the laboratory so that Tm ice measurements would be interpretable. Tm ice was measured from five different fluid inclusion assemblages. The values range from -1.8 to -0.5°C (Fig. 11.26).

Interpretation.—

The petrography indicates that the calcites in this study precipitated early, before significant burial, because they occur before compaction features. The distribution of the cathodoluminescent zonation pattern indicates that this cement did not precipitate beyond the bounds of the individual cyclothem in which it is contained. This suggests that the calcite precipitated before deposition of the next overlying cyclothem. The occurrence of primary all-liquid inclusions alongside two-phase inclusions with somewhat consistent ratios of liquid to vapor indicates that these cements precipitated from a low-temperature phreatic-zone fluid that was below about 50°C. The two-phase inclusions are from thermal reequilibration of originally all-liquid fluid inclusions. The all-liquid inclusions should retain the pristine record of salinity of the low-temperature fluid responsible for calcite cementation. Because the petrography indicates an early origin for the cements and because an early history of marine to freshwater conditions is predicted based on the cyclothem character, a seawater model will be assumed for interpreting the Tm ice data. The data indicate that calcite precipitated from a low-temperature phreatic zone environment in which salinity varied from 33.4 ppt to 9.3 ppt seawater equivalent. These cements clearly have precipitated from a low-temperature marine-freshwater mixing zone.

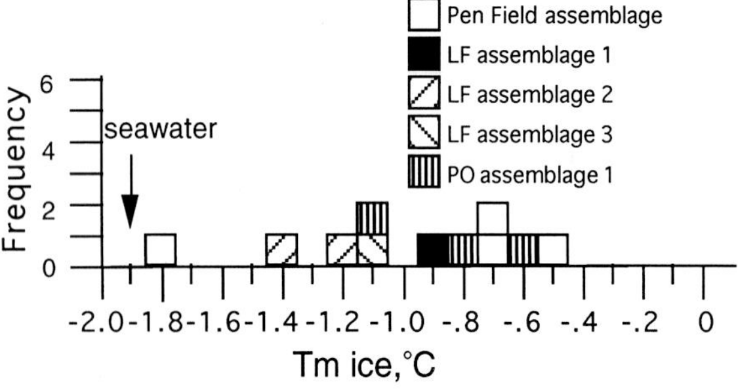

Fig. 11.26. *Tm ice data taken from all-liquid fluid inclusions, various localities from the Pennsylvanian of Kansas. Each symbol represents a separate fluid inclusion assemblage. Data indicate calcite precipitation from a mixing zone. Data from Phares (1991) and Wojcik (1991).*

Fluid Inclusions in Aragonitic Submarine Cement, Belize

Coarse, fibrous and botryoidal aragonite cements were collected from reef limestones, offshore Belize, at water depths of 65-120 m (Ginsburg and James, 1976). This aragonite is Holocene in age and formed from marine waters. The cements have never been subject to burial or burial heating and have only experienced a history of marine waters.

Objective.—

The objective of this case history is to present a schematic example to show that aragonite submarine cements do trap fluid inclusions of seawater. The goal is to aid in understanding of case histories to be presented later in this chapter.

Petrography.—

The aragonite consists of tightly packed fibers that form botryoids. Fluid inclusions are distributed throughout the botryoids. Inclusions range in size from 1 µm to 200 µm in size. The smaller inclusions are equidimensional and globular or polygonal in form. They are either randomly distributed or concentrated along linear zones parallel to aragonite fibers. The larger inclusions are typically elongate in shape and parallel to aragonite fibers. Most fluid inclusions are all-liquid.

Interpretation.—

Modern submarine cements do trap all-liquid fluid inclusions of seawater. Without recrystallization, one might predict that such seawater inclusions could be preserved in ancient sedimentary rocks.

Origin of Cambrian-Ordovician Calcite Cements, Llano Uplift

Three shallow marine truncation surfaces in limestones of the Wilberns formation, Texas, span the Cambrian-Ordovician boundary. The surfaces truncate grains and bladed cements providing timing on cement formation. Although maximum depth of burial is somewhat controversial, the cements were clearly heated during burial or hot brine migration to temperatures above 100°C. This study is covered in more detail in Johnson and Goldstein (1993).

Objective.—

The objective of this study is to determine the origin and original mineralogy of cements truncated along marine truncation surfaces in the Wilberns Formation. The study attempts to determine the temperature of cement precipitation, the salinity of the fluid from which the cement precipitated, the diagenetic environment, and the original mineralogy of the cement. In so doing, every piece of fluid inclusion data must be explained. The goal of this study is to illustrate the preservation of fluid inclusion evidence for low-temperature precipitation of calcite cement in the submarine environment, in a setting that has been subjected to significant heating and experienced pore fluids of many different compositions.

Petrography.—

The bladed calcite cements precipitated in pores of marine limestones and are truncated along knife-sharp erosion surfaces. The erosion surfaces are overlain by more marine limestone, so the age of the cements is well constrained as predating deposition of the overlying limestone. The cements are typically bladed low-Mg calcite containing 0.5-2.0 mol.% $MgCO_3$. The absence of microdolomite inclusions argues against closed-system recrystallization of high-Mg calcite. Fluid inclusions occur as randomly distributed cloudy blades and cloudy cores to blades that are surrounded by clearer bladed calcite zones (Fig. 11.27). The termination of the cloudy area at a former growth-zone boundary provides strong evidence of primary origin for fluid inclusions in this material. Secondary fluid inclusions also cut across the material. All-liquid primary fluid inclusions are abundant and occur along with liquid dominant two-phase inclusions that have relatively consistent ratios of liquid to vapor. Most of the all-liquid inclusions are not petrographically paired with vapor-rich inclusions, suggesting that most all-liquid inclusions do not result from necking down after a phase change. All-liquid inclusions are in the same size range as two-phase inclusions and do not nucleate vapor bubbles during cooling unless they have been stretched in the laboratory. Thus, all-liquid inclusions are not the result of *significant* metastability.

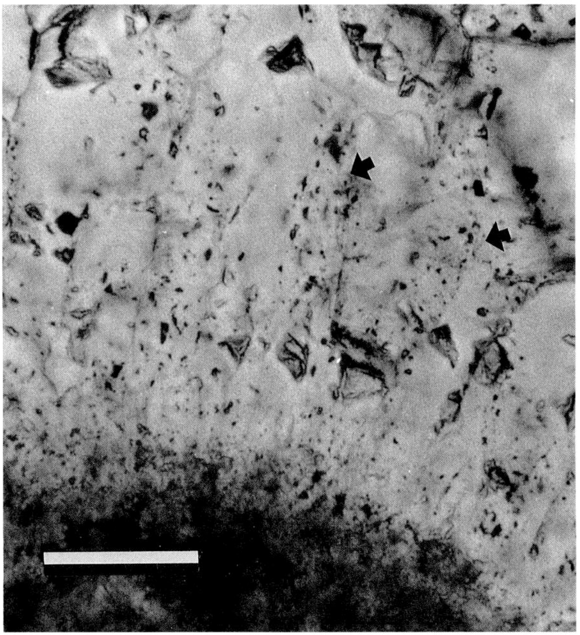

Fig. 11.27. *Transmitted light photomicrograph of bladed calcite cement from Cambrian of Texas. Just out of the field of view, the cements appear truncated along a planar erosion surface. Cloudy, blade-shaped areas within the calcite cement contain primary fluid inclusions. Bar scale is 50 µm. Photograph courtesy of W. J. Johnson. After Johnson and Goldstein (1993).*

Microthermometry.—

Most values for Tm ice of primary all-liquid inclusions are between -1.7 and -2.5°C with a mode at -1.9°C (Fig. 11.28). Rare primary all-liquid inclusions yield data immediately outside of this range, but these yield identical values to secondary all-liquid inclusions, so the primary vacuole may have been refilled by later fluids. The few primary all-liquid inclusions with significantly lower Tm ice yield the same values as two-phase secondary fluid inclusions. These rare inclusions appear to be associated with areas in which a few of the all-liquid inclusions are petrographically paired with vapor-rich inclusions, indicating that these higher salinity all-liquid inclusions represent leakage and refilling of primary cavities with later elevated temperature fluids, followed by necking down after a

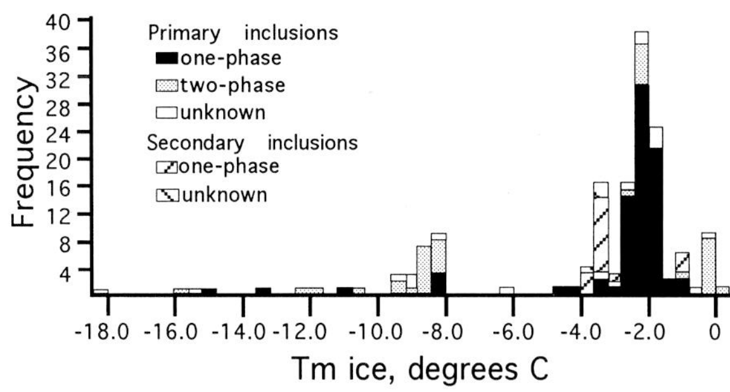

Fig. 11.28. *Tm ice data from primary and secondary fluid inclusions from the Cambrian of Texas. Before analysis, fluid inclusions were either all-liquid two-phase, or of unknown original phase ratio (presence or absence of bubble unknown). Data indicate calcite cementation from fluid of seawater salinity. Inclusion data outside of this range are the same composition as secondary fluid inclusions and can be explained by later leakage and refilling of primary vacuoles. After Johnson and Goldstein (1993).*

phase change. Thus, all of the all-liquid inclusions that occur outside of the -1.7 to -2.5°C range can be ascribed to leakage and refilling of primary fluid inclusions with fluids that were introduced after cementation.

Interpretation.—

The all-liquid inclusions that produce Tm ice within the -1.7 to -2.5°C range are those that were preserved from initial precipitation of the cement. They indicate that the cement precipitated below about 50°C from a phreatic environment. The two-phase inclusions and inclusions outside of this salinity range are unreliable because they result from leakage and refilling during microfracture formation and during thermal reequilibration. The Tm ice data can be interpreted using a seawater model because of the close association between cementation and the marine environment. Thus, the bladed calcite cement precipitated from a fluid ranging from 31 to 47 ppt seawater equivalent, with a mode at 35 ppt and a median at 39 ppt seawater equivalent. Modern normal seawater has a salinity of 35 ppt, and the range of values obtained are well within the range of *slightly* evaporated seawater. Thus, the calcite cement precipitated as low-Mg calcite from a low-temperature marine phreatic environment of Cambrian-Ordovician age. There is no evidence to suggest the cement has recrystallized.

Origin of Dolomite in Eocene Strata, Enewetak Atoll

Enewetak is an atoll located at about 11°N latitude in the Pacific Ocean, and consists of between 1200-1400 m of carbonate strata overlying a basaltic volcanic basement. Two deep wells penetrate the atoll margin and provide core samples from deep within the atoll. Today, the deeply-cased bore holes are filled with seawater and show tidal fluctuations that indicate relatively open circulation with seawater. Dolomite is not abundant in the recovered cores. As reported by Saller (1984a, b) in most core segments, dolomite abundance is less than 1%. However, below 1200 m some samples in a few Upper Eocene intervals may be completely dolomitized. Saller noted that the dolomite precipitated after development of brittle compaction features, and thus, must have formed after some burial. Saller and Koepnick (1990) have studied the $^{87}Sr/^{86}Sr$ ratios of samples from Enewetak. In general, the $^{87}Sr/^{86}Sr$ of dolomite is significantly higher than depositional carbonate at the same depth. Its composition matches depositional carbonate 950 to 1,260 m higher in the section. There is no postulated source of radiogenic Sr (basement is probably less than 0.7070); thus, the dolomite must have acquired its Sr and formed after burial of at least 950 m of overburden. So the dolomite formed fairly recently. With this in mind, Saller built a logical argument on the origin of the dolomite. The atoll margin today contains seawater circulating from tidal pumping and thermal convection. At the present burial depth, the dolomitized interval is at about 10°C to 20°C. The oxygen isotopic data are perfectly consistent with dolomite precipitation from seawater at these temperatures. Thus, Saller developed a model whereby dolomite was predicted to form from cold seawater below the approximate calcite saturation depth of 1000 m.

Objective.—

The objective of this study is to use fluid inclusions as a test of the hypothesis that the dolomite of Enewetak Atoll precipitated from cold, deeply circulating seawater. The Tm ice of primary fluid inclusions are a logical method by which the seawater model can be tested. If the model is shown to be inappropriate, fluid inclusions can be used to constrain the diagenetic environment. The goal of this study is to demonstrate the power of using Tm ice data to constrain the origin of dolomite. Although the seawater model has been predicted, this study forms the basis for demonstrating the applicability of the fluid inclusion technique to environments of low-temperature, hypersaline reflux. Later, it will form the basis for comparison to ancient examples that have been subject to burial heating.

Petrography.—

Dolomite crystals are generally 100-200 μm in diameter. They are euhedral where they partially replace the host limestone and less euhedral in completely dolomitized intervals. Calcite bioclasts have commonly dissolved in the dolomitized intervals. All-liquid fluid inclusions are abundant and are confined to cloudy growth-zone bounded cores (Fig. 11.29) and later inclusion-rich growth zones within the dolomite. This argues for a primary origin for the inclusions.

Microthermometry.—

The fluid inclusion Tm ice data from the Enewetak dolomite are presented in Figure 11.30. The data range from about -4.4°C to about -2.4°C.

Interpretation.—

The all-liquid inclusions indicate dolomite formation below about 50°C. The Tm ice data are interpreted using a seawater model because the atoll is currently bathed in seawater. Tm ice for normal seawater salinity would be about -1.9. The lowest salinity recorded by the Tm ice is about 44 ppt seawater equivalent, the highest salinity is about 85 ppt, and the mode is about

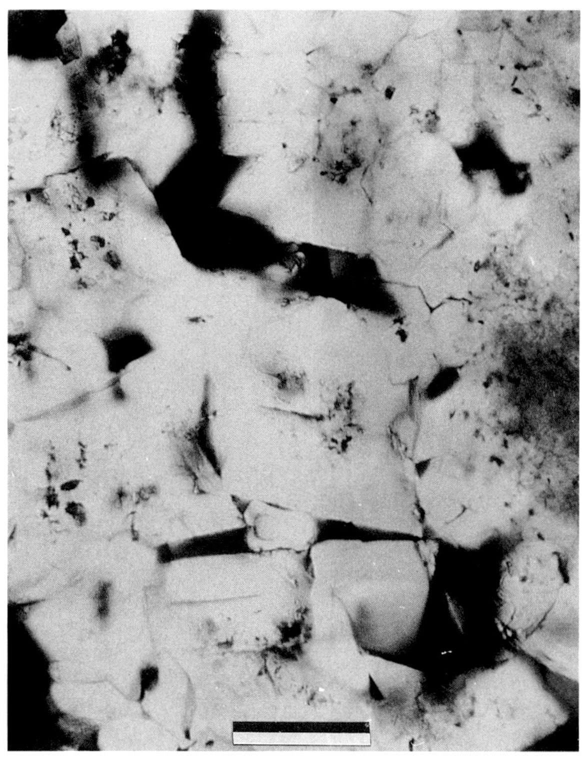

Fig. 11.29. Transmitted light photomicrograph of blue epoxy impregnated thin section of dolomite from Enewetak atoll. Dolomite crystals have cloudy cores which contain primary fluid inclusions. Bar scale is 100 μm.

60 ppt seawater equivalent. Thus, the Enewetak dolomite formed from brines that were approximately twice the salinity of normal seawater. The origin of the dolomite would have to be one that involves reflux of brines of salinity significantly higher than seawater. Evaporation would likely take place in the lagoon and would have been followed by reflux through the lagoon core followed by lateral discharge along permeable beds.

Origin of Early Post-Compactional Calcite Cement, Pennsylvanian, Lansing-Kansas City Groups

The Lansing and Kansas City groups occur in the subsurface of NW Kansas and SW Nebraska and consist of limestone-siliciclastic cyclothems. They are overlain by Upper Pennsylvanian and Lower Permian interbedded shales and carbonates (limestone and dolomite). Sandstones are common in the Lower Permian, and the Lower Permian strata also include anhydrite and halite evaporites. In the area of study, the Lansing and Kansas City groups include six to seven complete cyclothems. The data for this study and the interpretations herein are part of an M.S. thesis by Anderson (1989).

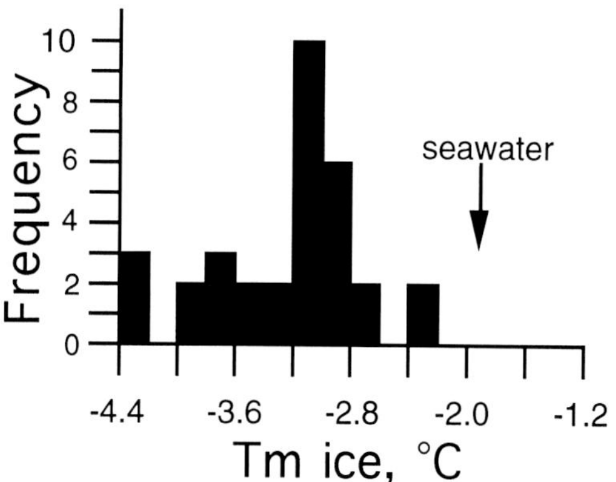

Fig. 11.30. Tm ice data from primary fluid inclusions in dolomite from Enewetak Atoll. Data indicate that dolomite precipitated from fluid that was higher than seawater salinity.

Objective.—
The objective of this study is to determine the origin of calcite cements that significantly reduce the porosity of limestone reservoir facies. The diagenetic environment will be constrained by determining the timing and distribution of the cement, the temperature at which it precipitated, and the compositions of the fluids responsible for its precipitation. The goal of presenting this study is to illustrate the preservation of fluid inclusion evidence of low-temperature precipitation of calcite cement from a hypersaline, refluxing brine.

Petrography.—
Evidence of compaction includes overly close packing and crushed grains that predate precipitation of the cements in question. The calcite cements generally are clear, equant cements that reduce the moldic, interparticle, and intraparticle pores. The #1 calcite cement is a nonluminescent, nonferroan, calcite cement zone that is the first phase of calcite cementation in almost all thin sections. It is the first generation of calcite that postdates compaction. The #2 calcite cement is a nonferroan calcite cement that postdates the nonluminescent #1 calcite cement. The luminescence of the #2 calcite cement is complex and subzoned, and these subzones range from moderately bright to moderately dull. Fluid inclusions occur in cloudy zones that are confined by the cathodoluminescent growth banding of the calcite (Fig. 11.31). Most of the primary fluid inclusions in these two cements are all-liquid and they occur along with liquid-rich two-phase fluid inclusions that have somewhat consistent ratios of liquid to vapor. The all-liquid inclusions are not petrographically paired with vapor-rich fluid inclusions and the proportion of two-phase inclusions is small, so necking down after a phase change is not a valid explanation for the all-liquid inclusions. The all-liquid

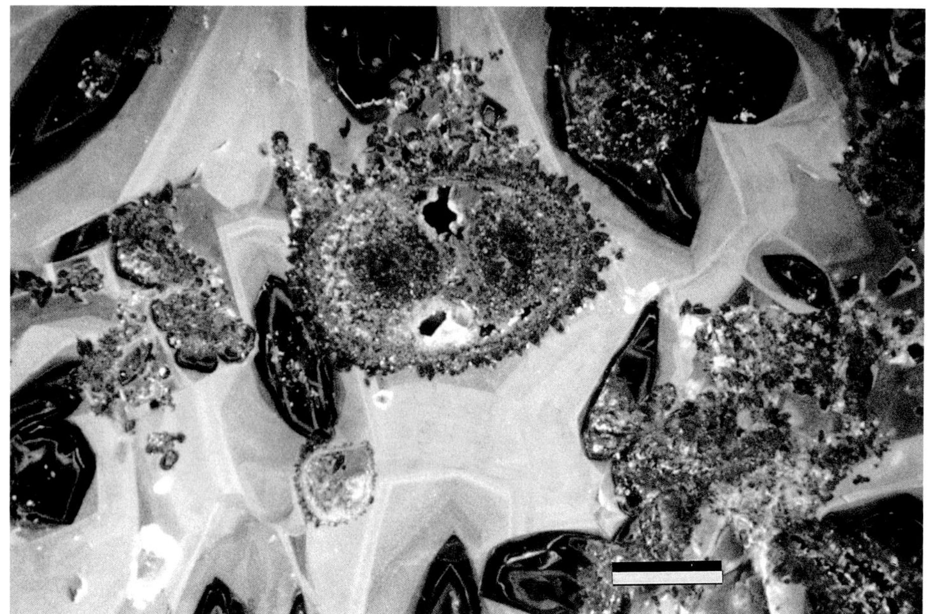

Fig. 11.31. Cathodoluminescence photomicrograph of calcite cement from Pennsylvanian Lansing and Kansas City Groups, Kansas. Primary fluid inclusions occur in growth zones that are easily defined by the compositional growth zonation observed using cathodoluminescence. Bar scale is 75 μm.

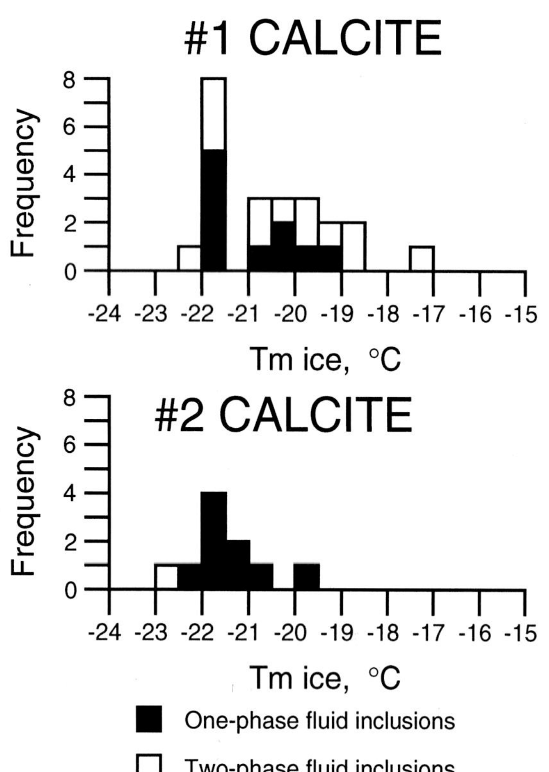

Fig. 11.32. *Tm ice data from primary fluid inclusions in two distinctive generations of early calcite from Pennsylvanian Lansing-Kansas groups, Kansas. Data from Anderson (1989).*

inclusions are in the same size range as the two-phase inclusions and all-liquid inclusions do not generate vapor bubbles during cooling unless stretched. Therefore, all-liquid inclusions do not result from significant metastability.

Microthermometry.—
The repeatable Te data from the all-liquid inclusions range from -42° to -52°C in calcite #1 and from -50° to -52°C in calcite #2. The Tm ice data for the one-phase fluid inclusions in calcite #1 cover a more narrow range than two-phase fluid inclusions and occur between -22.0°C and -19.1°C. The Tm ice for the one-phase fluid inclusions in calcite #2 cover a more narrow range than the two-phase fluid inclusions and are between -22.2°C and -19.8°C (Fig. 11.32).

Interpretation.—
The presence of all-liquid fluid inclusions indicates that the calcite precipitated from a fluid below about 50°C. The two-phase fluid inclusions are those that have reequilibrated during burial heating and are not to be trusted to provide reliable data on origin of the diagenetic phase. The eutectics indicate a likely H_2O-NaCl-$CaCl_2$ composition for the brine. But because the Tm hydrohalite were difficult to observe, the salinities will be interpreted assuming the H_2O-NaCl system. The Tm ice data indicate an approximate composition of 23 wt.% NaCl equivalent. The high salinity of these fluids suggest an evaporitic source. The Permian System in this area includes anhydrite and halite. The stratigraphic interval between the top of the Lansing

Group and the base of an evaporite in the Stone Corral Formation ranges from 300 to 500 m. The brines from which the calcite cements precipitated could have been generated and refluxed downward during deposition of Lower Permian evaporites and yielded low temperatures of formation. The Ca content of the brines from which calcite precipitated may be explained by dolomitization, albitization of plagioclase in the sandstones, and the exchange of sodium for calcium in smectite clays of the mudstones, as these brines refluxed down through the Wolfcampian and Virgilian strata (Fig. 11.33).

Origin of Mid-Cretaceous Dolomite, Valles Platform, Mexico

This study deals with the origin of dolomite from the mid-Cretaceous Tamabra Formation of Mexico. The dolomite occurs along the margin of the Valles Platform and is abundant in the outer margin of the platform, and the slope facies of the basal El Abra and Tamabra formations. The dolomite of this study is confined to the slope facies. The platform sequence is underlain by approximately 520 m (50,000 km^3) of anhydrite of the Lower Cretaceous Guaxcama Formation. In the Tamabra Formation, the dolomite replaces and cements resedimented slope deposits and fine basinal pelagics. The dolomite has $\partial^{13}C$ between +2 and +3 ‰ PDB. The $\partial^{18}O$, for the most part, is between -1 and 0 ‰ PDB. Such an oxygen isotopic composition would be consistent with precipitation between tropical surface temperatures and 50°C from a brine of seawater composition to several per mil more positive than seawater. This study was conducted by M.S. student Bryan Stephens and more details can be found in Stephens (1988).

Objective.—
The objective of this study is to interpret the diagenetic environment of dolomite formation by understanding the temperature and fluid composition responsible for its precipitation. The goal of presenting this study is to illustrate the degree to which all-liquid fluid inclusions have been preserved despite later histories of significant heating during burial.

Petrography.—
The matrix is replaced by planar, subhedral to euhedral dolomite crystals 75-200 µm in size and dolomite overgrowths protrude into pore space (Fig. 11.34). The dolomite is postdated by fracturing and precipitation of calcite, quartz, and fluorite. The dolomite contains a growth zone with a mixture of primary all-liquid and two-phase inclusions that are liquid-dominated and have relatively consistent ratios of

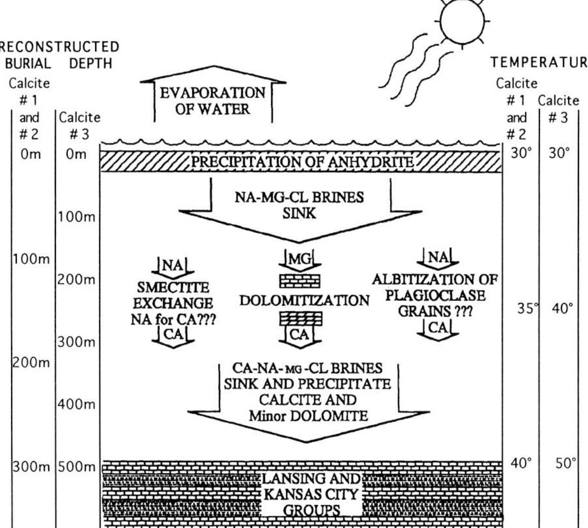

Fig. 11.33. *Diagrammatic illustration of origin of fluids that precipitated the early calcite (#1 and #2) and the later calcite (#3). Data from calcite #3 is described in the case history on page 178 of this chapter. Depths are reconstructed from assumed geothermal gradients and depth of Pennsylvanian below Permian evaporites. All calcites precipitated from Ca-rich brines that might have refluxed downward through the Pennsylvanian strata during deposition of Permian evaporites. Brines achieved a Ca-rich composition through any or all of the types of rock/water interaction illustrated. After Anderson (1989).*

Fig. 11.34. *Transmitted light photomicrograph of thin section from the Tamabra Formation. Notice cloudy and clear, growth-zoned dolomite "overgrowth" that extends into pore. Primary fluid inclusions occur in the overgrowth dolomite. The dolomite is followed by calcite cement and megaquartz cement (hexagonal cross sections). Bar scale is 200 µm.*

liquid to vapor. These inclusions are limited to thin growth zones defined by UV fluorescence. All-liquid inclusions are in the same size range as two-phase inclusions, are not petrographically paired with two-phase inclusions, and do not nucleate bubbles during cooling unless stretched. Thus, all-liquid inclusions do not result from necking down of two-phase fluid inclusions or *significant* metastability. Later mineral phases contain abundant primary and secondary two-phase fluid inclusions that have relatively consistent ratios of liquid to vapor.

Microthermometry.—

Freezing point depression (Tm ice) for all-liquid inclusions in dolomite are -12 to -17°C (Fig. 11.35). If one were to assume an H_2O-NaCl model, these would translate to a salinity of 16-20 wt.% NaCl equivalent. The Tm ice data from the two-phase inclusions in dolomite show a much wider spread in Tm ice. Because of the observed paragenetic relationships, the fluid inclusions in the calcite and fluorite must both postdate formation of the dolomite. Notice that dolomite, calcite, and fluorite all contain two-phase fluid inclusions that yield Tm ice within -9 to -4°C. Thus, the two-phase fluid inclusions in dolomite within this Tm ice range can easily be explained through leakage and refilling during the presence of later higher temperature fluids. This example also illustrates that the two-phase fluid inclusions that occur along with all-liquid inclusions are not to be trusted to give evidence of conditions of precipitation of the mineral in which they are enclosed. They can easily originate from thermal reequilibration. Homogenization temperatures of the two-phase fluid inclusions show quite a range of variability (Fig. 11.36), but yield data that indicate the

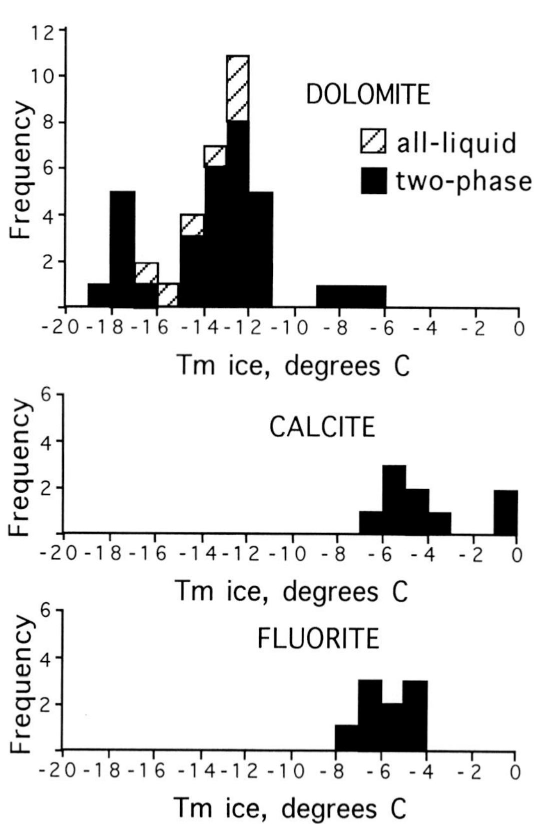

Fig. 11.35. Tm ice data from fluid inclusions from the Cretaceous Tamabra Formation. Notice that primary all-liquid fluid inclusions in dolomite span a narrow range that does not include the range of the later fluid inclusions. This illustrates that some of the two-phase inclusions in dolomite that appear primary have leaked and refilled during thermal reequilibration with later fluids. After Stephens (1988).

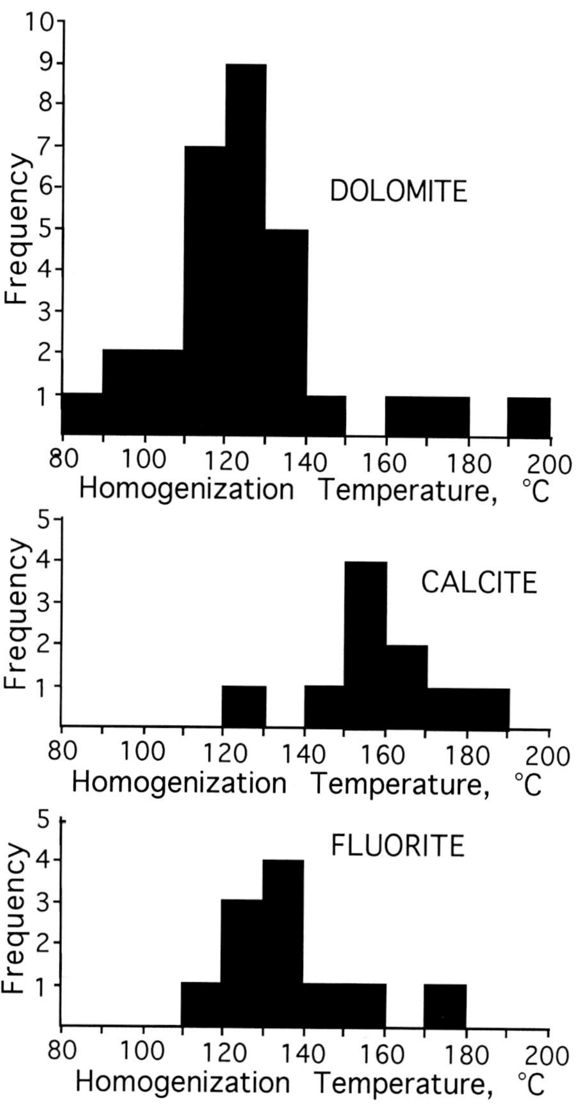

Fig. 11.36. Th data from two-phase fluid inclusions in the Cretaceous Tamabra Formation. Notice that rocks have been heated to at least 150°C. After Stephens (1988).

rocks had been heated to temperatures of at least about 150°C.

Interpretation.—

The one-phase inclusions in the dolomite provide evidence that the dolomite precipitated at a low temperature (below about 50°C) and formed from a fluid with a salinity between 16 and 20 wt.% NaCl equivalent, a salinity well within the field of gypsum precipitation. The most incredible thing to realize is that this low-temperature record is preserved despite later natural heating to at least 150°C. There are two logical models to explain the origin of dolomite that are both consistent with the geology, isotopic data, and fluid inclusion data. Either a brine was generated during deposition of some yet unfound evaporite deposited on the platform, and reflux took place (Golden Lane evaporites are known of appropriate age), or connate fluids within the Guaxcama evaporites were expelled to produce dolomitization.

Minimum Temperature of Formation of Halite Associated with Sylvite

The Devonian Prairie Formation is an extensive evaporite unit in the subsurface of western Canada. Halite dominates but potash salts are present in southern Saskatchewan. The mechanism of deposition of potash-bearing mineral sequences has remained in question for years. One method of dealing with this question is to determine the temperatures at which potash salts have formed. Fluid inclusions in halite associated with the potash salts provide the necessary temperature information to aid in evaluation of the origin of potash salts. This case history reproduces part of a more extensive study by Lowenstein and Spencer (1990).

Objective.—

The objective of this study is to refine the origin of halite in sylvite-bearing sequences of the Prairie Formation by using fluid inclusion data to constrain minimum entrapment temperature of fluid inclusions. The goal of presenting this study is to show one example of the application of fluid inclusion studies to evaporite sequences, and to illustrate that homogenization by daughter mineral dissolution can provide useful information on minimum entrapment temperature.

Petrography.—

The halite associated with potash salts in the Prairie Formation display polygonal mosaic textures of halite and sylvite or carnallite, halite and sylvite poikilotopically enclosed in sylvite, halite enclosed in poikilotopic carnallite, and interlayered halite and carnallite (Lowenstein and Spencer, 1990). Some of the halite contains fluid inclusion-rich bands that reflect growth zoning, and provide strong evidence for a primary origin. Some of these inclusions contain liquid and a single crystal of sylvite. Genetically related inclusions typically have the same ratio of sylvite and brine and the sylvite crystals are interpreted to be daughter crystals. Some inclusions contain only brine until they are cooled to a temperature at which they nucleate sylvite daughter crystals.

Microthermometry.—

The inclusions homogenize by dissolution of the sylvite daughter crystals. The dissolution temperatures of sylvite provide measures of minimum entrapment temperature. Closely associated fluid inclusions (not necessarily FIAs) yield sylvite dissolution temperatures that remain consistent within a closely related population (Fig. 11.37) lending support to an interpretation that the inclusions have not been altered significantly during burial.

Interpretation.—

Sylvite dissolution temperatures from the Prairie Formation indicate minimum entrapment temperatures as low as 5°C and up to about 80°C. See Lowenstein and Spencer (1990) for other implications.

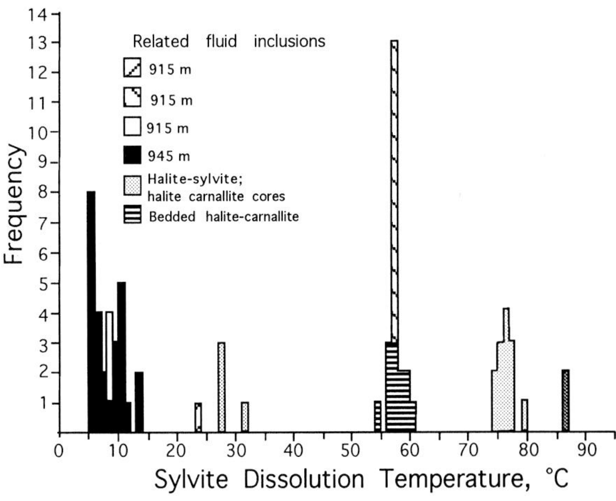

Fig. 11.37. *Sylvite dissolution temperatures from primary fluid inclusions in halite from the Prairie evaporite, western Canada. Each symbol represents a related population of fluid inclusions. Modified from Lowenstein and Spencer (1990).*

Calcite Cementation During Progressing Conditions, Permian Laborcita Formation, New Mexico

Limestones of the Permian (Wolfcampian) Laborcita Formation crop out in areas of the Sacramento Mountains, New Mexico. A significant amount of the porosity of these units is occluded by blocky calcite cement. The origin of this calcite cement has remained in question and has been ascribed to cementation in the meteoric phreatic realm (Cys and Mazzullo (1977) or from cementation in a deep-burial, semiclosed system (Major, 1985). Preliminary fluid inclusion results for this study are published in Lambert (1990).

Objective.—
The objective of this study is to identify the temperature and salinity of formation of the calcite to identify the diagenetic environment in which it precipitated. The goal of presenting this case history is to illustrate a method for interpreting fluid inclusion data in which the diagenetic environment is at an elevated temperature rather than the low-temperature environments previously presented. This case history illustrates a method for dealing with fluid inclusion data that come from a broad, primary growth zone that represents a range of geologic events and produces data that are highly variable.

Petrography.—
Large vugs and primary and moldic pores are reduced by coarse calcite cement. A late stage of this cement contains a thick, cloudy growth zone that overgrows an earlier stage of clear calcite, and is overgrown by a later stage of clear calcite. No compositional subzonation is recognizable for this thick cloudy growth zone. The primary fluid inclusions that are limited to this growth zone consist of two-phase fluid inclusions that appear to have consistent ratios of liquid to vapor. All-liquid inclusions are absent.

Microthermometry.—
Unfortunately, the fluid inclusion assemblage consists of inclusions from a broad, cloudy zone that is the width of hundreds of fluid inclusions. The Th and Tm ice data gathered from this broad FIA is highly variable. This can be explained either as data that has reequilibrated thermally during progressive heating, or data that represents a wide range of conditions during precipitation of the calcite. Without finer petrographic differentiation, it would be impossible to distinguish between these two possibilities, and thus, impossible to interpret the fluid inclusions. However, some closely associated fluid inclusions can be measured in the same crystal in transects from earlier (inner portion of zone) to later (outer portion of zone) parts of the broad growth zone. When the inner inclusions are compared to the outer inclusions, a consistent increase in Th (Fig. 11.38) and a consistent decrease in Tm ice (Fig. 11.39) is detected outward in the crystal.

Interpretation.—
Normally, one would not be able to use such variable Th and Tm ice data from a single FIA, but because there is evidence for a consistent change in Tm ice and Th, it is apparent that at least some of an original signature is preserved. Thus, there is strong evidence that the calcite precipitated during conditions of increasing temperature and increasing salinity through time. The degree to which the early Th data preserve the original fluid inclusion signature is unknown, because some thermal reequilibration from stretching cannot be ruled out for these data. The salinity increased from about 9.6 wt.% to about 18.6 wt.% NaCl equivalent based on the trend in Tm ice data.

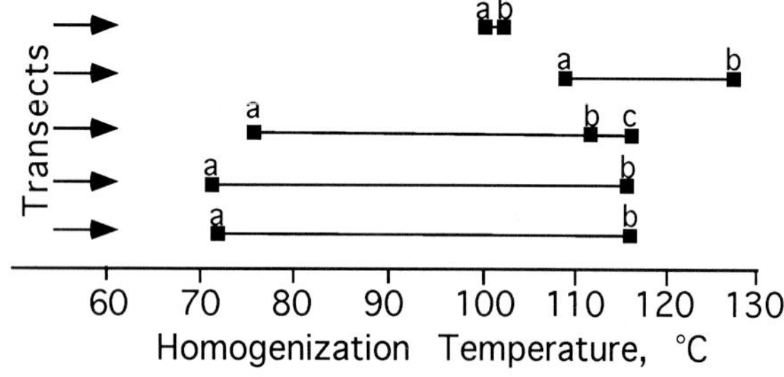

Fig. 11.38. *Th data from primary fluid inclusions in calcite, Permian Laborcita Formation. Inclusion data are linked with lines to represent transects across the crystal. From a to b represents earlier to later. Data indicate an increase in temperature during cementation. Modified from Lambert (1990).*

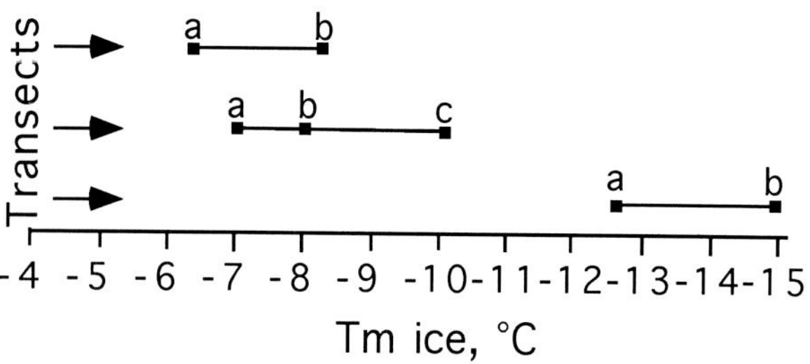

Fig. 11.39. *Tm ice data from primary fluid inclusions in calcite, Permian Laborcita Formation. Inclusion data are linked with lines to represent transects across the crystal. From a to b to c represents earlier to later. Data indicate an increase in salinity during cementation. Modified from Lambert (1990).*

Origin of Calcite and Thermal and Fluid History of Pennsylvanian Holder Formation, New Mexico

The Holder Formation is a Pennsylvanian unit (Virgilian) that crops out in the Sacramento Mountains of southern New Mexico and has petroleum reservoir analogs in the Permian basin. The limestones in this unit lack porosity because of extensive calcite cementation. It is important to understand the origin of this calcite to understand distribution of petroleum reservoirs in other areas. Furthermore, any evidence of later fluid migration will be helpful in determining the timing of charging of petroleum reservoirs relative to cementation. This study is summarized in more detail in Goldstein (1988).

Objective.—
The objective of this study is to determine the temperature of calcite precipitation, the salinity of the fluid from which it formed, and the later history of fluid composition and temperatures. The goal of presenting this study is to illustrate a method for determining conditions of cement precipitation for elevated-temperature calcite cementation by demonstrating the method of convergent Th-Tm ice trends in fluid inclusions from one fluid inclusion assemblage. This method will show how such data can be used to reconstruct the pre-thermal reequilibration fluid inclusion population and the later history of increase in temperature and salinity through time.

Petrography.—
The late-stage calcite cements post-date fractures, and post-date compaction features. By correlating the cathodoluminescent zonation of this cement, it appears that late calcite distribution is unrelated to early events of subaerial exposure in the section and that cement precipitation must post-date deposition of the Holder Formation. Primary fluid inclusions are confined to a cloudy growth zone that correlates well with the compositional growth zonation observable in cathodoluminescence (Fig. 11.40). The primary fluid inclusions in the late cement are all two-phase and appear to have relatively consistent ratios of liquid to vapor. There are also secondary aqueous fluid inclusions and secondary petroleum-filled fluid inclusions that fluoresce brightly with UV illumination and that cut across the late cement.

Microthermometry.—
The Th of primary aqueous fluid inclusions in the cloudy growth zone of the late cement show high variability from just over 70°C to just under 120°C (Fig. 11.41A). Having such variability in a single FIA is strong evidence for a significant range of geologic conditions, or thermal reequilibration of the original population. The Tm ice data also show significant variability, ranging from about -18°C to -5°C (Fig. 11.41B). This variability in a single fluid inclusion assemblage is also suggestive of a range of geologic conditions during cementation or thermal reequilibration of fluid inclusions. When a bivariate plot is constructed, two trends are defined and the trends converge on a Th between 70° and 80°C and a salinity of about -6°C (Fig. 11.42).

Interpretation.—
The petrography of the late cement suggests precipitation after significant burial. The fluid inclusion data from those late cements provide more specific evidence of this environment. The variability of the Th and Tm ice data from a single FIA suggests either a range of conditions during cement precipitation or thermal reequilibration of the original FIA. The hypothesis of a range of geologic conditions can be evaluated from the bivariate plot of fluid inclusion data (Fig. 11.42): in these data, significantly differing salinities would have to exist at the same temperature, or the temperature would had to have fluctuated wildly during a range of fluid compositions. This is a rather complex explanation that does not make much sense

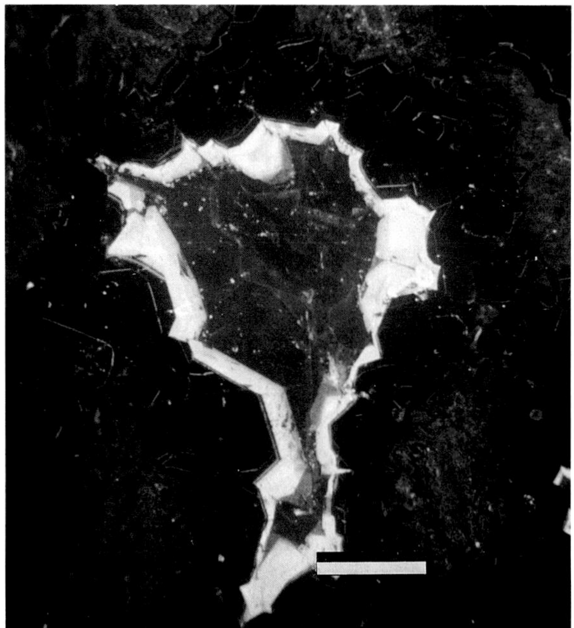

Fig. 11.40. *Cathodoluminescence photomicrograph of growth-zoned calcite from the Pennsylvanian Holder Formation, New Mexico. Primary fluid inclusions occur confined to some of the late compositional zones. Bar scale is 300 μm.*

geologically. The simplest hypothesis is that the range in data represent thermal reequilibration of an originally consistent FIA. Two trends are obvious in the data. One shows variable Th but no change in Tm ice, which could have been caused by stretching of the lower temperature fluid inclusions during thermal reequilibration. The other trend shows a decrease in Tm ice with increasing Th, could represent leakage and refilling of fluid inclusions as the rocks experienced pore fluids of higher temperature and higher salinity. The point of convergence at about 75°C Th and -6°C Tm ice indicates the position of the original FIA before reequilibration took place. Thus, the late calcite that contains fluid inclusions precipitated at a minimum temperature of 75°C, and from a fluid with salinity of about 9 wt.% NaCl equivalent. After the cement precipitated, the Holder Formation experienced conditions of increasing temperature and salinity approximated by the leakage and refilling trend (Fig. 11.42). Secondary fluid inclusions of oil indicate that some oil migrated through or into the system after the porosity had already been occluded by the late calcite cement.

Origin of Late Post-Compactional Calcite Cement, Pennsylvanian, Lansing-Kansas City Groups

The Lansing and Kansas City groups occur in the subsurface of NW Kansas and SW Nebraska. The limestones are cemented with early calcites that precipitated from brines at temperatures below about 50°C (see earlier case study on page 171 in this chapter). They are overlain by Upper Pennsylvanian and Lower Permian interbedded shales and carbonates (limestone and dolomite). The Lower Permian strata include anhydrite and halite. Much of the porosity is occluded by calcite cement that postdates the early calcite cement. This study deals with the origin of this late calcite cement. The data for this study and the interpretations herein are part of an M.S. thesis by Anderson (1989).

Objective.—
The objective of this study is to determine the origin of the late calcites (calcite #3) of the Lansing and Kansas City groups, and to determine the history of temperature and pore fluids experienced by the rocks after the calcite was precipitated. The goal of presenting this study is to illustrate an example of using convergent trends to determine the temperature and salinity of cement precipitation for a different setting than the one illustrated above, and to show that partially reequilibrated FIAs can be useful in determining the history of post cementation history of a unit.

Petrography.—
The calcite cements occur in moldic and primary pores. They postdate compaction features and overgrow the early cements (#1 and #2) discussed earlier in this chapter. Compositional growth zonation is apparent with cathodoluminescence illumination (Fig. 11.30). There is no relationship between the

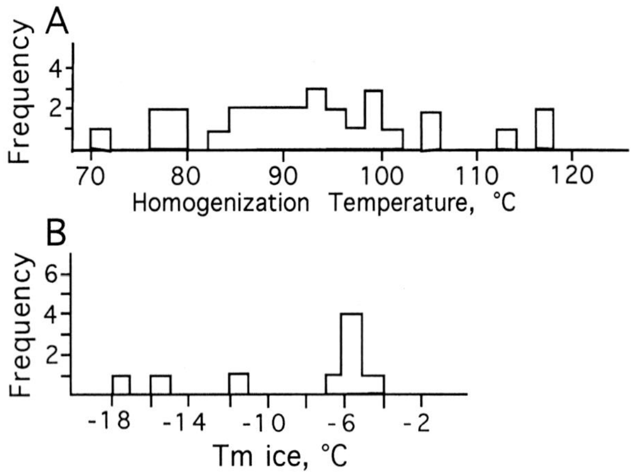

Fig. 11.41. *Histograms of data from primary fluid inclusions in a growth zone of calcite from the Pennsylvanian Holder Formation, New Mexico.* ***A)*** *Histograms showing Th data.* ***B)*** *Histograms showing Tm ice data. Variability of both Th and Tm ice data make it impossible to rule out thermal reequilibration of fluid inclusions. Modified from Goldstein (1988).*

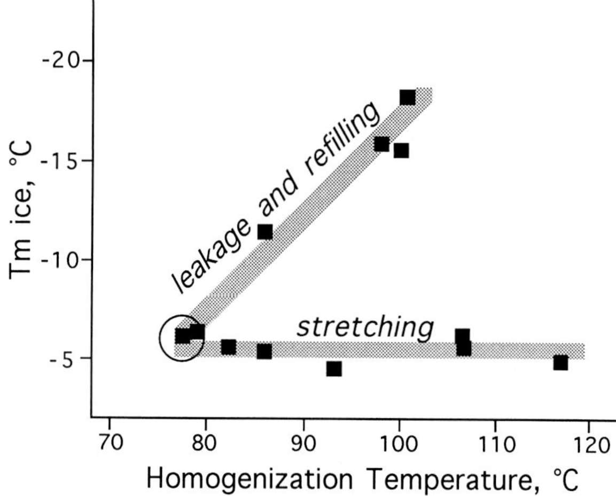

Fig. 11.42. *Bivariate plot of primary fluid inclusion data from a growth zone from the Pennsylvanian Holder Formation, New Mexico. Stretching trend and leakage and refilling trend are both caused by thermal reequilibration. They converge at the circle, which represents the original conditions of fluid inclusions entrapped before reequilibration. Modified from Goldstein (1988).*

cathodoluminescent zonation and the cyclothem boundaries. Fluid inclusions in the late cement are confined to cloudy cathodoluminescent growth bands and are classified as primary on that basis. The fluid inclusions are two-phase liquid-rich with apparently consistent ratios of liquid to vapor. Remember that earlier growth zones contain preserved all-liquid fluid inclusions.

Crushing.—
Samples containing two-phase primary fluid inclusions in the late cement can be crushed under kerosene to identify the presence and amount of organic gases. Upon crushing, all bubbles expanded as the inclusions were opened to one-atmosphere pressure. All dissolved into the kerosene soon after indicating the presence of organic gases. Calculations of internal pressure before crushing (from amount of expansion of the bubble) yield pressures ranging from 7 to 40 atmospheres with a mean value of 22 atmospheres.

Microthermometry.—
Each FIA yields Th (Fig. 11.43A) and Tm ice values (Fig. 11.43B) that are highly variable. No petrographic consistency or trends can be identified. Cross-plots of Th and Tm ice data, however, illustrate that data are constrained within two trends, a trend of variable Th and consistent Tm ice around -22.2°C, and a trend of increasing Th and increasing Tm ice (Fig. 11.44).

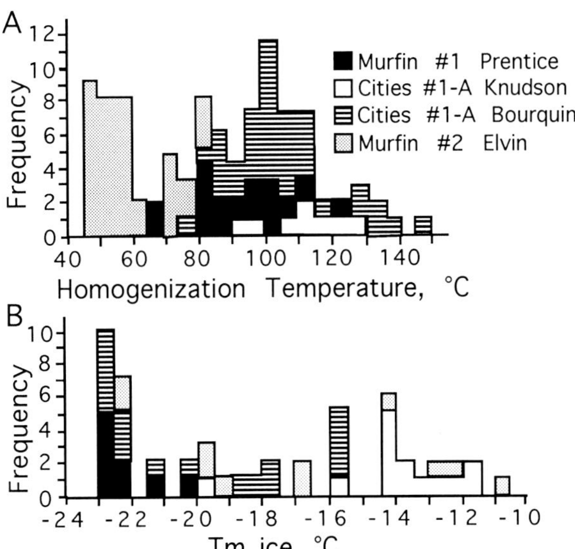

Fig. 11.43. *Histograms of data from primary fluid inclusions from calcite growth zone #3, Pennsylvanian Lansing-Kansas City group, Kansas.* ***A)*** *Histograms showing Th data.* ***B)*** *Histograms showing Tm ice data. Each symbol represents data from a different locality. The variability of the data makes it difficult to rule out thermal reequilibration. Modified from Anderson (1989).*

Eutectic melting temperatures around -52°C suggest that the brines are in the H_2O-NaCl-$CaCl_2$ model system.

Interpretation.—
The presence of only two-phase inclusions in the late calcite of this study, compared to the all-liquid fluid inclusions in the earlier growth zones, indicate that the lowest temperature part of the FIA in the late growth zone must have been preserved and could not represent inclusions that have been altered by thermal reequilibration. However, the variability of the Tm ice and Th data in individual FIAs indicate that at least partial thermal reequilibration is likely for some data from the late calcite. The bivariate plot illustrates a stretching trend and a leakage and refilling trend that converge at the lowest temperature, lowest Tm ice value. This point of convergence is indicative of the condition of original entrapment of the fluid inclusions. Thus, the late calcite cement precipitated at a minimum temperature of about 50°C and from a highly concentrated brine. The brine was a Ca-rich brine based on the Te values. From the burial history, the most reasonable explanation for the origin of the calcite cement (cement #3) is that it precipitated after burial and

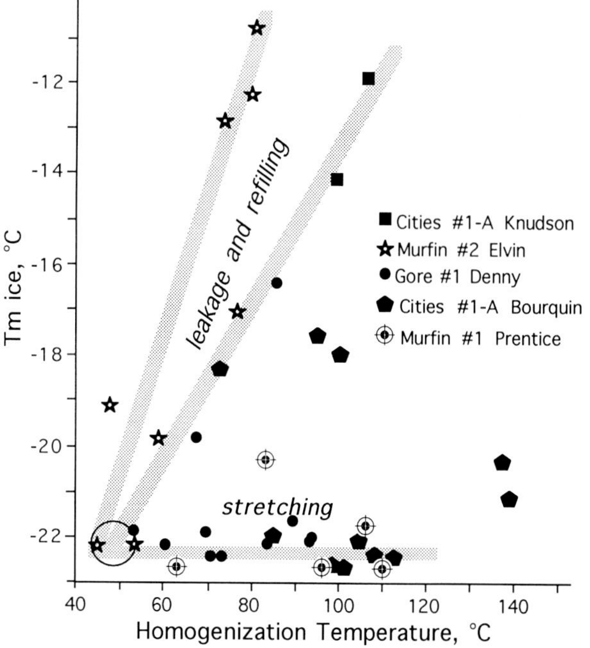

Fig. 11.44. *Bivariate plot of data from primary fluid inclusions in calcite cement zone #3, Pennsylvanian Lansing-Kansas City group, Kansas. Data define a stretching and two similar leakage and refilling trends which converge at the circle. The point of convergence represents the conditions at which the original inclusions were entrapped, before thermal reequilibration. Modified from Anderson (1989).*

during deposition of the overlying Permian evaporites. As the brines were generated at the surface in the Permian, they refluxed downward through the underlying strata, evolved in composition and were heated to at least 50°C (Fig. 11.33). Burial of only 500 m was required for this amount of heating to occur. This burial depth is consistent with the thickness of strata between the Lansing and Kansas City groups and the overlying Permian evaporites. The leakage and refilling trend shows that after precipitation of the calcite cement, the unit experienced a history of increasing temperature and decreasing salinity. Minimum amounts of total burial can be estimated from the fluid inclusion internal pressures in cement #3. The highest room temperature internal pressure is 40 bars which would correspond to a minimum burial depth of about 400 m.

Origin of Authigenic Quartz Cement, Upper Jurassic, North Sea

Quartz overgrowths and fracture-filling quartz cement are common components of North Sea Upper Jurassic sandstones. It is important to understand the timing of formation of these cements relative to thermal and fluid migration events to place diagenesis in the context of petroleum migration and generation. More complete documentation of this case history is available in Burley and others (1989) and Guscott and Burley (1993).

Objective.—
The objective of this study is to determine the temperature and diagenetic environment of quartz cementation. The goal of presenting this study is to illustrate that given consistent fluid inclusion data from individual FIAs, interpretation of fluid inclusion data from elevated-temperature diagenetic systems is extremely straightforward.

Petrography.—
Cathodoluminescence petrography of quartz cements illustrates that overgrowths are zoned and may be cross-cut by several generations of fracture-filling cements. Fluid inclusions are trapped along the detrital grain-overgrowth boundary and are interpreted to be the earliest assemblage of primary fluid inclusions entrapped. Two later fracture generations can be defined that are filled by quartz cement. These fractures contain primary fluid inclusions as well. Fluid inclusions are two-phase, of various sizes, and contain consistent ratios of liquid to vapor.

Microthermometry.—
Fluid inclusion data were gathered from three FIAs that can be put in sequence of entrapment using the petrographic observations. The first FIA entrapped consists of those inclusions trapped at the detrital grain-overgrowth boundary. The Th of these inclusions range from 99.2-101.5°C (Fig. 11.45). The Th of inclusions from the next fluid inclusion assemblage, fracture fill 1, range from 98.8-106.6°C (Fig. 11.45) and the final fracture fill ranges from 121.5-124.0°C (Fig. 11.45). Notice the narrow range of Th for each FIA, which provides evidence that fluid inclusions have not undergone thermal reequilibration and that they are faithful records of minimum entrapment temperature. The Tm ice from inclusions in fracture fill 1 range from -6.9 to -7.4°C and the Tm ice from fracture fill 2 range from -8.9 to -9.3°C.

Interpretation.—
The Th data are consistent within individual FIAs, so they can be interpreted without cause to suspect thermal reequilibration: all inclusions are representative of original conditions and none have reequilibrated. These data indicate that quartz was precipitating during increasing temperature from a minimum temperature of 100°C to a minimum temperature of about 125°C. There also was a record of increasing salinity through time from about 10.5 to about 13 wt.% NaCl equivalent.

Geothermometry and Geobarometry Using Fluid Inclusions in Quartz from the Central Alps

Fluid inclusions in quartz that formed in association with Alpine metamorphism can be used to determine the pressure and temperature of metamorphic events. Gas chromatography and mass spectrometry of fluid inclusions indicate many are dominated by methane. We have refrained from discussing this study in great detail but useful details are discussed in Mullis (1979).

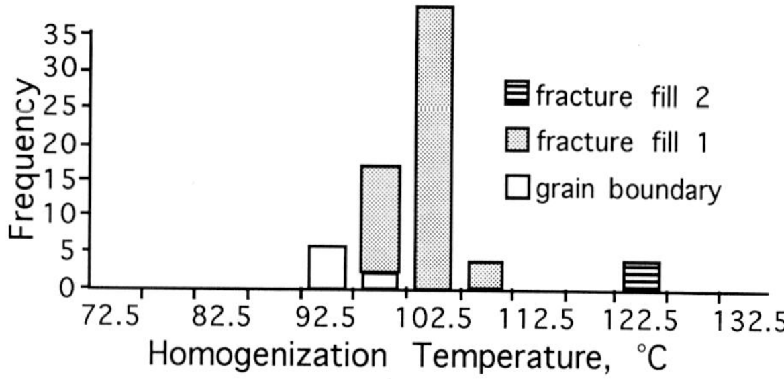

Fig. 11.45. *Fluid inclusion Th data from quartz overgrowth and two later fracture-filling quartz cements, Jurassic sandstone, North Sea. Notice that the Th data from each fluid inclusion assemblage is very consistent, suggesting it has not been subject to thermal reequilibration. Modified from Guscott and Burley (1993).*

Objective.—

The objective of this study is to use fluid inclusions to determine the pressure and temperature of metamorphism during formation of quartz. The goal of presenting this study is to illustrate briefly the importance of entrapment of immiscible fluids for yielding entrapment temperature and entrapment pressure of fluid inclusions.

Petrography.—

The quartz was collected from alpine mineral clefts. An early phase of quartz cement is recognized that may contain a "white stripe" corresponding to syntaxial or stretched crystal fibers, and this is followed by later quartz that may be prismatic and skeletal. Many fluid inclusions are trapped along growth zones that mark successive generations of quartz cement. Because their distribution parallels and defines the growth zonation, they are interpreted to be primary. One growth zone may contain water-rich fluid inclusions with consistent ratios of liquid to vapor whereas the next growth zone may contain all methane-dominated fluid inclusions, lacking an obvious liquid water phase. This close association of the two types of inclusions suggest that quartz was precipitating at or close to conditions of water-methane immiscibility.

Microthermometry.—

The fluid inclusion assemblages containing water dominated fluid inclusions yield Th data typically between 200 and 260°C. The inclusions that are dominated by methane yield Th below -82.5°C and down to about -108°C.

Interpretation.—

The homogenization temperature of the methane dominated inclusions can be used to define the isochore on which the separate methane phase was entrapped. The Th of the water-rich fluid inclusions would be approximately equal to true entrapment temperature assuming that they are the water-rich end member of an immiscible system. The pressure at which the methane isochore, determined from the methane Th, intersects the entrapment temperature yields the pressure of fluid inclusion entrapment. Thus the quartz crystals precipitated from metamorphic temperatures ranging from about 200 to 260°C and pressures ranging from about 1200 to about 3100 bars. If the water-rich inclusions were not actually trapped as part of an immiscible system, then the temperatures and pressures determined represent minimum pressures and temperatures of entrapment.

SUMMARY

The case histories presented in this chapter were chosen specifically to illustrate the principles and procedures for conducting fluid inclusion studies and for interpreting fluid inclusion data. In each of the cases, something was learned about some thermal and compositional aspect of the fluid history experienced in the diagenetic realm. Numerous other studies that combine the fluid inclusion technique with other analytical techniques or with other geological information that require significant interpretation (e.g., burial history reconstructions), can be found in the literature, but these examples were intentionally avoided so as to highlight the power of information gained from inclusions alone.

The authors predict that many readers will become very enthusiastic and optimistic about the potential use of fluid inclusions in helping to decipher diagenetic processes after studying the case histories in this chapter. Such a response will be appreciated, but readers should know that there have been numerous other instances in our careers where objectives could not be fulfilled because nature did not store the necessary inclusions in the selected samples. In your endeavors with fluid inclusions, patience and perseverance most probably will pay off.

Happy hunting!

Chapter 12

OTHER ANALYTICAL APPROACHES

INTRODUCTION

The previous chapters have dealt with the theory and practice of conducting fluid inclusion studies in the diagenetic realm by concentrating on petrographically based approaches that should be readily accessible to all researchers. This chapter concentrates on other analytical approaches that may be applied in a fluid inclusion study to constrain further the composition of fluid inclusions. A few laboratories conduct these types of analyses as the initial part of a fluid inclusion study, and such an approach may prove to be worthwhile for oil and gas exploration. However, in a fluid inclusion study that places fluid inclusion origins into a paragenetic and geologic context, and that involves microthermometric analysis, these analytical techniques will be applied only for specific purposes that arise during the course of a petrographically based study. As the application of these additional analyses requires equipment that will probably not be available in every laboratory, each technique will not be treated in detail: the utility of a range of analytical techniques is only summarized. Most of these techniques are still in developmental stages, so successful utilization will probably require commitment of time and sufficient funding for continued development.

Of the analytical approaches available, no single technique can provide a complete analysis of all of the components in a fluid inclusion. Each technique has its strengths and weaknesses in terms of components that may be analyzed and the concentrations and amounts required. A single fluid inclusion may need to be studied by a variety of analytical techniques to answer a compositional question posed. Thus, it is important to know something of the capabilities of the entire arsenal of analytical tools and approaches that can be used.

The analytical approaches presented below can be divided into those in which the analysis takes place without opening the fluid inclusion vacuole (nondestructive) and those in which the fluid inclusion is opened for the analysis (destructive). Also, some of the techniques can be applied to a single fluid inclusion, and others require analysis of a bulk sample, utilizing large numbers of fluid inclusions for an analysis. The most complete summaries of analytical approaches have been compiled by Roedder (1984, 1990). This chapter summarizes our view of the most important approaches, their application, strengths, and limitations, as well as references that will enable a reader to learn about the fundamental and procedural aspects of any technique from other sources.

NONDESTRUCTIVE TECHNIQUES

Nondestructive techniques typically involve spectroscopic analyses that require illumination of a fluid inclusion with simultaneous measurement of the spectrum of absorption or emission from the fluid inclusion.

Ultraviolet Fluorescence Emission Spectroscopy

Ultraviolet fluorescence emission spectroscopy is a relatively simple technique that utilizes an ultraviolet source for illumination of fluid inclusions through a microscope. Oil-filled fluid inclusions commonly fluoresce under such UV illumination (typically 365 nm) because of the presence of aromatic hydrocarbons and nitrogen-, sulfur-, and oxygen-bearing compounds (Bertrand and others, 1985; Hagemann and Hollerbach, 1985). The spectrum of emitted light can be quantitatively documented with a spectrophotometer, or qualitatively determined by visual observation of the color of the fluorescing light, and reflects compositional differences of petroleum inclusions.

The measured spectra are then employed to distinguish between oil-filled inclusions of significantly different compositions, and roughly to assess their relative differences in maturity. The timing and origin of various generations of petroleum fluid inclusions have been successfully determined using UV spectroscopy (Burruss and others, 1980; Burruss and Goldstein, 1980; Burruss, 1981; Burruss and others, 1985; McLimans, 1987; Guilhaumou and others, 1989, 1990; Bodnar, 1990; Kihle, 1993). Comparisons between fluorescence emission spectra of oil-filled inclusions with API gravity and maturity of *related* oils show that there is a tendency for petroleum fluorescence to change from yellow to blue colors with increasing maturity (McLimans, 1987). Such a correlation is best when comparing oil-filled fluid inclusions from within one genetically related suite of oils, and will lose some applicability when comparing oils from different basins. Similar comparisons have been reported for fluorescence emission spectra and API gravity (Tsui, 1990). McLimans (1987) also reported that the emission spectra measured upon excitation with a wide range of wavelengths from 270-366 nm, produced with a pulsed laser as a source, contain more compositional information than the spectra obtained with excitation at only one wavelength, and that fluorescence emission lifetimes determined using the pulsed laser are highly correlated to API gravity.

The main limitations for application of this technique are linked to the ability to accumulate a signal from the sample. The best spectral signals will result from the most intensely emitted light, but some oil-filled inclusions lack bright fluorescence either due to the composition of the oil or the small size of the inclusion measured. Also, scattered background fluorescence from the enclosing mineral or from the mounting medium can cause difficulties. Even though the technique of illuminating an oil-filled inclusion with intense UV light and measuring its fluorescence emission spectrum has proven to be useful, it remains a crude method for gaining information on compositions of oils in fluid inclusions.

Infrared Absorption Microspectroscopy

Infrared absorption microspectroscopy, also known as micro-FTIR (micro-fourier transform infra-red spectrometry), is a technique that uses an optical microscope to sample spectroscopic information from a single fluid inclusion (Wopenka and others, 1990). The instrument is equipped with a source for polychromatic IR-radiation (1.1-200 μm) and a detector which measures the resultant spectrum. Micro-FTIR has the capability to yield more detailed spectra than UV spectroscopy and such may be useful in future applications. Currently, however, a severe limitation is that inclusions must be very large (200 μm^2 in cross-sectional area) for usable spectra to be obtained. Other problems involve interference of atmospheric H_2O and CO_2, and strong absorption of infrared light by the enclosing mineral.

In the past micro-FTIR has been employed to determine CO_2/H_2O ratios (Vry and others, 1987; A. T. Anderson and others, 1989), and to identify and quantify organic species in fluid inclusions such as methane and higher hydrocarbons (Barres and others, 1987; Guilhaumou and others, 1989, 1990; O'Grady and others, 1989; Pironon and Barres, 1990, 1992). It may be possible for internal pressures within inclusions to be determined from shifts in shape and position of IR absorption bands from gaseous components such as CO_2 and CH_4 (Dahan and Couty, 1987). At present, most use of the micro-FTIR technique is limited to identification of water and some gaseous and organic components in relatively large fluid inclusions.

Synchrotron X-Ray Fluorescence Microprobe

The synchrotron X-ray fluorescence microprobe employs a high energy (25 KeV) beam of X-rays of about 10-15 μm diameter to irradiate a single or group of fluid inclusions located with an optical microscope. The excited mineral and fluid give off a spectrum of characteristic X-rays that is detected by standard EDS and WDS techniques. The apparatus is in air so that some of the emitted X-ray spectrum is absorbed by the air. Also, X-rays emitted from the inclusion are absorbed by the enclosing mineral, and the enclosing mineral's spectrum may interfere with that of the inclusion, so the capabilities are dependent on the host mineral.

Vanko and others (1992, 1993), and Cline and others (1993) have reported analyses of individual aqueous fluid inclusions in quartz using the synchrotron. They have found that brine inclusions near the surface of the host quartz routinely yield X-ray spectra for Cl, K, and Ca. Elements such as Rb, Sr, Br, and other heavy elements also may be detected easily. Typically, light elements such as Cl, K, S, and Ca are difficult, but possible to detect. The spectra can be quantified by determining ratios internal to the fluid inclusions, such as K/Cl ratio, or by measuring the inclusion path length and the host mineral path length through which the beam traveled to develop a correction factor for the X-ray peak intensity. So far, minimum detection limits for aqueous fluid inclusions in quartz are about 1000 ppm for Cl, 700 ppm for K, 700 ppm for Ca, 200 ppm for Mn, 250 ppm for Fe, 100 ppm for Cu, 150 ppm for Zn, 150 ppm for Pb, 50 ppm for Br, and 500-1000 ppm for Mo.

The synchrotron is applied most easily to relatively large fluid inclusions of uniform shape, so that the beam path length can be determined in order to quantify concentrations. The sample must either be mounted to a silica glass base or not mounted at all to yield valid analyses, because thin section glass and epoxies typically contain contaminants that interfere with the fluid inclusion analyses.

Laser Raman Microprobe

The laser Raman microprobe is an instrument that focuses a laser through an optical microscope into a single fluid inclusion. The laser beam is highly focused so that spatial resolution is probably about 3 μm, which means that one may probe individual phases within fluid inclusions as well as the entire inclusion. Scattered Raman radiation is detected with a spectrometer and characteristic spectra are attained from many polyatomic, solid, fluid, or dissolved species (Wopenka and others, 1990). Solid daughter minerals in fluid inclusions have been identified by laser Raman (Rosasco and others, 1975; McLimans, 1987; Pasteris and Wanamaker, 1988). When fluid inclusions have been frozen for the analysis, salt-hydrate species and gas-hydrate species (clathrates) can also be identified (Dubessey and others, 1982, 1992; Seitz and others, 1987; Schiffries, 1990). The most common application has been the identification and quantification of polyatomic gases such as CH_4, CO_2, H_2S, N_2, H_2, O_2 and others. By shifts in the Raman peak position, partial pressures of these various gases can be determined (Fabre and Couty, 1986; Wopenka and Pasteris, 1987; Pasteris and others, 1988; Dubessy and others, 1988; Chou and others, 1990; Wilkins and Jenatton, 1991). Recent work has illustrated several complications with this approach for mixed gas systems (Seitz and others, 1993a, b). Polynuclear species such as $SO_4^=$, NO_3^-, $CO_3^=$, PO_4^{-3} dissolved in the aqueous phase may also be detectable. Among these, $SO_4^=$ can be commonly detected and compared to the water peaks to quantify its concentration in the aqueous phase (Cunningham and others, 1977; Rosasco and Roedder, 1979; Dubessey and others, 1982; 1983; 1992; Higgins and Stein, 1986). In addition, there is a possibility that the O-H stretching-peak of water can be used to determine salinity at room temperature (Mernagh and Wilde, 1989).

The Raman microprobe is basically an optical technique, which means that its performance and detection limits are very much controlled by a combination of the optics of the fluid inclusion sample being studied, its mineral host, and the efficiency of Raman scattering of the sample. Pasteris and others (1988) evaluated many of the practical controls on use of the technique. The Raman signal scattered from a particular polynuclear species in a fluid inclusion is a function of its Raman-scattering efficiency, the degree of polarization of the spectrum, and the amount of the species encountered by the laser. The amount of Raman radiation making it back through the spectrometer and to a detector is an optical problem. Clarity of the mineral is important for escape of the Raman radiation. Some minerals and oil inclusions will absorb much of the laser light and essentially "cook" the sample and the inclusion. Background fluorescence of the enclosing mineral, from hydrocarbons in the inclusions, or from organics in the mineral, as well as fluorescence of the sample mounting medium may obscure the Raman signal. The optical anisotropy of the mineral medium also affects the scattering of the Raman radiation and the polarization direction. One of the most important factors affecting the use of this technique is the morphological character of the fluid inclusions themselves. Poorly-shaped inclusions typically have a focusing or defocusing effect which prevents efficient gathering of the Raman signal or even efficient illumination with the laser. Typically, the largest fluid inclusions with the highest concentrations of the species analyzed yield the best signal. Thus, detection limits are actually sample specific. Even with these ambiguities, the laser Raman technique is a highly valuable one with great potential for analyzing polynuclear species in fluid inclusions nondestructively.

Proton Microprobe

The proton microprobe bombards a mineral sample with a focused beam of protons. If the protons penetrate the mineral host, they may interact with the species in inclusion fluids and cause emission of X-rays (PIXE or proton-induced X-ray emission) or of gamma rays (PIGE or proton induced gamma ray emission) that yield characteristic spectra for various elements from a single fluid inclusion. The PIXE technique is best used for analysis of elements with atomic number greater than about 30 because of problems with host mineral absorption of lower energy X-rays. It may prove sensitive to elements in ppm concentrations (Horn and Traxel, 1987). Lighter elements can be analyzed by the PIGE technique (Anderson and others, 1987; 1989). Both techniques have potential for producing quantitative and qualitative analyses of ions dissolved in inclusion fluids without destruction of the vacuole.

Nuclear Magnetic Resonance

Nuclear magnetic resonance is a bulk, but nondestructive technique for identifying water in fluid inclusions (Poty and others, 1987). Pironon and others (1992) used the technique to show differences between aqueous and synthetic hydrocarbon fluid inclusions. They also inferred a possible spatial resolution as fine as 20 μm. Other applications have included detection of ^{23}Na and ^{35}Cl in fluid inclusions in minerals (Kohn and others, 1988).

DESTRUCTIVE TECHNIQUES

Daughter Minerals

In the sedimentary realm, daughter minerals are most likely to be encountered in evaporite minerals. Fortunately, inclusions in evaporite phases commonly can be very large (>50 μm), so that it is feasible to extract a daughter mineral from a single fluid inclusion for identification. Daughter minerals also simply may be exposed for analysis by opening the fluid inclusion. An evaporite mineral containing fluid inclusions with daughter crystals may be crushed, commonly under an immersion medium to free daughter crystals from fluid inclusions. Also, a needle probe or drill may be used to open a fluid inclusion so that the mineral may be extracted with a microsyringe or a fine needle. The extracted mineral can be analyzed using EDS (energy dispersive system on an SEM or electron microprobe), electron diffraction, or can be analyzed using single crystal X-ray diffraction (Zolensky and Bodnar, 1982; Blasch and Coveney, 1988). If a daughter mineral has not been extracted, but has simply been exposed through removal of the mineral roof of a fluid inclusion, it can be identified using EDS in an SEM or electron microprobe (Le Bel, 1976; Metzger and others, 1977; Anthony and others, 1984). Daughter mineral identification is used to constrain fluid inclusion composition.

EDS of Salts From Fluids in Inclusions

The salts that have been extracted from a single fluid inclusion can be analyzed qualitatively and semiquantitatively using EDS. The extraction method involves heating fluid inclusions in a polished plate to the point of thermal decrepitation. Then EDS analysis of salts remaining on the surface of the plate after evaporation of escaping inclusion liquids is performed (Kozlowski, 1978; Haynes and Kesler, 1987; Haynes and others, 1987, 1988). This technique is a useful one for determining cation ratios of the major salts in fluid inclusions and may yield data on Cl, S, Na, Ca, and K, but is limited by inherent heterogeneities in the salts precipitated, loss of some constituents during decrepitation, and interferences from host minerals containing species similar to those that could be present in the fluid inclusions. A variant on this scheme is the analysis of frozen fluid inclusions using the EDS (Kelly and Burgio, 1983).

Cation and Anion Analyses of Extracted Bulk Samples or of Individual Inclusions

Release of aqueous fluids from inclusions can be followed by cation and anion analyses by various techniques to produce quantitative and semiquantitative analyses of inclusion solutes. However, the release procedures are not trivial. Probably the most effective extraction procedure is the direct removal of aqueous fluid from a single fluid inclusion. In this technique, the host mineral is drilled and fluids are removed from huge fluid inclusions using a microsyringe (Lazar and Holland, 1988; Stein and Krumhansl, 1988). There are great advantages of this technique in that it removes fluid from a single fluid inclusion of known origin, avoids contact with mineral surfaces that could cause preferential adsorption of some ions or contamination, and can be used to obtain a record of salinity by measuring the volume of the fluid extracted into the microsyringe. The great disadvantage of this extraction technique is that it can be used only with very large fluid inclusions.

The other method for release and extraction of aqueous ions from fluid inclusions involves bulk crushing or decrepitation and leaching of large amounts of mineral sample containing thousands of fluid inclusions. The crushing and decrepitation opens the fluid inclusions, and leaching in various solvents extracts the cations and anions. The greatest weakness of this technique is that it samples thousands of fluid inclusions at the same time. For most samples then, it may be impossible to know the various contributions of fluid inclusions of various origins and hence compositions. A few methods have been applied to eliminate problems with preferential adsorption of certain ions, to improve the release of the inclusion salts, and to reduce contamination from the solid phase. For most crush and leach procedures, the sample is precleaned and is then crushed in a clean agate mortar and pestle or within a stainless steel tube. Crushing can be performed in the presence of a solvent such as water or alcohol to facilitate separation of the ions in the inclusion fluids from the host mineral. Sometimes the solvent is introduced after the crush and must redissolve the minerals precipitated. The first leachate removed contains ions from the inclusions plus contamination from the mineral phase. The sample is then flushed again with a solvent to remove any remaining ions from the inclusions, and then the sample is flushed again with solvent to correct for contamination from dissolution of the host mincral phase. Various components have been added to the leaching fluid to facilitate effective leaching of inclusion contents and to prevent adsorption. Details of many crushing and leaching procedures can be found in Norman (1987), Bottrell and Yardley (1987, 1988), Bottrell and others (1988), Changkakoti and others (1988), Aulstead and others (1988), Rankin and others (1992), Bennet and Barker, (1992), Banks and Yardley (1992), and Channer and Spooner (1992). The leachate contains cations and anions from the inclusions, may contain some contamination from the solid phase, and may have been altered somewhat by differential adsorption of various ions. Although crush and leach is a powerful tool for extracting components from fluid inclusions, it is not without its uncertainties.

Analyses of fluid inclusion extracts can be accomplished by various techniques depending on the ions of interest that have been extracted. Common methods for analysis of cations have been atomic absorption spectrophotometry (Aulstead and others, 1988), DCP (Stein and Krumhansl, 1988), ICP-AES (Stein and Krumhansl, 1988; Banks and Yardley, 1992; Rankin and others, 1992), ICP-MS, neutron activation analysis (Norman and others, 1987), and ion chromatography (Lazar and Holland, 1988). Anions in leachates have been analyzed by neutron activation (Sabouraud, 1974), ion chromatography (Lazar and Holland, 1988; Channer and Spooner, 1992).

The analysis of fluid inclusion extracts is an extremely active field; the procedures can be used for analyzing the cations and anions in aqueous fluid inclusions quantitatively.

Analysis of Solutes Released Directly Into Instrumentation

There has been significant recent work on analyzing fluid inclusion solutes by directly releasing the fluid into an instrument using intense laser light or some other method. For many of these procedures, fractionation that possibly could occur upon fluid release is unknown, so the analyses may not be representative of the inclusion composition. Also, contamination from volatilization of the host mineral and loss of some solutes during volatilization are plausible sources of error for the higher energy extractive techniques that involve a blast with a laser or sputtering away of the mineral with an ion beam. The degree to which these methods can provide quantitative information will not be known until calibration procedures and instrument limitations are established using synthetic inclusions of known compositions.

Simple thermal decrepitation into a carrier gas feeding into an ICP or ICP-MS has been used (Thompson and others, 1980; Alderton and others, 1982; Viets and others, 1985), but most techniques involve focusing a laser or ion beam on a single fluid inclusion to extract its contents. Laser ablation directly into an ICP or ICP-MS has been tried (Chenery and Rankin, 1989; Horn and Tye, 1989; Canals and others, 1992; Rankin and others, 1992; Shepherd and Chenery, 1993). Current work suggests that sufficient material is generated by laser ablation of single 10-15 μm fluid inclusions into an ICP-MS to analyze major and minor elements. A similar approach using laser produced plasma emission spectroscopy (LPES) has produced Mg and Ca analyses of fluid inclusions (Boiron and others, 1992). Laser probe mass spectrometry (LPMS) is a technique that uses a laser for inclusion release and partial ionization, and then accelerates the ions through a mass spectrometer for analysis (Deloule and Eloy, 1982). Secondary Ion Mass Spectrometry (SIMS) uses an ion beam to sputter through the host mineral and release the fluid inclusion contents into a mass spectrometer (Diamond and others, 1990) and has great potential for the analysis of fluid inclusion solutes.

Finally, another technique that has great potential is the laser microprobe noble gas mass spectrometric (LMNGMS) technique. This technique uses a laser to excavate the contents of a fluid inclusion and to analyze its contents using a noble gas mass spectrometer. Abundance and isotopic ratios of noble gases Ar, Kr, and Xe can be analyzed and Cl, Br, I and K can be analyzed from their conversion to noble gases after neutron irradiation (before sample release), yielding accuracy and precision for elemental abundance and isotope ratios of 5-10%. In principle, Ca, Ba, U, Se and Te also are detectable. The approach has had some success in analyzing single fluid inclusions with diameters as small as 20-50 μm (Böhlke and others, 1989; Böhlke and Irwin, 1992).

Isotopic Composition of Inclusion Fluids

There is a significant body of literature dealing with the extraction of fluid and the quantitative H, O, C, and S stable isotopic analysis of inclusion fluids. Most of these focus on bulk extraction techniques and suffer from the possibility that more than one population of fluid inclusions is being sampled. Some of these studies have yet to evaluate the effect of hydrogen diffusion or the effects of isotopic exchange between the mineral host and fluid. However, these studies still have provided some extremely useful data on the stable isotopic composition of ancient fluids (Knauth and Beeunas, 1986).

One of the most exciting applications has been the use of radiometric dating of inclusion fluids. Most of these approaches use bulk sampling techniques followed by mass spectrometric analysis. However, some isotopic systems have been analyzed in single fluid inclusions using a laser probe (Böhlke and Irwin, 1992). The Pb/Pb system has been applied (Changkakoti and others, 1988). Bannon and others (1987) reported K/Ar dates for fluid inclusions and McLimans and others (1992) and Brannon and others (1992) recently reported success in the Rb/Sr system.

The determination of the isotopic composition of solutes and fluids in inclusions presents some tremendously exciting prospects and undoubtedly advances will be made in the future.

Analyses of Gas Composition

The analysis of the gases present in fluid inclusions has received significant attention over the years. It was even the subject of an entire issue of the Journal of Geochemical Exploration in 1991 (Volume 42, no. 1). Methods of extraction as well as methods of analysis are numerous. For most approaches, the sampling strategy involves bulk analyses, but some procedures involve attempts to acquire data from individual but unselected fluid inclusions or from individual petrographically selected fluid inclusions. Analyses are by capacitance manometer, gas chromatography (GC), mass spectrometry (MS), or GC-MS.

Many methods of extraction have been tried and are currently in use. A bulk sample may be crushed to release the gas contents of fluid inclusions. This may be done by simply ball milling the sample (Roedder, 1972), by crushing the sample in an evacuated stainless steel tube (Kreulen and Schuiling, 1982), or by bulk crushing the sample in any number of cleverly designed crushing devices. Some bulk extraction systems use thermal decrepitation of the sample to release gaseous components (Kesler and others, 1986). Other techniques use a more selective approach in which attempts are made to analyze gases from individual fluid inclusions or from selected fields of view. Burruss (1987b) described a crushing cell that fits directly into the injection port of a gas chromatograph and provided data from gases and from oils from 1 milligram samples of fluid inclusion-bearing minerals. Some researchers have devised stages that breach a single, petrographically selected fluid inclusion with a needle probe or with a diamond to release the gas content of a single fluid inclusion. Others have used lasers to blast a single fluid inclusion or group of fluid inclusions into submission. Also, a process of decrepitation has been employed, whereby a bulk sample is inserted into a heating cell and as the temperature is raised, and as individual inclusions decrepitate, the gas from each single burst is analyzed using a quadrupole mass spectrometer (Barker and Smith, 1986). Many workers have concentrated on designing systems that have the correct composition of materials and are kept at the correct temperatures, or flooded with the proper carrier gas, to prevent preferential adsorption of certain components before the gases are analyzed.

Once the compounds have been extracted, they may be analyzed by one of many techniques semiquantitatively and perhaps quantitatively. Compounds that can be detected include N_2, Ar, CO, CH_4, CO_2, C_2H_4, C_2H_2, COS, C_3H_6 $C_3H_8C_3H_4$, H_2O, SO_2, O_2, H_2S, SO_2, NH_3 HCl, HCN, H_2 and many higher hydrocarbons. Gas chromatography has proven a useful technique that provides detailed data on gases and fingerprints of organic compounds in oils (Burruss, 1987b). GC-MS analyses of oils extracted from fluid inclusions provide a very important tool in petroleum exploration. They may be used for biomarker analysis, evaluated for thermal maturity, and used to develop fingerprints for oil-source rock correlation. Bray and others (1991) have offered a useful summary of procedures and techniques for analyzing a wide range of gaseous compounds using GC. Quadrupole mass spectrometry and GC-MS work have proven to be equally popular analytical tools for analyzing the gaseous components extracted from fluid inclusions (Barker and Smith, 1986; Landis and others, 1987; Graney and others, 1991). Although these techniques have been actively used for quite some time, calibration of these instruments and determining just how representative analyses are of the fluid inclusion contents are still active areas of research.

REFERENCES CITED

ABEGG, F.E., 1990, Fluid-inclusion analysis of dolomite in lithoclasts from the Morgan Creek Limestone (Upper Cambrian) of central Texas: The Compass, v. 67, p. 135-146.

ADAMS, L.H., AND GIBSON, R.E., 1930, The melting curve of sodium chloride dihydrate. An experimental study of an incongruent melting at pressures up to twelve thousand atmospheres: Journal of the American Chemical Society, v. 52, p. 4252-4264.

ALDERTON, D.H.M., THOMPSON, M., RANKIN, A.H., AND CHRYSSOULIS, S.L., 1982, Developments of the ICP-linked decrepitation technique for the analysis of fluid inclusions in quartz: Chemical Geology, v. 37, p. 203-213.

ANDERSON, A.J., CLARK, A.H., MAX, P., PALMER, G.R., MACARTHUR, J.D., BODNAR, R.J., AND ROEDDER, E., 1987, *In situ* elemental analysis of fluid inclusions using PIXE and PIGME (Abstract): GAC/MAC Program and Abstracts, v. 12, p. 21.

ANDERSON, A.J., CLARK, A.H., MAX, P., PALMER, G.R., MACARTHUR, J.D., AND ROEDDER, E., 1989, Proton-induced X-ray and gamma-ray emission analysis of unopened fluid inclusions: Economic Geology, v. 84, p. 924-939.

ANDERSON, A.J., SMITH, D.M. AND BODNAR, R.J., 1992, An adaptation of the spindle stage for geometric analysis of fluid inclusions with applications (Abstract): PACROFI IV, Fourth Biennial Pan-American Conference on Research on Fluid Inclusions, Program and Abstracts, Lake Arrowhead, CA, v. 14, p. 10

ANDERSON, A.T., JR., NEWMAN, S., WILLIAMS, S.N., DRUITT, T.H., SKIRIUS, C., AND STOLPER, E., 1989, H_2O, CO_2, Cl, and gas in Plinian and ash-flow Bishop rhyolite: Geology, v. 17, p. 221-225.

ANDERSON, J.E., 1989, Diagenesis of the Lansing and Kansas City groups (Upper Pennsylvanian), northwestern Kansas and southwestern Nebraska, Unpublished M.S. Thesis, University of Kansas, Lawrence, 259 p.

ANTHONY, E.Y., REYNOLDS, T.J., AND BEANE, R.E., 1984, Identification of daughter minerals in fluid inclusions using scanning electron microscopy and energy dispersive analysis: American Mineralogist, v. 69, p. 1053-1057.

ARMSTRONG, A.K., SNAVELY, P.D., AND ADDICOTT, W.D., 1980, Porosity evolution of Upper Miocene Reefs, Almeria Province, Southern Spain: American Association of Petroleum Geologists Bulletin, v. 64, p. 188-208.

AULSTEAD, K.L., SPENCER, R.J., AND KROUSE, H.R., 1988, Fluid inclusion and isotopic evidence on dolomitization, Devonian of Western Canada: Geochimica et Cosmochimica Acta, v. 52, p. 1027-1035.

BAKKER, R.J., AND JANSEN, J.B.H., 1990, Preferential water leakage from fluid inclusions by means of mobile dislocations: Nature, v. 345, p. 58-60.

BAKKER, R.J., AND JANSEN, J.B.H., 1991, Experimental post-entrapment water loss from synthetic CO_2-H_2O inclusions in natural quartz: Geochimica et Cosmochimica Acta, v. 55, p. 2215-2230.

BANKS, D.A., AND YARDLEY, B.W.D., 1992, Crush-leach analysis of fluid inclusions in small natural and synthetic samples: Geochimica et Cosmochimica Acta, v. 56, p. 245-248.

BANNON, M.P., TURNER, G., AND KELLEY, S.P., 1987, ^{40}Ar-^{39}Ar and thermometric analysis of fluid inclusions in quartz from mineralization associated with the Cornubian Batholith, SW England (Abstract): ECROFI, European Current Research on Fluid Inclusions, IX Symposium, Oporto, p. 7-8.

BARKER, C.E., 1992, Sample Temperatures Reached During Cathodoluminescence Observations: Preliminary Measurements Using Reequilibrated Fluid Inclusions: Society for Luminescent Microscopy and Spectroscopy Newsletter (unpaginated).

BARKER, C.E., AND GOLDSTEIN, R.H., 1990, Fluid inclusion technique for determining maximum temperature and its comparison to the vitrinite reflectance geothermometer: Geology, v. 18, p. 1003-1006.

BARKER, C.E., GOLDSTEIN, R.H., HATCH, J.R., WALTON, A.W., AND WOJCIK, K.M., 1992, Burial history and thermal maturation of Pennsylvanian rocks, Cherokee basin, Southeastern Kansas, *in* Johnson and Cardott, eds., Source Rocks in the Southern Midcontinent, Oklahoma Geological Survey Circular 93 p. 299-310.

BARKER, C.E., AND HALLEY, R.B., 1988, Fluid inclusions in vadose cement with consistent vapor to liquid ratios, Pleistocene Miami Limestone, southeastern Florida: Geochimica et Cosmochimica Acta, v. 52, p. 1019-1025.

BARKER, C.E., AND KOPP, O.C., 1991, Luminescence Microscopy and Spectroscopy: Qualitative and Quantitative Applications: Society of Economic Paleontologists and Mineralogists Short Course, v. 25, 195 p.

BARKER, C.E., AND PAWLEWICZ, M.J., 1986, The correlation of vitrinite reflectance with maximum temperature in humic organic matter, *in* Buntebarth, G., and Stegena, L., eds., Paleogeothermics: Berlin, Springer Verlag, p. 79-93.

BARKER, C.E., AND REYNOLDS, T.J., 1984, Preparing doubly polished sections of temperature sensitive sedimentary rocks: Journal of Sedimentary Petrology, v. 54, p. 635-636.

BARKER, C., AND SMITH, M.P., 1986, Mass spectrometric determination of gases in individual fluid inclusions in natural minerals: Analytical Chemistry, v. 58, p. 1330-1333.

BARRES, O., BURNEAU, A., DUBESSY, J., AND PAGEL, M., 1987, Application of micro-FT-IR spectroscopy to individual hydrocarbon fluid inclusion analysis: Applied Spectroscopy, v. 41, p. 1000-1008.

BEIN, A.S.D., HOVORKA, R.S.F., AND ROEDDER, E., 1991, Fluid inclusions in bedded Permian halite, Palo Duro Basin, Texas: Evidence for modification of seawater in evaporite brine-pools and subsequent early diagenesis: Journal of Sedimentary Petrology, v. 61, p. 1-14.

BENNETT, D.G., AND BARKER, A.J., 1992, High salinity fluids: The result of retrograde metamorphism in thrust zones: Geochimica et Cosmochimica Acta, v. 56, p. 81-95.

BERTRAND, P., PITTION, J., AND BERNARD, C., 1985, Fluorescence of sedimentary organic matter in relation to its chemical composition: Organic Geochemistry, v. 10, p. 641-647.

BLACIC, J.D., 1975, Plastic-deformation mechanisms in quartz: The effect of water: Tectonophysics, v. 27, p. 271-294.

BLACIC, J.D., 1981, Water diffusion in quartz at high pressure: Tectonic implications: Geophysics Research Letters, v. 8, p. 721-723.

BLASCH, S.R., AND COVENEY, R.M., JR., 1988, Goethite-bearing brine inclusions, petroleum inclusions, and the geochemical conditions of ore deposition at the Jumbo mine, Kansas: Geochimica et Cosmochimica Acta, v. 52, p. 1007-1017.

BLOUNT, C.W., AND PRICE, L.C., 1982, Solubility of methane in water under natural conditions: A laboratory study: United States Department of Energy Report DOE/ET/12145-1, DE82 017680.

BODNAR, R.J., 1983, A method of calculating fluid inclusion volumes based on vapor bubble diameters and P-V-T-X properties of inclusion fluids: Economic Geology, v. 78, p. 535-542.

BODNAR, R.J., 1990, Petroleum migration in the Miocene Monterey Formation, California, USA: Constraints from fluid-inclusion studies: Mineralogical Magazine, v. 54, p. 295-403.

BODNAR, R.J., 1992a, Revised equation and table for freezing point depressions of H_2O-salt fluid inclusions (Abstract): PACROFI IV, Fourth Biennial Pan-American Conference on Research on Fluid Inclusions, Program and Abstracts, Lake Arrowhead, CA, v. 14, p. 15.

BODNAR, R.J., 1992b, The system H_2O-NaCl (Abstract): PACROFI IV, Fourth Biennial Pan-American Conference on Research on Fluid Inclusions, Program and Abstracts, Lake Arrowhead, CA, v. 4, p. 108-111.

BODNAR, R.J., AND BETHKE, P.M., 1984, Systematics of stretching of fluid inclusions; fluorite and sphalerite at 1 atmosphere confining pressure: Economic Geology, v. 79, p. 141-161.

BODNAR, R.J., BINNS, P.R., AND HALL, D.L., 1989, Synthetic fluid inclusions VI. Quantitative evaluation of the decrepitation behavior of fluid inclusions in quartz at one atmosphere confining pressure: Journal of Metamorphic Geology, v. 7, p. 229-242.

BODNAR, R.J., REYNOLDS, T. J., AND KUEHN, C.A., 1985, Fluid-inclusion systematics in epithermal systems, *in* Berger, B. R., and Bethke, P. M., eds., Geology and Geochemistry of Epithermal Systems: Society of Economic Geologists, Reviews in Economic Geology, v. 2, p. 73-97.

BODNAR, R.J., AND STERNER, S.M., 1985, Synthetic fluid inclusions in natural quartz. II. Application to PVT studies: Geochimica et Cosmochimica Acta, v. 49, p. 1855-1859.

BÖHLKE, J.K., AND IRWIN, J.J., 1992, Laser microprobe analyses of noble gas isotopes and halogens in fluid inclusions: Analyses of microstandards and synthetic inclusions in quartz: Geochimica et Cosmochimica Acta, v. 56, p. 187-201.

BÖHLKE, J.K., KIRSCHBAUM, C., AND IRWIN, J., 1989, Simultaneous analyses of noble-gas isotopes and halogens in fluid inclusions in neutron-irradiated quartz veins by use of a laser-microprobe noble-gas mass spectrometer: United States Geological Survey Bulletin, v. 1890, p. 61-88.

BOIRON, M.C., DUBESSY, J., BRIAND, A., MAUCHIEN, P., AND ALLE, P., 1992, Analysis of mono-atomic ions in individual fluid inclusions: A comparative study using L.P.E.S. and S.I.M.S. (Abstract): PACROFI IV, Pan-American Conference on Research on Fluid Inclusions, Program and Abstracts, Lake Arrowhead, CA, v. 4, p. 17-18.

BOLES, J.R., 1978, Active ankerite cementation in the subsurface of Eocene of Southwest Texas: Contributions of Mineralogy and Petrology, v. 68, p. 13-22.

BOTTRELL, S.H., AND YARDLEY, B.W.D., 1987, A modified crush-leach method for the analysis of fluid inclusion electrolytes (Abstract): ECROFI, European Current Research on Fluid Inclusions, IX Symposium, Oporto, p. 15-16.

BOTTRELL, S.H., AND YARDLEY, B.W.D., 1988, The composition of a primary granite-derived ore fluid from S. W. England, determined by fluid inclusion analysis: Geochimica et Cosmochimica Acta, v. 52, p. 585-588.

BOTTRELL, S.H., YARDLEY, B.W.D., AND BUCKLEY, F., 1988, A modified crush-leach method for the analysis of fluid inclusion electrolytes: Bulletin de Minéralogie, v. 111, p. 279-290.

BRANNON, J.C., PODOSEK, F.A., AND McLIMANS, R.K., 1992, Alleghenian age of the Upper Mississippi Valley zinc-lead deposit determined by Rb-Sr dating of sphalerite: Nature, v. 356, p. 509-511.

BRANTLEY, S.L., 1992, The effect of fluid chemistry on quartz microcrack lifetimes: The Earth and Planetary Science Express, p. 145-156.

BRANTLEY, S.L., EVANS, B., HICKMAN, S.H., AND CERAR, D.A., 1990, Healing of microcracks in quartz: Implications for fluid flow: Geology, v. 18, p. 136-139.

BRAY, C.J., SPOONER, E.T.C., AND THOMAS, A.V., 1991, Fluid inclusion volatile analysis by heated crushing, on-line gas chromatography; applications to Archean fluids: Journal of Geochemical Exploration, v. 42, p. 167-193.

BURLEY, S.D., MULLIS, J., AND MATTER, A., 1989, Timing diagenesis in the Tartan reservoir (UK North Sea): Constraints from combined cathodoluminescence microscopy and fluid inclusion studies: Marine and Petroleum Geology, v. 6, p. 98-120.

BURRUSS, R.C., 1981, Hydrocarbon fluid inclusions in studies of sedimentary diagenesis, *in* Hollister, L. S., and Crawford, M. L., eds., Short Course in Fluid Inclusions: Application to Petrology: Mineralogical Association of Canada Short Course Handbook, v. 6, p. 138-156.

BURRUSS, R.C., 1987a, Paleotemperatures from fluid inclusions: Advances in theory and technique, *in* Naeser, N. D., and McCulloh, T. H., Thermal History of Sedimentary Basins, Methods and Case Histories, American Association of Petroleum Geologists Special Publication 41, p. 121-131.

BURRUSS, R.C., 1987b, Crushing-cell, capillary column gas chromatography of petroleum fluid inclusions: Method and application to petroleum source rocks, reservoirs, and low temperature hydrothermal ores (Abstract): American Current Research on Fluid Inclusions, Socorro, NM, Program and Abstracts (unpaginated).

BURRUSS, R.C., 1991, Practical aspects of fluorescence microscopy of petroleum fluid inclusions, *in* Barker, C. E., and Kopp, O. C., eds., Luminescence Microscopy and Spectroscopy: Society of Economic Paleontologists and Mineralogists Short Course, v. 25, p. 1-7.

BURRUSS, R.C., 1992, Phase behavior in petroleum-water (brine) systems applied to fluid inclusion studies (Abstract): PACROFI IV, Pan-American Conference on Research on Fluid Inclusions, Program and Abstracts, Lake Arrowhead, CA, v. 4, p. 116-118.

BURRUSS, R.C., CERCONE, K.R., AND HARRIS, P.M., 1985, Timing of hydrocarbon migration: evidence from fluid inclusions in calcite cements, tectonics, and burial history, *in* Schneidermann, N., and Harris, P. M., eds., Carbonate Cements: Society of Economic Paleontologists and Mineralogists Special Publication 26, p. 277-289.

BURRUSS, R.C., AND GOLDSTEIN, R.H., 1980, Time and temperature of hydrocarbon migration: fluid inclusion evidence from the Fayetteville Formation, N. W. Arkansas (Abstract): Geological Society of America Abstracts with Program, v. 12, p. 396.

BURRUSS, R.C., AND HOLLISTER, L.S., 1979, Evidence from fluid inclusions for a paleogeothermal gradient at the geothermal test wells sites, Los Alamos, New Mexico: Journal of Volcanology and Geothermal Research, v. 5, p. 163-177.

BURRUSS, R.C., AND REYNOLDS, T.J., 1993, Nucleation, Growth, and Dissociation of Methane Gas Hydrate: Microscope Observations in Natural Fluid Inclusions in Fluorite: United States Geological Survey Open-file Report 93-388 (video tape).

BURRUSS, R.C., TOTH, D.J., AND GOLDSTEIN, R.H., 1980, Fluorescence microscopy of hydrocarbon fluid inclusions: Relative timing of hydrocarbon migration events in the Arkoma basin, N. W. Arkansas (Abstract): EOS, v. 61, p. 400.

CANALS, A., GUILHAUMOU, N., RAMSEY, M.H., COLES, B.W., AND ROSENBERG, E., 1992, Laser ablation ICP-AES and X-ray micro-probe for analysis of individual fluid inclusion: The case of halite from Mulhouse basin, (Abstract): PACROFI IV, Pan-American Conference on Research on Fluid Inclusions, Program and Abstracts, Lake Arrowhead, CA, v. 4, p. 19.

CAROTHERS, W.W., AND KHARAKA, Y.K., 1978, Aliphatic acid anions in oil field waters—Implications for origin of natural gas: American Association of Petroleum Geologists Bulletin, v. 62, p. 2441-2453.

CAROTHERS, W.W., AND KHARAKA, Y.K., 1980, Stable carbon isotopes in oil-field waters and the origin of CO_2: Geochimica et Cosmochimica Acta, v. 44, p. 323-332.

CARPENTER, A.B., 1978, Origin and chemical evolution of brines in sedimentary basins: Oklahoma Geological Survey Circular, v. 79, p. 60-77.

CASAS, E., LOWENSTEIN, T.K., SPENCER, R.J., AND PENGXI, Z., 1992, Carnallite mineralization in the nonmarine, Qaidam basin, China: Evidence for the early diagenetic origin of potash evaporites: Journal of Sedimentary Petrology, v. 62, p. 881-898.

CHANGKAKOTI, A., GRAY, J., KRSTIC, D., CUMMING, G.L., AND MORTON, R.D., 1988, Determination of radiogenic isotopes (Rb/Sr, Sm/Nd and Pb/Pb) in fluid inclusion waters: An example from the Bluebell Pb-Zn deposit, British Columbia, Canada: Geochimica et Cosmochimica Acta, v. 52, p. 961-967.

CHANNER, D.M.D., AND SPOONER, E.T.C., 1992, Analysis of fluid inclusion leachates from quartz by ion chromatography: Geochimica et Cosmochimica Acta, v. 56, p. 249-259.

CHENERY, S.R.N., AND RANKIN, A.H., 1989, The use of laser ablation microprobe (LAMP) attached to an inductively coupled plasma spectrometer for the elemental analysis of individual fluid inclusions (Abstract): ECROFI, European Current Research on Fluid Inclusions, X Symposium, London, p. 21.

CHOU, I-M., PASTERIS, J.D., AND SEITZ, J.C., 1990, High-density volatiles in the system C-O-H-N for the calibration of a laser Raman microprobe: Geochimica et Cosmochimica Acta, v. 54, p. 535-543.

CLINE, J.S., VANKO, D.A., GHAZI, A.M., SUTTON, S.R., AND ROEDDER, E., 1993, Synchrotron X-ray fluorescence analysis of dense brine and lower salinity inclusions from the Questa porphyry molybdenum deposit, Questa, New Mexico, U.S.A.: ECROFI XII, Twelfth Biennial Symposium, European Current Research on Fluid Inclusions, Warsaw-Cracow, p. 39-41.

COLLINS, A.G., 1975, Geochemistry of Oilfield Brines: Amsterdam, Elsevier Scientific Publishing Company, 496. p.

COLLINS, P.L.F., 1979, Gas hydrates in CO_2-bearing fluid inclusions and the use of freezing data for estimation of salinity: Economic Geology, v. 74, p. 1435-1444.

COMINGS, B.D., AND CERCONE, K.R., 1986, Experimental contamination of fluid inclusions in calcite: Society of Economic Paleontologists and Mineralogists Abstracts with Programs, v. 3, p. 24.

CRAWFORD, M.L., 1981, Phase equilibria in aqueous fluid inclusions, *in* Hollister, L.S., and Crawford, M.L., eds., Fluid Inclusions: Applications to Petrology: Mineralogical Association of Canada Short Course Handbook 6, p. 75-100.

CUNNINGHAM, K.M., GOLDBERG, M.C., AND WEINER, E.R., 1977, Investigation of detection limits for solutes in water measured by laser Raman spectrometry: Analytical Chemistry, v. 49, p. 70-75.

CYS, J., AND MAZZULLO, S., 1977, Biohermal submarine cements, Laborcita Formation (Permian), northern Sacramento Mountains, New Mexico, *in* Butler, J., ed., Geology of the Sacramento Mountains, Otero County, New Mexico: West Texas Geological Society Publication 1977-68, p. 39-51.

DAHAN, N., AND COUTY, R., 1987, Infrared microspectroscopy of hydrocarbon fluid inclusions in fluorite. Effect of heating under confining pressure (400°C-400 bars): Compte Rendue Académie des Sciences, Paris, v. 305, p. 687-589.

DAVIS, D.W., LOWENSTEIN, T.K., AND SPENCER, R.J., 1990, Melting behavior of fluid inclusions in laboratory-grown halite crystals in the systems NaCl-H_2O, NaCl-KCl-H_2O, NaCl-$MgCl_2$-H_2O, and NaCl-$CaCl_2$-H_2O: Geochimica et Cosmochimica Acta, v. 54, p. 591-601.

DELOULE, E., AND ELOY, J.F., 1982, Improvements of laser probe mass spectrometry for the chemical analysis of fluid inclusions in ores: Chemical Geology, v. 37, p. 191-202.

DIAMOND, L.W., MARSHALL, D.D., JACKMAN, J.A., AND SKIPPEN, G.B., 1990, Elemental analysis of individual fluid inclusions in minerals by secondary ion mass spectrometry (SIMS): Application to cation ratios of fluid inclusions in an Archaean mesothermal gold-quartz vein: Geochimica et Cosmochimica Acta, v. 54, p. 545-552.

DICKEY, P.A., 1969, Increasing concentration of subsurface brines with depth: Chemical Geology, v. 4, p. 361-370.

DICKSON, J.A.D., 1965, A modified staining technique for carbonates in thin section: Nature, v. 205, p. 587.

DIX, D.R., AND JACKSON, M.P.A., 1982, Lithology, Microstructures, Fluid Inclusions and Geochemistry of Rock Salt and of the Cap-rock Contact in Oakwood Dome, East Texas: Significance for Nuclear Waste Storage: Texas Bureau of Economic Geology Reports of Investigations, v. 120, 63 p.

DOLOMIEU, DEODAT DE, 1792, Sur de l'huile de pétrole dans le cristal de roche et les fluides élastiques tirés du quartz: Observations sur la Physique, v. 40, p. 318-319.

DUAN, Z., MOLLER, N., GREENBERG, J., AND WEARE, J.H., 1992, The prediction of methane solubility in natural waters to high ionic strength from 0 to 250°C and from 0 to 1600 bar: Geochimica et Cosmochimica Acta, v. 56, p. 1451-1460.

DUBESSY, J., AUDEOUD, D., WIKINS, R., AND KOSZTOLANYI, C., 1982, The use of the Raman microprobe MOLE in the determination of the electrolytes dissolved in the aqueous phase of fluid inclusions: Chemical Geology, v. 37, p. 137-150.

DUBESSY, J., BOIRON, M., MOISSETTE, A., MONNIN, C., AND SRETENSKAYA, N., 1992, Determinations of water, hydrates and pH in fluid inclusions by micro-Raman spectrometry: European Journal of Mineralogy, v. 4, p. 885-894.

DUBESSY, J., GEISLER, D., KOSZTOLANYI, C., AND VERNET, M., 1983, The determination of sulphate in fluid inclusions using the M.O.L.E. Raman microprobe. Application to a Keuper halite and geochemical consequences: Geochimica et Cosmochimica Acta, v. 47, p. 1-10.

DUBESSY, J., PAGEL, M., BENY, J-M., CHRISTENSEN, H., HICKEL, B., KOSZTOLANYI, C., AND POTY, B., 1988, Radiolysis evidenced by H_2-O_2- and H_2-bearing fluid inclusions in three uranium deposits: Geochimica et Cosmochimica Acta, v. 52, p. 1155-1167.

FABRE, D., AND COUTY, R., 1986, Etude, par spectroscopie Raman, du methane comprime jusqu'a 3 kbar. Application a la mesure de pression dans les inclusions fluides contenues dans les minéraux: Compte Rendue Académie des Sciences, Paris, v. 303 p. 1305-1308.

FISHER, J.R., 1976, The volumetric properties of H_2O—a graphical portrayal: Journal of Research, United States Geological Survey, v. 4, p. 189-193.

FOLK, R.L., 1965, Some aspects of recrystallization in ancient limestones, *in* Pray, L. C. and Murray, R. C., eds., Dolomitization and Limestone Diagenesis: A Symposium: Society of Economic Paleontologists and Mineralogists Special Publication, v. 13, p. 14-48.

FUJINO, K., LEWIS, E.L., AND PERKIN, R.G., 1974, The freezing point of seawater at pressures up to 100 bars: Journal of Geophysical Research, v. 79, p. 1792-1797.

GAFFEY, S.J., 1988, Water in skeletal carbonates: Journal of Sedimentary Petrology, v. 58, p. 397-414.

GAFFEY, S.J., 1990, Skeletal versus nonbiogenic carbonates-UV-Visible-Near IR (0.3-2.7 mm) reflectance properties, *in* Coyne, L. M., McKeever, S. W., and Blake, D. F., eds., Spectroscopic Characterization of Minerals and Their Surfaces: Washington, D. C., American Chemical Society Symposium Series No. 415, p. 94-116.

GERLACH, H. AND HELLER, S., 1966, Concerning artifically produced fluid inclusions in rock salt crystals: Deutsche Gesellschaft für Geologische Wissenschaften, Reihe B, Mineralogische Lagerstättenforschung, v. 11, p. 195-214 (in German).

GINSBURG, R.N., AND JAMES, N.P., 1976, Submarine botryoidal aragonite in Holocene reef limestones, Belize: Geology, v. 4, p. 431-436.

GOLDSTEIN, R.H., 1986a, Reequilibration of fluid inclusions in low-temperature calcium-carbonate cement: Geology, v. 14, p. 792-795.

GOLDSTEIN, R.H., 1986b, Integrative Carbonate Diagenesis Studies: Fluid Inclusions in Calcium Carbonate Cement: Paleosols and Cement Stratigraphy of Late Pennslylvanian Cyclic Strata, New Mexico: Unpublished Ph.D. Dissertation, University of Wisconsin, Madison, 343 p.

GOLDSTEIN, R.H., 1988, Cement stratigraphy of Pennsylvanian Holder Formation, Sacramento Mountains, New Mexico: American Association of Petroleum Geologists Bulletin, v. 72, p. 425-438.

GOLDSTEIN, R.H., 1990, Petrographic and geochemical evidence for origin of paleospeleothems, New Mexico: Implications for the application of fluid inclusions to studies of diagenesis: Journal of Sedimentary Petrology, v. 60, p. 282-292.

GOLDSTEIN, R.H., 1993, Fluid inclusions as microfabrics: a petrographic method to determine diagenetic history, *in* Rezak, R. and Lavoi, D., eds., Carbonate Microfabrics, Frontiers in Sedimentary Geology: New York, Springer-Verlag, p. 279-290.

GOLDSTEIN, R.H., FRANSEEN, E.K., AND MILLS, M.S., 1990, Diagenesis associated with subaerial exposure of Miocene strata, southeastern Spain: Implications for sea-level change and preservation of low-temperature fluid inclusions in calcite cement: Geochimica et Cosmochimica Acta, v. 54, p. 699-704.

GOLDSTEIN, R.H., STEPHENS, B.P., AND LEHRMANN, D.J., 1991, Fluid inclusions elucidate conditions of dolomitization in Eocene of Enewetak Atoll and Mid-Cretaceous Valles Platform of Mexico: Dolomieu Conference on Carbonate Platforms and Dolomitization Abstracts, Ortisei, Italy, p. 92-93.

GRANEY, J.R., KESLER, S.E., AND JONES, H.D., 1991, Application of gas analysis of jasperoid inclusion fluids to exploration for micron gold deposits: Journal of Geochemical Exploration, v. 42, p. 91-106.

GRATIER, J.P., AND JENATON, L., 1984, Deformation by solution-deposition, and re-equilibration of crystals depending on temperature, internal pressure and stress: Journal of Structural Geology, v. 6, p. 189-200.

GREGG, J.M., AND SHELTON, K.L., 1990, Dolomitization and dolomite neomorphism in the back reef facies of the Bonneterre and Davis Formations (Cambrian), southeast Missouri: Journal of Sedimentary Petrology, v. 60, p. 549-562.

GUILHAUMOU, N., SZYDLOWSKI, N., AND PRADIER, B., 1989, Characterization of hydrocarbon fluid inclusions by infra red and fluorescence microspectrometry (Abstract): ECROFI, European Current Research on Fluid Inclusions, X Symposium, London, v. 40 (unpaginated).

GUILHAUMOU, N., SZYDLOWSKI, N., AND PRADIER, B., 1990, Characterization of hydrocarbon fluid inclusions by infra-red and fluorescence microspectrometry: Mineralogical Magazine, v. 54, p. 311-324.

GUSCOTT, S.C., AND BURLEY, S.D., 1993, A systematic approach to reconstructing palaeofluid evolution from fluid inclusions in authigenic quartz overgrowths, *in* Parnell, J., and others, eds., Conference Proceedings, Geofluids, v. 93, p. 323-328.

HAAS, J.L., JR., 1978, An Empirical Equation with Tables of Smoothed Solubilities of Methane in Water and Aqueous Sodium Chloride Solutions up to 25 Weight Percent, 360°C, and 138 MPa.: United States Geological Survey Open-File Report 78-1004, 41 p.

HAGEMANN, H.W., AND HOLLERBACH, A., 1985, The fluorescence behavior of crude oils with respect to their thermal maturation and degradation: Organic Geochemistry, v. 10. p. 473-480.

HALL, D.L., BODNAR, R.J., AND CRAIG, J.R., 1991, Evidence for postentrapment diffusion of hydrogen into peak metamorphic fluid inclusions from the massive sulfide deposits at Ducktown, Tennessee: American Mineralogist, v. 76, p. 1344-1355.

HALL, D.L., AND STERNER, S.M., 1992, The effect of a vapor phase on fluid inclusion salinity determinations: Experimental evidence from the system NaCl-H_2O (Abstract): PACROFI IV, Pan-American Conference on Research on Fluid Inclusions, Program and Abstracts, Lake Arrowhead, CA, v. 4, p. 38.

HALL, D.L., STERNER, S.M., AND BODNAR, R.J., 1988, Freezing point depression of NaCl-KCl-H_2O solutions: Economic Geology, v. 83, p. 197-202.

HALL, D.L., STERNER, S.M., AND BODNAR, R.J., 1989, Experimental evidence for hydrogen diffusion into fluid inclusions in quartz (Abstract): Geological Society of America Abstracts with Programs, v. 21, p. A-358.

HALL, D.L., STERNER, S.M., AND WHEELER, J.R., 1993, One-atmosphere decrepitation behavior of synthetic fluid inclusions in natural calcite: Implications for preservation of calcite-hosted inclusions during burial: ECROFI XII, Twelfth Biennial Symposium, European Current Research on Fluid Inclusions, Warsaw-Cracow, p. 91-92.

HALL, D.L., AND WHEELER, J.R., 1992, Fluid composition and the decrepitation behavior of synthetic fluid inclusions in quartz (Abstract): PACROFI IV, Pan-American Conference on Research on Fluid Inclusions, Program and Abstracts, Lake Arrowhead, CA, v. 4, p. 39.

HANOR, J.S., 1980, Dissolved methane in sedimentary brines: potential effect on the PVT properties of fluid inclusions: Economic Geology, v. 75, p. 603-609.

HANOR, J.S., 1984, Variation in the chemical composition of oil-field brines with depth of northern Louisiana and southern Arkansas: Implications for mechanisms and rates of mass transport and diagenetic reaction: Transactions—Gulf Coast Association of Geological Societies, v. 34, p. 55-61.

HASZELDINE, R.S., SAMSON, I.M., AND CORNFORD, C., 1984, Dating diagenesis in a petroleum basin, a new fluid inclusion method: Nature, v. 307, p. 354-357.

HAYNES, F.M., 1988, Fluid-inclusion evidence of basinal brines in Archean basement, Thunder Bay Pb-Zn-Ba district, Ontario, Canada: Canadian Journal of Earth Sciences, v. 25, p. 1884-1894.

HAYNES, F.M., AND KESLER, S.E., 1987, Chemical evolution of brines during Mississippi Valley-type mineralization: Evidence from East Tennessee and Pine Point: Economic Geology, v. 82, p. 53-71.

HAYNES, F.M., STERNER, S.M. AND BODNAR, R.J., 1987, Chemical analysis of fluid inclusions by SEM/EDA: Methodology and results from synthetic inclusions in natural quartz (Abstract): American Current Research on Fluid Inclusions, Socorro, NM, Program and Abstracts (unpaginated).

HAYNES, F.M., STERNER, S.M., AND BODNAR, R.J., 1988, Synthetic fluid inclusions in natural quartz. IV. Chemical analyses of fluid inclusions by SEM/EDA: Evaluation of method: Geochimica et Cosmochimica Acta, v. 52, p. 969-977.

HECKEL, P.H., 1980, Paleogeography of eustatic model for deposition of midcontinent Upper Pennsylvanian cyclothems, *in* Fouch, T. D., and Magathan, E. R., eds., Paleozoic Paleogeography of the West-central United States: Rocky Mountain Section, Society of Economic Paleontologists and Mineralogists, Rocky Mountain Paleogeography Symposium, v. 1, p. 197-216.

HENNIKER, J.C., 1949, The depth of the surface zone of a liquid: Reviews in Modern Physics, v. 21, p. 322-341.

HIGGINS, K.L., AND STEIN, C.L., 1986, Micro-Raman spectroscopy of fluid inclusions in a hopper crystal in halite, *in* Romig, A. D., Jr., and Chambers, W. F., eds., Microbeam Analysis: San Francisco, San Francisco Press, Inc., p. 31-34.

HORITA, J., FRIEDMAN, T.J., LAZAR, B., AND HOLLAND, H.D., 1991, The composition of Permian seawater: Geochimica et Cosmochimica Acta, v. 55, p. 417-432.

HORN, E.E, AND TRAXEL, K., 1987, Investigations of individual fluid inclusions with the Heidelberg proton microprobe—A nondestructive analytical method: Chemical Geology, v. 61, p. 29-35.

HORN, E.E., AND TYE, C.T., 1989, Analysis of fluid inclusions in minerals by VG laser ablation ICP-MS (Abstract): PACROFI, Second Pan-American Conference on Research on Fluid Inclusions, Program and Abstracts, Blacksburg, VA, v. 2., p. 32.

HORSFIELD, B., AND McLIMANS, R.K., 1984, Geothermometry and geochemistry and aqueous and oil-bearing fluid inclusions from Fateh field, Dubai: Organic Geochemistry, v. 6, p. 733-740.

HUNT, J.M., 1979, Petroleum Geochemistry and Geology: San Francisco, Freeman, 617 p.

ITARD, Y., CHAMPENOIS, M., CHEILLETZ, A., AND RAMBOZ, C.C., 1989, Volume estimation of fluid inclusions using an interactive image analyser (Abstract): ECROFI, European Current Research on Fluid Inclusions, X Symposium, London, p. 54.

JAMES, N.P., AND BONE, Y., 1992, Synsedimentary cemented calcarenite layers in Oligo-Miocene cool-water shelf limestones, Eucla platform, Southern Australia: Journal of Sedimentary Petrology, v. 62, p. 860-872.

JANSSEN-VAN ROSMALEN, R., AND BENNEMA, P., 1977, The role of hydrodynamics and supersaturation in the formation of liquid inclusions in KDP: Journal of Crystal Growth, v. 42, p. 224-227.

JOHNSON, W.J., AND GOLDSTEIN, R.H., 1993, Cambrian sea water preserved as inclusions in marine low-magnesium calcite cement: Nature, v. 362, p. 335-337.

JONES, P.H., 1976, Natural Gas Resources of the Geopressured Zones in the Northern Gulf of Mexico Basin, *in* Natural Gas from Unconventional Geologic Sources: National Research Council, National Academy of Sciences, p. 17-33.

KALYUZHNYI, V.A., 1971, The refilling of liquid inclusions in minerals and its genetic significance: L'vov. Gos. Univ. Mineral. Sbornik, v. 25, p. 124-131.

KANNO, H., AND ANGELL, C.A., 1977, Homogeneous nucleation and glass formation in aqueous alkali halide solutions at high pressures: The Journal of Physical Chemistry, v. 81, p. 2639-2643.

KEKULAWALA, K.R.S.S., PATERSON, M.S., AND BOLAND, J. N., 1981, An experimental study of the role of water in quartz deformation, *in* Mechanical Behavior of Crustal Rocks, Geophysics Monograph 24, American Geophysical Union, p. 49-60.

KELLY, W.C., AND BURGIO, P.A., 1983, Cryogenic scanning electron microscopy of fluid inclusions in ore and gangue minerals: Economic Geology, v. 78, p. 1262-1267.

KENDALL, A.C., AND BROUGHTON, P.L., 1978, Origin of fabrics in speleothems composed of columnar calcite crystals: Journal of Sedimentary Petrology, v. 48, p. 519-538.

KESLER, S.E., HAYNES, P.S., CREECH, M.Z., AND GORMAN, J. A., 1986, Application of fluid inclusion and rock-gas analysis in mineral exploration: Journal of Geochemical Exploration, v. 25, p. 201-215.

KHARAKA, Y.K., CAROTHERS, W.W., AND ROSENBAUER, R. J., 1983, Thermal decarboxylation of acetic acid: Implications for origin of natural gas: Geochimica et Cosmochimica Acta, v. 47, p. 397-402.

KHARAKA, Y.K., LAW, L.M., CAROTHERS, W.W., AND GOERLITZ, D.F., 1986, Role of organic species dissolved in formation waters from sedimentary basins in mineral diagenesis, *in* Guatier, D. L., Roles of Organic Matter in Sediment Diagenesis: Society of Economic Paleontologists and Mineralogists Special Publication 38, p. 111-122.

KIHLE, J., 1993, Non-destructive fingerprinting of single hydrocarbon fluid inclusions: micro-luminescence spectroscopy; state of the art and prospects for tomorrow: Organic Geochemistry Poster Sessions from the Sixteenth International Meeting on Organic Geochemistry, p. 784-790.

KIHLE, J., AND JOHANSEN, H., 1994, Low-temperature isothermal trapping of hydrocarbon fluid inclusions in synthetic crystals of KH_2PO_4: Geochimica et Cosmochimica Acta, v. 58, p. 1193-1202.

KLOSTERMAN, M.J., 1981, Applications of fluid inclusion techniques to burial diagenesis in carbonate rock sequences: *in* Applied Carbonate Research Program Technical Series Contributions, v. 7, Baton Rouge, Louisiana State University, 102 p.

KNAUTH, L.P., AND BEEUNAS, M.A., 1986, Isotope geochemistry of fluid inclusions in Permian halite with implications for the isotopic history of ocean water and the origin of saline formation waters: Geochimica et Cosmochimica Acta, v. 50, p. 419-433.

KOHN, S.C., DUPREE, R., AND FARNAN, I., 1988, Volatiles in silicate glasses: A magic angle spinning NMR study (Abstract): Terra Cognita, v. 8, p. 66-69.

KOZLOWSKI, A., 1978, Pneumatolytic and hydrothermal activity in the Karkonosze-Izera block: Acta Geologica Polonica, v. 28, p. 171-222.

KRAMER, J.R., 1969, Subsurface brines and mineral equilibria: Chemical Geology, v. 4, p. 37-50.

KREULEN, R., AND SCHUILING, R.D., 1982, N_2-CH_4-CO_2 fluids during formation of the Dome d l'Agout, France: Geochimica et Cosmochimica Acta, v. 46, p. 193-203.

LACAZETTE, A., 1990, Application of linear elastic fracture mechanics to the quantitative evaluation of fluid-inclusion decrepitation: Geology, v. 18, p. 782-785.

LACAZETTE, A., 1991, Natural hydraulic fracturing in the Bald Eagle Sandstone in central Pennsylvania and the Ithaca Siltstone at Watkins Glen, New York: Ph.D. Dissertation, The Pennsylvania State University, State College, 225 p.

LAMBERT, M.W., 1990, Fluid-inclusion evidence for deep burial origin of late diagenetic spar in Martinez Mound bioherm, Lower Wolfcampian Laborcita Formation, New Mexico: The Compass, v. 67, p. 130-134.

LAND, L.S., 1973, Contemporaneous dolomitization of Middle Pleistocene reefs by meteoric water, North Jamaica: Bulletin of Marine Science, v. 23, p. 64-92.

LANDIS, G.P., HOFSTRA, A.H., LEACH, D.L., AND RYE, R.O., 1987, Quantitative analysis of fluid-inclusion gases—Applications to studies of ore deposits (Abstract): United States Geological Survey Circular 995, p. 38-39.

LAWLER, J.P., AND CRAWFORD, M.L., 1983, Stretching of fluid inclusions resulting from a low-temperature microthermometric technique: Economic Geology, v. 78, p. 527-529.

LAZAR, B., AND HOLLAND, H.D., 1988, Analysis of fluid inclusions in halite: Geochimica et Cosmochimica Acta, v. 52, p. 485-490.

LE BEL, L., 1976, Preliminary note on the mineralogy of solid phases in quartz phenocryst inclusions in the porphyry copper from Cerro Verde/Santa Rosa, S. Peru: Bull. Soc. Vaudoise Sci. Nat., v. 73, p. 201-208 (in French).

LEMMLEIN, G.G., 1951, The fissure-healing process in crystals and change in cavity shape in secondary liquid inclusions: Akademiya Nauk SSSR Doklady, v. 78, p. 685-688 (in Russian).

LEMMLEIN, G.G., AND KLIYA, M.O., 1952, Distinctive features of the healing of a crack in a crystal under conditions of declining temperature. Akademiya Nauk SSSR Doklady, v. 87, p. 957-960 (in Russian; translated in International Geologic Review, v. 2, p. 125-128, 1960).

LEROY, J., 1979, Contribution a l'etalonnage de la pression interne des inclusions fluides lors de leur décrepitation: Bulletin de Minéralogie, v. 120, p. 584-593.

LOHMANN, K.C., 1978, Closed system diagenesis of high magnesium calcite and cements: Geological Society of America, Abstracts with Programs, v. 10, p. 446.

LOWENSTEIN, T.K., AND SPENCER, R.J., 1990, Syndepositional origin of potash evaporites: Petrographic and fluid inclusion evidence: American Journal of Science, v. 290, p. 1-42.

LYMAN, J. AND FLEMING, R.H., 1940, The composition of seawater: Journal of Marine Research, v. 3, p. 134-146.

MAJOR, R., 1985, Isotopic evidence for burial diagenesis of a Permian (Wolfcampian) phylloid algal bioherm (Abstract): Society of Economic Paleontologists and Mineralogists Annual Midyear Meeting Abstracts, v. 2, p. 58.

MAVROGENES, J.A., AND BODNAR, R.J., 1994, Hydrogen movement into and out of fluid inclusions in quartz: Experimental evidence and geologic implications: Geochimica et Cosmochimica Acta, v. 58, p. 141-148.

McGEE, K.A., SUSAK, N.J., SUTTON, A.J., AND HAAS, J.L., JR., 1981, The Solubility of Methane in Sodium Chloride Brines: United States Geological Survey Open-File Report 81, 41 p.

McLAREN, A.C., COOK, R.F., HYDE, S. T., AND TOBIN, R.C., 1983, The mechanism of the formation and growth of water bubbles and associated dislocation loops in synthetic quartz: Physics and Chemistry of Minerals, v. 9, p. 79-94.

McLIMANS, R.K., 1987, The application of fluid inclusions to migration of oil and diagenesis in petroleum reservoirs: Applied Geochemistry, v. 2, p. 585-603.

McLIMANS, R.K., BRANNON, J.C., AND PODOSEK, F.A., 1992, Upper Mississippi Valley Zn-Pb district: Geochemistry of ore fluids and age of mineralization as determined from studies of fluid inclusions (Abstract): PACROFI IV, Pan-American Conference on Research on Fluid Inclusions, Program and Abstracts, Lake Arrowhead, CA, v. 4, p. 59.

McNEIL, B., AND MORRIS, E., 1992, The preparation of double-polished fluid inclusion wafers from friable, water-sensitive material: Mineralogical Magazine, v. 56, p. 120-122.

MERNAGH, T.P., AND WILDE, A.R., 1989, The use of the laser Raman microprobe for the determination of salinity in fluid inclusions: Geochimica et Cosmochimica Acta, v. 53, p. 765-771.

METZGER, F.W., KELLY, W.C., NESBITT, B.E., AND ESSENE, E.J., 1977, Scanning electron microscopy of daughter minerals in fluid inclusions: Economic Geology, v. 72, p. 141-152.

MEUNIER, J.D., 1989, Assessment of low-temperature fluid inclusions in calcite using microthermometry: Economic Geology, v. 84, p. 167-170.

MEYERS, W.J., AND LOHMANN, K.C., 1985, Isotope geochemistry of regionally extensive calcite cement zones and marine components in Mississippian limestones, New Mexico, *in* Schneidermann, N., and Harris, P. M., eds., Carbonate Cements: Society of Economic Paleontologists and Mineralogists Special Publication 36, p. 223-239.

MILLIKEN, K.L., LAND, L.S., AND LOUCKS, R.G., 1981, History of burial diageneis determined from isotopic geochemistry, Frio Formation, Brazoria County, Texas: American Association of Petroleum Geologists Bulletin, v. 65, p. 1397-1413.

MOORE, C.H., AND DRUCKMAN, Y., 1981, Burial diagenesis and porosity evolution, Upper Jurassic Smackover, Arkansas and Louisiana: American Association of Petroleum Geologists Bulletin, v. 65, p. 597-628.

MOORE, J.N., AND ADAMS, M.C., 1988, Evolution of the thermal cap in two wells from the Salton Sea geothermal system, California: Geothermics, v. 17, p. 695-710.

MORGAN, G.B., CHOU, I-M., PASTERIS, J.D., AND OLSEN, S.N., 1993, Re-equilibration of CO_2 fluid inclusions at controlled hydrogen fugacities: Journal of Metamorphic Geology, v. 11, p. 155-164.

MULLIS, J., 1979, The system methane-water as a geologic thermometer and barometer from the external part of the central Alps: Bulletin de Minéralogie, v. 102, p. 526-536.

NARR, W.M., AND BURRUSS, R.C., 1984, Origin of reservoir fractures in Little Knife Field, North Dakota: American Association of Petroleum Geologists Bulletin, v. 68, p. 1087-1100.

NEDKITNE, T., KARLSON, D.A., BJORLYKKE, K., AND LARTER, S., 1993, Relationship between reservoir diagenetic evolution and petroleum emplacement in the Ula Field, North Sea: Marine and Petroleum Geology, v. 10, p. 225-270.

NELSON, K.H., AND THOMPSON, T.G., 1954, Deposition of salts from sea water by frigid concentration: Journal of Marine Research, v. 13, p. 166-182.

NELSON, R.C., 1973, Fluid inclusions as a clue to diagenesis of carbonate rocks (Abstract): Geological Society of America Abstracts with Programs, v. 5, p. 748.

NICHOLS, F.A., AND MULLINS, W.W., 1965, Morphological changes of a surface of revolution due to capillarity-induced surface diffusion: Journal of Applied Physics, v. 36, p. 1826-1835.

NORMAN, D.I., 1987, Analysis of Rb-Sr and Sm-Nd in fluid inclusion waters (Abstract): American Current Research on Fluid Inclusions, Socorro, NM, Program and Abstracts, (unpaginated).

NORMAN, D., KYLE, P., SEGALSTAD, T., AND WALDER, I., 1987, Mobility of trace elements in thermal waters in granite terranes (Abstract): NATO Advanced Research Workshop. Fluid movements, element transport, and the composition of the deep crust, Lindas, Norway (unpaginated).

O'GRADY, M.R., BODNAR, R.J., HELLGETH, J.W., CONROY, C.M., TAYLOR, L.T., AND KNIGHT, C.L., 1989, Fourier-transform infrared (FTIR) microspectrometry of individual petroleum fluid inclusions in geological samples, *in* Russell, P. E., ed., Microbeam Analysis—1989, San Francisco, San Francisco Press Inc., p. 579-582.

O'HEARN, T.C., 1985, A fluid inclusion study of diagenetic mineral phases, Upper Jurassic Smackover Formation, Southwest Arkansas and Northeast Texas, M.S. Thesis, Louisiana State University, Baton Rouge.

OAKES, C.S., BODNAR, R.J., AND SIMONSON, J.M., 1990, The system $NaCl\text{-}CaCl_2\text{-}H_2O$. I. The ice liquidus at 1 atm total pressure: Geochimica et Cosmochimica Acta, v. 54, p. 603-610.

OAKES, C.S., SHEETS, R.W., AND BODNAR, R.J., 1992, $(NaCl+CaCl_2)$ {aq}: Phase equilibria and volumetric properties (Abstract), PACROFI IV, Pan-American Conference on Research on Fluid Inclusions, Program and Abstracts, Lake Arrowhead, CA, v. 4, p. 128-132.

OSBORNE, M., AND HASZELDINE, S., 1993, Evidence for resetting of fluid inclusion temperatures from quartz cements in oilfields: Marine and Petroleum Geology, v. 10, p. 271-278.

PAGEL, M., WALGENWITZ, F., AND DUBESSY, J., 1986, Fluid inclusions in oil and gas-bearing sedimentary formations, *in* Burrus, J., ed., Thermal Modeling in Sedimentary Basins, Collection Colloques et Seminaires, v. 44, p. 565-583.

PASTERIS, J.D., AND WANAMAKER, B.J., 1988, Laser Raman microprobe analysis of experimentally re-equilibrated fluid inclusions in olivine: Some implications for mantle fluids: American Mineralogist, v. 73, p. 1074-1088.

PASTERIS, J.D., WOPENKA, B., AND SEITZ, J.C., 1988, Practical aspects of quantitative laser Raman microprobe spectroscopy for the study of fluid inclusions: Geochimica et Cosmochimica Acta, v. 52, p. 979-988.

PECHER, A., 1981, Experimental decrepitation and reequilibration of fluid inclusions in synthetic quartz: Tectonophysics, v. 78, p. 567-583.

PECHER, A., AND BOUILLIER, A., 1984, Evolution a pression et température elévées d'inclusion fluides dans un quartz synthétique: Bulletin de Minéralogie, v. 107, p. 139-153.

PENG, D.Y., AND ROBINSON, D.B., 1976, A new two constant equation of state: Industrial Engineering Chemical Fundamentals, v. 15, p. 59-64.

PETER, J.M., PELTONEN, P., SCOTT, S.D., SIMONEIT, B.R.T., AND KAWKA, O.E., 1991, ^{14}C ages of hydrothermal petroleum and carbonate in Guaymas Basin, Gulf of California: Implications for oil generation, expulsion, and migration: Geology, v. 19, p. 253-256.

PETRICHENKO, O.I., 1973, Methods of Study of Inclusions in Minerals of Saline Deposits. "Naukova Dumka" Pub. House, Kiev, 92 p. (in Ukranian; translated in Fluid Inclusion Research—Proceedings of COFFI, v. 12, 1979).

PHARES, R.A., 1991, Characterization and reservoir performance of the Lansing-Kansas City "I" and "J" zones (Upper Pennsylvanian) in the Pen Oil Field, Graham County, Kansas: Unpublished M.S. Thesis, University of Kansas, Lawrence, 445 p.

PICHAVANT, M., RAMBOZ, C., AND WEISBROD, A., 1982, Fluid immiscibility in natural processes: use and misuse of fluid inclusion data, I. Phase equilibria analysis—a theoretical and geometrical approach: Chemical Geology, v. 37, p. 1-27.

PIRONON, J., AND BARRES, O., 1990, Semi-quantitative FT-IR microanalysis limits: Evidence from synthetic hydrocarbon fluid inclusions in sylvite: Geochimica et Cosmochimica Acta, v. 54, p. 509-518.

PIRONON, J., AND BARRES, O., 1992, Influence of brine-hydrocarbon interactions on FT-IR microspectroscopic analyses of intracrystalline liquid inclusions: Geochimica et Cosmochimica Acta, v. 56, p. 169-174.

PIRONON, J., DEREPPE, J-M., AND MOREAU, C., 1992, ^1H NMR analysis of fluid content in rocks (Abstract): PACROFI IV, Pan-American Conference on Research on Fluid Inclusions, Program and Abstracts, Lake Arrowhead, CA, v. 4, p. 65.

POTTER, R.W., 1977, Pressure corrections for fluid inclusion homogenization temperatures based on the volumetric properties of the system $NaCl-H_2O$: United States Geological Survey Journal of Research, v. 5, p. 603-607.

POTTER, R.W. II, AND BROWN, D.L., 1975, The volumetric properties of aqueous sodium chloride solutions from 0°C to 500°C at pressures up to 2000 bars based on a regression of the available literature data: United States Geological Survey Open-File Report 75-636, 31 p.

POTTER, R.W. II, AND BROWN, D.L., 1977, The volumetric properties of aqueous sodium chloride solutions from 0° to 500°C at pressures up to 2000 bars based on a regression of available data in the literature: Preliminary steam tables for NaCl solutions: United States Geological Survey Bulletin 1421-C, p. C1-C36.

POTTER, R.W. II, AND CLYNNE, M.A., 1978, Pressure correction for fluid inclusion homogenization temperatures (Abstract): International Association on the Genesis of Ore Deposits, 5th Symposium, Program and Abstracts, Alta, UT, p. 146.

POTTER, R.W. II, CLYNNE, M. A., AND BROWN, D. L., 1978, Freezing point depression of aqueous sodium chloride solutions: Economic Geology, v. 73, p. 284-285.

POTY, B., DEREPPE, J.M., LANDAIS, P., AND PIRONON, J., 1987, Use of ^1H NMR for discrimination of solutions having different proton concentrations. Applications to fluid inclusions (Abstract): ECROFI, European Current Research on Fluid Inclusions, IX Symposium, Oporto, p. 101-102.

PRAY, L.C., 1961, Geology of the Sacramento Mountains Escarpment, Otero County, New Mexico: New Mexico Bureau of Mines and Mineral Resources Bulletin, v. 35, 144 p.

PREZBINDOWSKI, D.R. AND LARESE, R.E., 1987, Experimental stretching of fluid inclusions in calcite—Implications for diagenetic studies: Geology, v. 15, p. 333-336.

PREZBINDOWSKI, D.R. AND TAPP, J.B., 1991, Dynamics of fluid inclusion alteration in sedimentary rocks: a review and discussion: Organic Geochemistry, v. 17, p. 131-142.

RAMSEYER, K., FISCHER, J., MATTER, A., EBBERHARDT, P., AND GEISS, J., 1989, A cathodoluminescence microscope for low luminescence: Journal of Sedimentary Petrology, v. 59, p. 619-622.

RANKIN, A.H., RAMSEY, M.H., COLES, B., VAN LANGEVELDE, F., AND THOMAS, C.R., 1992, The composition of hypersaline, iron-rich granitic fluids based on laser-ICP and Synchrotron-XRF microprobe analysis of individual fluid inclusions in topaz, Mole Granite, eastern Australia: Geochimica et Cosmochimica Acta, v. 56, p. 67-79.

REEDER, R.J. AND WARD, W.B., 1985, Possible stretching mechanisms in fluid inclusions in calcite: Geological Society of America Abstracts with Programs, v. 17, p. 696-697.

ROEDDER, E., 1962, Ancient fluids in crystals: Scientific American, v. 207, p. 38-47.

ROEDDER, E., 1967, Metastable superheated ice in liquid-water inclusions under high negative pressure: Science, v. 155, p. 1413-1417.

ROEDDER, E., 1970, Application of an improved crushing stage to studies of gases in fluid inclusions: Schweizerische Mineralogische und Petrographische Mitteilungen, v. 50, p. 41-58.

ROEDDER, E., 1971, Metastability in fluid inclusions: Society of Mining Geology of Japan, Special Issue 3, (Proc. IMA-IAGOD Meetings '70, IAGOD Vol.), p. 327-334.

ROEDDER, E., 1972, Composition of Fluid Inclusions: United States Geological Survey Professional Paper 440JJ, 164 p.

ROEDDER, E., 1979, Fluid inclusion evidence on the environment of sedimentary diagenesis, review, *in* Scholle, P. A. and Schlunger, P .R., eds., Aspects of Diagenesis: Society of Economic Paleontologists and Mineralogists Special Publication 26, p. 89-107.

ROEDDER, E., 1984, Fluid Inclusions: Mineralogical Society of America, Reviews in Mineralogy, v. 12, 644 p.

ROEDDER, E., 1990, Fluid inclusion analysis—Prologue and epilogue: Geochimica et Cosmochimica Acta, v. 54, p. 495-507.

ROEDDER, E., 1992, Optical microscopy identification of the phases in fluid inclusions in minerals: Microscope, v. 40., p. 59-79.

ROEDDER, E., AND BELKIN, H.E., 1979, Application of studies of fluid inclusions in Permian Salado salt, New Mexico, to problems of siting the Waste Isolation Pilot Plant, *in* McCarthy, G. J., ed., Scientific Basis for Nuclear Waste Management, Volume 1, New York, Plenum Press, p. 313-321.

ROEDDER, E., AND BELKIN, H.E., 1980, Thermal gradient migration of fluid inclusions in single crystals of salt from the Waste Isolation Pilot Plant Site (WIPP), *in* Northrup, C. J. M., ed., Scientific Basis for Nuclear Waste Management, Volume 2, New York, Plenum Press, p. 453-464.

ROEDDER, E., AND BELKIN, H.E., 1981, Petrographic study of fluid inclusions in salt core samples from Asse mine, Federal Republic of Germany, United States Geological Survey Open-File Report 81-1128, 32 p.

ROEDDER, E., AND HOWARD, K.W., 1988, Taolin Zn-Pb fluorite deposit, Peoples Republic of China: an example of some problems in fluid inclusion research in mineral deposits: Journal of the Geological Society of London, v. 145, p. 163-174.

ROSASCO, C.J., ETZ, E.S., AND CASSATT, W.A., 1975, The analysis of discrete fine particles by Raman spectroscopy: Applied Spectroscopy, v. 19, p. 396-404.

ROSASCO, G.J., AND ROEDDER, E., 1979, Application of a new Raman microprobe spectrometer to nondestructive analysis of sulfate and other ions in individual phases in fluid inclusions in minerals: Geochimica et Coschimica Acta, v. 43, p. 1907-1915.

ROWAN, E.L., BODNAR, R.J., AND BETHKE, P.M., 1985, Stretching of fluid inclusions in fluorite at confining pressures up to 1 kilobar: United States Geological Survey Open-File Report 85-471, 34 p.

SABOURAUD-ROSSET, C., 1969, Expériences sur les inclusions hypersalinés ($NaCl-H_2O$ et $KCl-H_2O$). Diagnose de la halite et de la sylvite intracristalline: Compte Rendue Académie des Sciences, Paris, v. 268, p. 1671-1674.

SABOURAUD-ROSSET, C., 1972, Microcryoscopie des inclusions liquides du gypse et salinité des milieux générateurs: Revue de Geographie Physique et de Geologie Dynamique, v. 14, p. 133-144.

SABOURAUD-ROSSET, C., 1974, Détermination par activation neutronique des rapports Cl/Br des inclusions fluides de divers gypses. Correlation avec les données de la microcryoscopie et interpretations génétiques: Sedimentology, v. 21, p. 415-431.

SABOURAUD-ROSSET, C., 1976, Solid and Liquid Inclusions in Gypsum: These d'Etat, Université Paris Sud, Centre d'Orsay, 173 p.

SALLER, A.H., 1984a, Petrologic and geochemical constraints on the origin of subsurface dolomite, Enewetak Atoll: An example of dolomitization by normal seawater: Geology, v. 12, p. 217-220.

SALLER, A.H., 1984b, Diagenesis of Cenozoic limestones on Enewetak Atoll: Ph.D. Dissertation, Louisiana State University, Baton Rouge, 362 p.

SALLER, A.H., AND KOEPNICK, R.B., 1990, Eocene to early Miocene growth of Enewetak Atoll: Insight from strontium-isotope data: Geological Society of America Bulletin, v. 102, p. 381-390.

SCHIFFRIES, C.M., 1990, Liquid-absent aqueous fluid inclusions and phase equilibria in the system $CaCl_2$-$NaCl-H_2O$: Geochimica et Cosmochimica Acta, v. 54, p. 611-619.

SEITZ, J.C., AND PASTERIS, J.D., 1990, Theoretical and practical aspects of differential partitioning of gases by clathrate hydrates in fluid inclusions: Geochimica et Cosmochimica Acta, v. 54, p. 631-639.

SEITZ, J.C., PASTERIS, J.D., AND CHOU, I-M, 1993a, Raman spectroscopic characterization of gas mixtures. I. Quantitative composition and pressure determination of CH_4, N_2, and their mixtures: American Journal of Science, v. 293, p. 297-321.

SEITZ, J.C., PASTERIS, J.D., AND MORGAN, G. B. VI, 1993b, Quantitative analysis of mixed volatile fluids by Raman microprobe spectroscopy: A cautionary note on spectral resolution and peak shape: Applied Spectroscopy, v. 47, p. 816-820.

SEITZ, J.C., PASTERIS, J.D., AND WOPENKA, B., 1987, Characterization of CO_2-CH_4-H_2O fluid inclusions by microthermometry and laser Raman microprobe spectroscopy: Inferences for clathrate and fluid equilibria: Geochimica et Cosmochimica Acta, v. 51, p. 1651-1664.

SELLY, R.C., 1985, Elements of Petroleum Geology: New York, W. H. Freeman and Company, 449 p.

SHELTON, K.L., BAUER, R.M., AND GREGG, J.M., 1992, Fluid inclusion studies of regionally extensive epigenetic dolomites, Bonneterre Dolomite (Cambrian), southeast Missouri: Evidence of multiple fluids during dolomitization and lead-zinc mineralization: Geological Society of America Bulletin, v. 104, p. 675-683.

SHELTON, K.L, AND ORVILLE, P.M., 1980, Formation of synthetic fluid inclusions in natural quartz: American Mineralogist, v. 65, p. 1233-1236.

SHEPHERD, T.J., AND CHENERY, S.R., 1993, Chemical characterisation of single inclusions by laser ablation microprobe-inductively coupled plasma-mass spectrometry (LAMP-ICP-MS): ECROFI XII, Twelfth Biennial Symposium, European Current Research on Fluid Inclusions, Warsaw-Cracow, p. 194.

SHEPHERD, T.J., RANKIN, A.H. AND ALDERTON, D.H. M., 1985, A Practical Guide to Fluid Inclusion Studies: Glasgow, Blackie and Son, 239 p.

SKINNER, B.J., 1966, Thermal expansion, *in* Clark, S. P., Jr., ed., Handbook of Physical Constants, Revised Edition, Geological Society of America Memoirs, v. 97, p. 75-96.

SMITH, D.L., AND EVANS, B., 1984, Diffusional crack healing in quartz: Journal of Geophysical Research, v. 89, p. 4125-4135.

SORBY, H.C., 1858, On the microscopic structure of crystals, indicating the origin of minerals and rocks: Geological Society of London Quarterly Journal, v. 14, p. 453-500.

SPENCER, R.J., MOLLER, N., AND WEARE, J.H., 1990, The prediction of mineral solubilities in natural waters: A chemical equilibrium model for the Na-K-Ca-Mg-Cl-SO_4-H_2O system at temperatures below 25°C: Geochimica et Cosmochimica Acta, v. 54, p. 575-590.

STEIN, C.L., AND KRUMHANSL, J.L., 1988, A model for the evolution of brines in salt from the lower Salado Formation, southeastern New Mexico: Geochimica et Cosmochimica Acta, v. 52, p. 1037-1046.

STEPHENS, B.P., 1988, Origin of dolomite in the Tamabra Formation (Mid-Cretaceous) east-central Mexico: Unpublished M.S. Thesis, University of Kansas, Lawrence, 157 p.

STERNER, S.M., AND BODNAR, R.J., 1984, Synthetic fluid inclusions in natural quartz I. Compositional types synthesized and applications to experimental geochemistry: Geochimica et Cosmochimica Acta, v. 48, p. 2659-2668.

STERNER, S.M., HALL, D.L., AND BODNAR, R.J., 1988, Synthetic fluid inclusions. V. Solubility relations in the system $NaCl-KCl-H_2O$ under vapor-saturated conditions: Geochimica et Cosmochimica Acta, v. 52, p. 989-1005.

STOESSEL, R.K., AND BYRNE, P.A., 1982, Salting out of methane in single-salt solutions at 25°C and below 800 psi: Geochimica et Cosmochimica Acta, v. 46, p. 1327-1332.

SWANENBERG, H.E.C., 1980, Fluid inclusions in high-grade metamorphic rocks from S. W. Norway: Geologica Ultraiectina, University Utrecht, v. 20, 147 p.

THOMPSON, M., RANKIN, A.H., WALTON, S.J., HALLS, C., AND FOO, B.N., 1980, The analysis of fluid inclusion decrepitate by inductively-coupled plasma atomic emission spectroscopy: an exploratory study: Chemical Geology, v. 30, p. 121-133.

TILLEY, B.J., NESBITT, B.E., AND LONGSTAFFE, F.J., 1989, Thermal history of Alberta deep basin: Comparative study of fluid inclusion and vitrinite reflectance data: The American Association of Petroleum Geologists Bulletin, v. 73, p. 1206-1222.

TOBIN, R.C., 1991, Diagenesis, thermal maturation and burial history of the Upper Cambrian Bonneterre dolomite, southeastern Missouri: an interpretation of thermal history from petrographic and fluid inclusion evidence: Organic Geochemistry, v. 17, p. 143-151.

TSUI, T.F., 1990, Characterizing fluid inclusion oils via UV fluorescence microspectrophotometry—A method for projecting oil quality and constraining oil migration history (Abstract): American Association of Petroleum Geologists Bulletin, v. 74, p. 781.

TURGARINOV, A.I., AND VERNADSKY, V.I., 1970, Dependence of the decrepitation temperature of minerals on their gas-liquid inclusions and hardness: Akademiya Nauk SSSR Doklady, v. 195, p. 112-114.

ULRICH, M.R. AND BODNAR, R.J., 1984, Systematics of stretching of fluid inclusions in barite at 1 atm confining pressure: Geological Society of America Astracts with Programs, v. 16, p. 680.

ULRICH, M.R. AND BODNAR, R.J., 1988, Systematics of stretching of fluid inclusions II: barite at 1 atm confining pressure: Economic Geology, v. 83, p. 1037-1046.

VANKO, D.A., BODNAR, R.J., AND STERNER, S.M., 1988, Synthetic fluid inclusions: VIII. Vapor-saturated halite solubility in part of the sytem $NaCl-CaCl_2-H_2O$, with application to fluid inclusions from oceanic hydrothermal systems: Geochimica et Cosmochimica Acta, v. 52, p. 2451.

VANKO, D.A., GHAZI, A.M., SUTTON, S.R., AND CLINE, J., 1993, Synchrotron X-ray fluorescence microprobe applied to the analysis of fluid inclusions: ECROFI XII, Twelfth Biennial Symposium, European Current Research on Fluid Inclusions, Warsaw-Cracow, p. 230-232.

VANKO, D.A., SUTTON, S.R., RIVERS, M.L., BODNAR, R.J., AND ROEDDER, E., 1992, Synchrotron XRF microprobe analysis of fluid inclusions (Abstract): PACROFI IV, Pan-American Conference on Research on Fluid Inclusions, Program and Abstracts, Lake Arrowhead, CA, v. 4, p. 83.

VIETS, J.G., LEACH, D.L., MEIER, A.L., ROSE, S.C., AND ROWAN, E.L., 1985, Application of inductively coupled plasma mass spectrometry of the analysis of fluids extracted from Mississippi Valley-type deposits of the mid-continent, USA (Abstract): Geological Society of America Abstracts with Programs, v. 17, p. 740.

VISSER, W., 1982, Maximum diagenetic temperature in a petroleum source-rock from Venezuela by fluid inclusion geothermometry: Chemical Geology, v. 83, p. 95-101.

VITYK, M., DEMIHOV, Y., AND KROUSE, H.R., 1993, Preservation of $\partial^{18}O$ values of fluid inclusion's water in quartz over geological time in an epithermal environment: ECROFI XII, Twelfth Biennial Symposium, European Current Research on Fluid Inclusions, Warsaw-Cracow, p. 239-241.

VRY, J., BROWN, P.E., AND BEAUCHAINE, J., 1987, Application of micro-FTIR spectroscopy to the study of fluid inclusions (Abstract): EOS, v. 68, p. 1538.

WAGNER, P.D., AND MATTHEWS, R.K., 1982, Porosity preservation in the Upper Smackover (Jurassic) carbonate grainstone, Walker Creek Field, Arkansas: Response of paleophreatic lenses to burial processes—Reply: Journal of Sedimentary Petrology, v. 52, p. 24-25.

WALKER, G., AND BURLEY, S.D., 1991, Luminescence petrography and spectroscopic studies of diagenetic minerals, *in* Barker, C. E., and Kopp, O., eds., Luminescence Microscopy: Quantitative and Qualitative Aspects: Society of Economic Paleontology and Mineralogy Short Course Notes, v. 11, p. 83-96.

WALLACE, R.H., KRAEMER, T.F., TAYLOR, R.E., AND WESSELMAN, J.B., 1978, Assessment of geopressured-geothermal resources in the northern Gulf of Mexico basin: United States Geological Survey Circular 790, p. 132-155.

WELSCH, H., 1973, Die Systeme Xenon-Wasser und Methan-Wasser bei hohen Drucken und Temperaturen: Ph.D. Thesis, Institut für Physikalische Chemie, Universität Karlsruhe, Karlsruhe.

WOJCIK, K.M., 1991, Diagenesis of Pennsylvanian Sandstones and Limestones, Cherokee Basin, Southeastern Kansas: Importance of Regional Fluid Flow: Unpublished Ph.D. Thesis, University of Kansas, Lawrence, 349 p.

WOJCIK, K.M., GOLDSTEIN, R.H., AND WALTON, A.W., 1994, History of diagenetic fluids in a distant foreland area, Middle and Upper Pennsylvanian, Cherokee basin, Kansas, USA: Fluid inclusion evidence: Geochimica et Cosmochimica Acta, v. 58, p.1175-1191.

WOJCIK, K.M., McKIBBEN, M.A., GOLDSTEIN, R.H., AND WALTON, A.W., 1992, Diagenesis, thermal history, and fluid migration, Middle and Upper Pennsylvanian rocks, southeastern Kansas, *in* Johnson and Cardott, eds., Source Rocks in the Southern Midcontinent, Oklahoma Geological Survey Circular 93, p 144-159.

WOPENKA, B., AND PASTERIS, J.D., 1987, Raman intensities and detection limits of geochemically relevant gas mixtures for a laser Raman microprobe: Analytical Chemistry, v. 59, p. 2165-2170.

WOPENKA, B., PASTERIS, J.D., AND FREEMAN, J.J., 1990, Analysis of individual fluid inclusions by Fourier transform infrared and Raman microspectroscopy: Geochimica et Cosmochimica Acta, v. 54, p. 519-533.

YANATIEVA, O.K., 1946, Polythermal solubilities in the systems $CaCl_2-MgCl_2-H_2O$ and $CaCl_2-NaCl-H_2O$: Zhurnal Prikladnox Khimii, v. 19, p. 709-722.

YPMA, P.J.M., 1963, Rejuvenation of ore deposits as exemplified by the Belledonne Metalliferous Province: Ph.D. Dissertation, University of Leiden, Leiden, The Netherlands, 213 p.

ZHANG, Y., AND FRANTZ, J.D., 1987, Determination of the homogenization temperatures and densities of supercritical fluids in the system $NaCl-KCl-CaCl_2-H_2O$ using synthetic fluid inclusions: Chemical Geology, v. 64, p. 335-350.

ZOLENSKY, M.E., AND BODNAR, R.J., 1982, Identification of fluid inclusion daughter crystals using Gandolfi X-ray technique: American Mineralogist, v. 67, p. 137-141.